Modeling the 3D Conformation of Genomes

Series in Computational Biophysics

Series Editor
Nikolay Dokholyan

Molecular Modeling at the Atomic Scale
Methods and Applications in Quantitative Biology
Ruhong Zhou

Coarse-Grained Modeling of Biomolecules
Garegin A. Papoian

Computational Approaches to Protein Dynamics
From Quantum to Coarse-Grained Methods
Monika Fuxreiter

Modeling the 3D Conformation of Genomes
Guido Tiana, Luca Giorgetti

For more information about this series, please visit:
[www.crcpress.com/Series-in-Computational-Biophysics/book-series/CRCSERCOMBIO]

Modeling the 3D Conformation of Genomes

Edited By
Tiana Guido
Luca Giorgetti

CRC Press
Taylor & Francis Group
Boca Raton London New York

CRC Press is an imprint of the
Taylor & Francis Group, an **informa** business

CRC Press
Taylor & Francis Group
6000 Broken Sound Parkway NW, Suite 300
Boca Raton, FL 33487-2742

First issued in paperback 2020

ISBN-13: 978-1-138-50079-2 (hbk)
ISBN-13: 978-0-367-78045-6 (pbk)

Library of Congress Cataloging-in-Publication Data

Names: Tiana, G. (Guido), author.
Title: Modeling the 3D conformation of genomes / Guido Tiana, Luca Giorgetti.
Description: Boca Raton : Taylor & Francis, 2018. | Series: Series in computational biophysics ; 4 | Includes bibliographical references.
Identifiers: LCCN 2018030735 | ISBN 9781138500792 (hardback : alk. paper)
Subjects: LCSH: Genomes--Data processing. | Genomics--Technological innovations.
Classification: LCC QH447 .T53 2018 | DDC 572.8/6--dc23
LC record available at https://lccn.loc.gov/2018030735

**Visit the Taylor & Francis Web site at
http://www.taylorandfrancis.com**

**and the CRC Press Web site at
http://www.crcpress.com**

Contents

Preface

Characterizing the three-dimensional organization of chromosomes, as well as its mechanistic determinants, are central topics in contemporary science. Besides the curiosity towards fundamental questions such as how meters of DNA are folded inside every cell nucleus in our body, the quest for a comprehensive characterization of chromosome structure is animated by the urge to better understand gene expression. Indeed, many genes in mammalian genomes are controlled by regulatory DNA sequences such as *transcriptional enhancers*, which can be located at very large genomic distances from their target genes. Although the exact molecular details of their functional interactions are only partially understood, a large body of experimental evidence suggests that enhancers control transcription by physically contacting their target genes and looping out intervening DNA. Thus, it is crucial to understand how chromosomes are folded, which molecular mechanisms control their structure, and how chromosome architecture evolves in time, especially in large and complex genomes such as ours, where the vast majority of DNA sequence does not encode protein-coding genes. These questions are intrinsically quantitative, and lie at the interface between molecular biology and biophysics.

The last two decades have witnessed a revolution in our understanding of chromosome structure, which has been fueled by the development and refinement of a class of experimental techniques known as chromosome conformation capture (3C) and its derivatives such as 4C, 5C and Hi-C. In 3C and its derivatives, biochemical manipulation of fixed cell populations allows to measure population-averaged *contact probabilities* within chromosomes, which can be plotted in the form of two-dimensional matrices describing the contact propensities of the chromatin fiber. Several different 3C-based techniques have allowed spectacular discoveries, such as the existence of complex, highly non-random patterns of interactions across mammalian chromosomes that span several orders of magnitude in genomic length and range from 'loops' connecting DNA loci separated by few tens of kilobases, all the way up to huge, multi-megabase 'compartments' reflecting the association of transcribed and repressed parts of the genome, themselves subdivided in topologically associating domains (TADs) corresponding to sub-megabase domains of preferential interactions of the chromatin fiber. Many of these structures have been validated using independent methods, and notably using single-cell approaches that measure distances between genomic locations such as DNA fluorescence in situ hybridization (DNA FISH). However, the mechanisms that drive the formation of such a complex patter of interactions are still only poorly understood.

A fascinating aspect of chromosomal contact probabilities measured in 3C-based experiments is that they obey the same power-law scaling rules that can be often encountered in statistical physics, and in its application to polymers. In fact, even since the advent of 3C, polymer physics has played a key role in interpreting the experimental data. Polymer models have been extensively used to test hypotheses concerning the mechanisms that give rise to the observed experimental phenomenology, and for solving the inverse problem of determining the three-dimensional shape of the chromatin fiber that gives rise to the observed contact probabilities. This volume aims at giving an overview of the computational methods that have been developed to study chromosome structure, and have been motivated by the ever-growing amount of experimental data based on 3C methods as well as single-cell techniques such as DNA FISH.

In Chapter 1, Job Dekker reviews the technical and conceptual bases of 3C and its derivative techniques such as 5C and Hi-C, which were developed in his laboratory. This 'experimental' chapter is accessible to non-biologists and nicely describes how 3C-based methods laid the foundation for the current understanding of chromosome architecture, and eventually enabled to build and test physical models of chromosome folding.

The remaining chapters in this volume are divided into two groups. Chapters 2–9 describe models that follow a 'bottom-up' approach, where explicit hypotheses regarding the biophysical mechanisms driving chromosomal interactions are made, and model predictions are compared with experiments in order to test the validity of the underlying hypotheses. Chapters 10–14 instead describe 'top-down' modeling approaches, which start from the experimental data to derive models describing various properties of chromosome folding. This partition is convenient, but obviously only partially accurate. The cell nucleus is such a complex system that describing chromosome organization and dynamics in terms of purely *ab initio* models seems totally unrealistic. All current mechanistic, bottom-up chromosome folding models are markedly inspired by available experimental data, and Hi-C data in particular. On the other hand, top-down modeling strategies contain nontrivial assumptions concerning the interpretation of 3C-based data in terms of physical distances and/or contact probabilities, which in turn depend on more or less implicit hypotheses concerning chromosome folding mechanisms.

The first nine chapters on bottom-up approaches emphasize and combine different physical ingredients in order to reproduce the experimental data (namely the presence of loops, TADs and compartments in Hi-C data), predict the outcome of new experiments and learn the basic rules that control chromosomes in cell nuclei. In Chapter 2, Cédric Vaillant and Daniel Jost describe a model based on direct interaction between genomic locations, which depend on local chromatin modifications, which is able to accurately predict the outcome of Hi-C experiments in *Drosophila* based on the physics of block co-polymers. Mario Nicodemi and coworkers (Chapter 3) focus instead on the role of diffusing molecules that mediate interactions between chromosomal loci, mimicking the effect of nuclear proteins that might promote direct looping across chromosomes. In Chapter 4, Leonid Mirny and colleagues discuss the highly influential loop-extrusion model, which incorporates hypotheses on how the DNA-binding proteins CTCF and cohesin promote the formation of out-of-equilibrium, ATP-driven interactions across mammalian genomes. A combination of the diffusing-molecule and loop-extrusion models (in a version where the loop extruding factors diffuse in an ATP-independent manner) is discussed in Chapter 5 by Davide Marenduzzo and coworkers.

Irrespective of the mechanisms that drive loops and higher-order structures in a site-specific manner, a ubiquitous phenomenon that is likely to impact the three-dimensional folding of genomic DNA is torsional stress (known as super-coiling) generated by active biological processes and notably transcription through RNA polymerases. In Chapter 6, Andrzej Stasiak and coworkers show that models describing supercoiling can predict the formation of chromosomal domains such as TADs, and can also be integrated with loop extrusion.

Chapters 7 and 8 focus on the temporal dynamics of chromosome folding. Starting from hypotheses on the physical mechanism controlling the structure of chromosomes, Andrea Papale and Angelo Rosa discuss the dynamics of a poly-mer model subject to topological constraints (Chapter 7). A dynamic polymer model controlled by local interactions and physical confinement is described by Assaf Amitai and David Holcman in Chapter 8, along with its application to study the dynamics of chromosomal loci in budding yeast. Finally, Chapter 9 describes a polymer model designed to describe the dynamics of bacterial genomes, and how its predictions can be extended to higher organisms.

Chapters in the second part of the book describe 'data-driven' models that use different strategies for interpreting experimental 3C-based data in terms of physical conformations of the chromatin fiber. Marco Di Stefano and Marc Marti-Renom review in Chapter 10 how to derive three-dimensional models of chromosomes by implementing spatial restraints derived from Hi-C data. Frank Alber and coworkers discuss in Chapter 11 how it is possible to integrate experi-mental data generated using multiple experimental techniques to build models of chromosome structure.

We discuss in Chapter 12 a maximum-entropy approach allowing to extract the full equilibrium ensemble of conformations giving rise to 5C or Hi-C data at the TAD level, and to make predictions regarding statistical and dynamical properties of chromosome conformation, which can be validated experimen-tally. A similar maximum-entropy approach developed by Peter Wolynes, José Onuchic and coworkers is described in Chapter 13, with a focus on the energy landscape of the model. Finally, Christian Micheletti and coworkers review in Chapter 14 a restraint-based approach allowing to extract the structure of entire chromosomes from Hi-C dataset.

The chapters in this volume give a comprehensive overview of the state-of-the-art of computational research in the area of chromosome conformation and nuclear structure, and testify to how theoretical work is instrumental in reaching a mechanistic, quantitative understanding of the basic rules that shape biology at the very heart of cellular processes. We are deeply grateful to all the authors who have invested their time and thought in making this collection possible.

Tiana Guido
Luca Giorgetti

Note to Readers: For access to figures in full color format, please visit the book's home page at the publisher's website: www.crcpress.com/9781138500792

Editor

Guido Tiana, PhD, is Associate Professor of Theoretical Biophysics at the University of Milan. He obtained a PhD at the Niels Bohr Institute (Copenhagen) in 2000 and since then has worked on the physics of complex systems of biological interest, such as proteins, DNA, RNA, chromosomes and genetic networks. The methods come from the realm of statistical mechanics, making heavy use of computational tools and some experimental work.

Luca Giorgetti, PhD, is a group leader at the Friedrich Miescher Institute for Biomedical Research in Basel. He obtained his PhD at the European Institute of Oncology (IEO) and University of Milan followed by a postdoctoral training at the Curie Institute in Paris. He is an expert in combining physical modeling and experimental research in chromosome conformation and transcriptional regulation.

Contributors

A. Amitai
Department of Chemical
Engineering
Massachusetts Institute of
Technology
Cambridge, MA
and
Institute for Medical Engineering
and Science
Massachusetts Institute of
Technology
Cambridge, MA
and
Ragon Institute of MGH
MIT and Harvard
Cambridge, MA

Andrea Esposito
Berlin Institute for Medical Systems
Biology
Max-Delbrück Centre (MDC) for
Molecular Medicine
Berlin, Germany

Andrea M. Chiariello
Dipartimento di Fisica
Università di Napoli Federico II, and
INFN Napoli
Complesso Universitario di Monte
Sant'Angelo
Naples, Italy

Andrea Papale
Sissa (Scuola Internazionale
Superiore di Studi Avanzati)
Trieste, Italy

Andrzej Stasiak
Center for Integrative Genomics
University of Lausanne
Lausanne, Switzerland
and
SIB Swiss Institute of Bioinformatics
Lausanne, Switzerland

Angelo Rvosay
Sissa (Scuola Internazionale
Superiore di Studi Avanzati)
Trieste, Italy

Anton Goloborodko
Institute for Medical Engineering
and Science, and Department of
Physics
4DN MIT-UMass Center for Physics
and Structure of the Genome
Massachusetts Institute of
Technology
Cambridge, MA

C. A. Brackley
SUPA, School of Physics &
Astronomy
University of Edinburgh
Edinburgh, UK

Carlo Annunziatella
Dipartimento di Fisica
Università di Napoli Federico II, and
INFN Napoli
Complesso Universitario di Monte
Sant'Angelo
Naples, Italy

Cristian Micheletti
SISSA, International School for
Advanced Studies
Trieste, Italy

Cédric Vaillant
University of Lyon, ENS de Lyon,
University of Claude Bernard, CNRS,
Laboratoire de Physique
Lyon, France

D. Holcman
Ecole Normale Superieure
Paris, France

D. Marenduzzo
SUPA, School of Physics &
Astronomy
University of Edinburgh
Edinburgh, UK

D. Michieletto
SUPA, School of Physics &
Astronomy
University of Edinburgh
Edinburgh, UK

Daniel Jost
University of Grenoble Alpes, CNRS,
Grenoble INP, TIMC-IMAG
Grenoble, France

Dusan Racko
Center for Integrative Genomics
University of Lausanne
Lausanne, Switzerland
and
SIB Swiss Institute of Bioinformatics
Lausanne, Switzerland
and
Polymer Institute of the Slovak
Academy of Sciences
Bratislava, Slovakia

Eivind Hovig
Department of Informatics
University of Oslo
Oslo, Norway

Eivind Hovig
Department of Tumor Biology
Institute for Cancer Research
Oslo University Hospital
Oslo, Norway

Eivind Hovig
Institute of Cancer Genetics and
Informatics
Oslo, Norway

**Fabrizio Benedetti, Andrzej
Stasiak**
Center for Integrative Genomics
University of Lausanne
Lausanne, Switzerland
and
Vital-IT, SIB Swiss Institute of
Bioinformatics
Lausanne, Switzerland

**Guido Polles, Nan Hua, Asli
Yildirim, Frank Alber**
Department of Biological Sciences
Molecular and Computational
Biology, the Bridge Institute
University of Southern California
Los Angeles, CA

Guido Polles, Nan Hua, Asli Yildirim, Frank Alber
Department of Biological Sciences
Molecular and Computational
Biology, the Bridge Institute
University of Southern California
Los Angeles, CA

Guido Tiana
Department of Physics and Center
for Complexity and Biosystems
Universita degli Studi di Milano and
INFN
Milan, Italy

J. Johnson
SUPA, School of Physics &
Astronomy
University of Edinburgh
Edinburgh, UK

Job Dekker
Department of Biochemistry and
Molecular Pharmacology
Howard Hughes Medical Institute,
University of Massachusetts Medical
School
Worcester, MA

Jonas Paulsen
Institute of Basic Medical Sciences
University of Oslo
Oslo, Norway

José N. Onuchic, Peter G. Wolynes
Department of Physics &
Astronomy, Department of
Chemistry and Department of
Biosciences
Rice University
Houston, TX

Julien Dorier
Vital-IT, SIB Swiss Institute of
Bioinformatics
Lausanne, Switzerland
and
Center for Integrative Genomics
University of Lausanne
Lausanne, Switzerland

Leonid A. Mirny
Institute for Medical Engineering
and Science and Department of
Physics
4DN MIT-UMass Center for Physics
and Structure of the Genome
Massachusetts Institute of
Technology
Cambridge, MA

Luca Fiorillo
Dipartimento di Fisica
Università di Napoli Federico II, and
INFN Napoli
Complesso Universitario di Monte
Sant'Angelo
Naples, Italy

Luca Giorgetti
Friedrich Miescher Institute for
Biomedical Research
Basel, Switzerland

M. C. Pereira
SUPA, School of Physics &
Astronomy
University of Edinburgh
Edinburgh, UK

Marc A. Marti-Renom
Gene Regulation, Stem Cells and
Cancer Programme
Centre for Genomic Regulation
(CRG)
The Barcelona Institute of Science
and Technology (BIST)
Barcelona, Spain

Marc A. Marti-Renom
Institució Catalana de Recerca i
Estudis Avançats (ICREA)
Barcelona, Spain

Marco Cosentino Lagomarsino
CNRS, UMR 7238
Paris, France
Marco Cosentino Lagomarsino
FIRC Institute of Molecular
Oncology (IFOM)
Milan, Italy

Marco Di Stefano, Marc A. Marti-Renom
Department of Experimental and Health Sciences
Universitat Pompeu Fabra
Barcelona, Spain

Marco Di Stefano, Marc A. Marti-Renom
Structural Genomics Group, CNAG-CRG
The Barcelona Institute of Science and Technology (BIST)
Barcelona, Spain

Marco Di Stefano
Department of Experimental and Health Sciences
Universitat Pompeu Fabra
Barcelona, Spain

Marco Di Stefano
Gene Regulation, Stem Cells and Cancer Programme
Centre for Genomic Regulation (CRG), The Barcelona Institute of Science and Technology (BIST)
Barcelona, Spain

Marco Di Stefano
Structural Genomics Group, CNAG-CRG
The Barcelona Institute of Science and Technology (BIST)
Barcelona, Spain

Marco Gherardi, Marco Cosentino Lagomarsino
Computational and Quantitative Biology
Sorbonne Universités, UPMC Univ
Paris, France

Marco Gherardi
Dipartimento di Fisica
Università degli Studi di Milano and INFN
Milan, Italy

Mario Nicodemi
Berlin Institute of Health (BIH)
Max-Delbrück Centre (MDC) for Molecular Medicine
Berlin, Germany

Michele Di Pierro, Ryan R. Cheng, Bin Zhang, José N. Onuchic, Peter G. Wolynes
Center for Theoretical Biological Physics
Rice University
Houston, TX

Remus Thei Dame
Leiden Institute of Chemistry and Centre for Microbial Cell Biology
Leiden University
Leiden, The Netherlands

Simona Bianco
Dipartimento di Fisica
Università di Napoli Federico II, and INFN Napoli
Complesso Universitario di Monte Sant'Angelo
Naples, Italy

Vittore Scolari
CNRS, UMR 3525
Paris, France

Vittore Scolari
Spatial Regulation of Genomes, Genomes & Genetics Department
Institut Pasteur
Paris, France

1

Chromosome Folding: Contributions of Chromosome Conformation Capture and Polymer Physics

JOB DEKKER

1.1 INTRODUCTION

It is now hard to imagine that there was a time when there was a debate about the extent to which the genome is organized in any specific way inside the interphase cell nucleus or nucleoid. One reason for this debate is that in some experiments chromatin can appear highly structured, e.g., forming hierarchies of increasingly folded and thicker fibers (1), while in other experiments very little structure is detected in what appears to be an ocean of nucleosomes (2). Further, fluorescence in situ hybridization experiments to localize specific loci reveals tremendous cell-to-cell variability in sub-nuclear position and distance between any pair of loci. However, these experiments also show some clear trends. For instance, chromosomes occupy their own territories (3) with some limited intermingling at their borders (4). The positions of chromosomes can vary greatly between individual cells, but some chromosomes and chromosomal domains tend to be more often at the periphery of the nucleus while others are more often near the center of the nucleus (5, 6). Further, in general, euchromatic loci have been observed to associate with other euchromatic loci, and heterochromatic loci to interact with other heterochromatic loci (3, 7). Combined, these and many other findings suggest that nuclear organization is variable and stochastic on the one hand, while also guided by some common principles and that this relates to whether chromatin is (transcriptionally) active or inactive (8). Further, chromosomes are obviously organized in some defined manner during mitosis when the classic rod-shaped structures are observed. However, even in that case, how chromatin is folded during mitosis (and meiosis) has been debated for many years.

Over the last two decades, there has been enormous progress in our understanding and appreciation of the spatial organization of genomes and how it mediates or modulates the many functions of chromosomes such as the regulation of gene expression, the repair and replication of DNA, and chromosome condensation and transmission to daughter cells. This has been driven by improved imaging techniques, including super-resolution and live cell approaches but most particularly by two developments that, as outlined in the following, are significantly interlinked: First, the development of molecular genomic approaches for mapping the structure of chromosomes, mainly based on chromosome conformation capture (3C) technology; and, second, the development and application of insights from the field of polymer physics to the problem of chromosome folding.

The debate has now moved to mechanistic interpretations of structural features observed with different technologies, their dynamical properties, how variable these structures are between otherwise identical cells, and how chromosome architecture instructs or influences any of the genome's functions. Here I outline the conception of 3C, it's development over the last two decades, and how it has stimulated interactions between cell biologists, molecular biologists, (polymer) physicists, mathematicians, and computational biologists. This rich interdisciplinary interface is now producing remarkable new insights into the chromosome folding problem, discussed in this book.

1.2 CHROMOSOME CONFORMATION CAPTURE

The introduction of chromosome conformation capture in 2002 (9) has revolutionized the study of the spatial organization of genomes by allowing mapping of three-dimensional chromosome structure at increasingly high resolution directly to sequence and at the scale of complete genomes. The key new concept that motivated the subsequent development of 3C was the idea that when a matrix of many or all interaction frequencies between and among loci located along a chromosome could be measured, the three-dimensional organization of that chromosome could be inferred. 3C is used to detect the frequency of interaction of any pair of genomic loci, and when combined with deep DNA sequencing

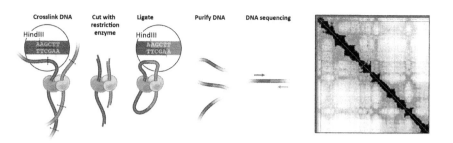

Figure 1.1 Schematic outline of the key steps of 3C-based assays. Left: Chromatin is crosslinked, and then digested and religated. (adapted from (23)). Ligation products are then sequenced. 3C variants such as 4C, 5C, and Capture C differ in how ligation products are detected, or include steps to label digested ends (Hi-C), or include a step to selectively purify fragments bound by specific proteins (ChIA-PET, HiChIP). Right: 3C-based assays are used to obtain matrices of interaction frequencies that can be depicted as heatmaps. This example shows an interaction matrix for human chromosome 21 in HeLa S3 cells. The color intensity reflects relative interaction frequencies.

can generate genome-wide all-by-all interaction frequency matrices (Figure 1.1). Such matrices have now been shown to indeed allow the derivation of models of the spatial organization of chromosomes and even help gain insights into the dynamics and mechanisms of their folding.

In 3C cells are fixed with formaldehyde (Figure 1.1). This essentially freezes the spatial arrangement of chromosomes in place. The spatial proximity between loci is then determined by fragmenting the chromatin with a restriction enzyme, while the chromatin remains frozen in place. Spatially proximal DNA ends are subsequently re-ligated and DNA is purified. This assay produces a large collection of unique DNA ligation products that each represents a spatial co-location event in one of the cells in the population. Given that DNA molecules are easily identified and characterized by PCR or DNA sequencing 3C reduces the difficult problem of determining relative spatial positions of loci inside cells to the much simpler process of DNA sequence analysis.

3C combines a number of molecular steps that had previously been used separately, e.g., proximity ligation had been used for many years to detect protein-induced bending and looping of DNA in vitro and in vivo (10–12). The innovation of 3C lies in the fact that it was designed to be unbiased and able to detect any spatial proximity, even when not mediated by a specific factor, so that it can detect all the spatial proximities irrespective of the specific mechanisms that brought the loci near each other. The concept on which it is based is also innovative, asserting that dense matrices of interaction frequencies reveal the principles of chromosome folding.

Initially, PCR was used to read and quantify specific ligation product formation events, e.g., to determine whether specific loci would interact with each other more frequently than expected. In the original 3C paper, previously known specific interactions between yeast centromeres, between telomeres, and between homologous chromosomes could be detected, thereby validating the approach (9). Further, the first (albeit sparse) matrix of interaction frequencies for a complete chromosome was presented and this matrix was used, through polymer physics theory and mathematical optimization to infer the population average folding of this chromosome. Of note, this very early work already employed polymer models for analysis of 3C data. Polymer theory and model building to interpret matrices of chromatin interaction frequencies has become a field of intense study and is now contributing to fundamental new insights into how chromosomes fold, as is outlined throughout the chapters in this book.

3C was rapidly adopted and in several landmark publications it helped demonstrate that enhancers could loop to their target genes, e.g., in the beta-globin and alpha-globin loci, in a tissue-specific manner and it was found that such specific looping interactions depend on specific transcription factors (13–19). Although such locus-specific studies to map interactions between and among specific elements in loci of interest continue to this day, further innovations in 3C technology now enable genome-wide and unbiased mapping of chromatin interaction matrices.

1.3 3C VARIANTS TO OBTAIN GENOME-SCALE AND HIGH-RESOLUTION CHROMATIN INTERACTION MATRICES

The size of genomes, and thus the number of possible chromatin interactions, makes 3C analysis using PCR impractical. Since the initial development of 3C, many experimental adaptations have been introduced that allow the mapping of chromatin interactions at a genome-wide scale, with increased resolution, in cell populations and in single cells. All these variants follow the basic 3C protocol of crosslinking chromatin, DNA fragmentation and relegation of DNA ends that are in close spatial proximity. They differ in the method for ligation product detection.

The first two 3C variants, 4C (20, 21) and 5C (22) were published in 2006. In 4C, inverse PCR is used to amplify all loci that interact with a single locus of interest. The amplified DNA is then analyzed by deep sequencing. 4C thus allows quantifying the genome-wide interaction profile of a specific genomic element, e.g., a gene promoter. Even though 4C does not produce a matrix of interaction frequencies and is, therefore, less suited for inference of chromosome folding, 4C profiles can be analyzed, e.g., by polymer models or statistical approaches, to determine whether such a promoter interacts significantly more frequently than expected with any other specific DNA elements, e.g., distal enhancers.

5C employs large sets of primers, one designed for each end of a restriction fragment to detect dense matrices of interaction frequencies between all loci throughout chromosomal domains that can be up to several Mb. 5C relies on highly multiplexed ligation-mediated amplification (LMA) to detect 3C ligation products. Pairs of 5C primers are designed to anneal immediately adjacent to each other across 3C ligation junctions. Annealed 5C primers can then be ligated and PCR amplified. Amplified DNA is analyzed by deep sequencing.

The advantage of 5C and LMA is that it allows multiplexing with thousands of primers so that matrices of all pair-wise interactions (up to several million) throughout large chromosomal domains can be detected and quantified in one reaction.

4C and 5C do not allow detection of all pair-wise interactions throughout genomes. The Hi-C variant solves this limitation by enabling unbiased detection of any pair-wise interaction (23). The main adaption in Hi-C is that DNA ends formed after chromatin fragmentation are labeled with biotinylated nucleotides prior to their relegation. This allows specific purification of DNA ligation junctions using streptavidin-coated beads. DNA is then directly sequenced. Given that genome-wide interaction maps are extremely large (10^{14} possible pairwise interactions for the human genome digested in 250 base pair fragments) Hi-C requires extremely deep sequencing, and even the most deeply sequenced datasets require binning interaction matrices at 1–10 Kb. Currently, Hi-C interaction maps are produced that are based on billions of chromatin interactions (24, 25) and these reveal a richness of features such as chromosomal domains and chromatin loops (see section "Insights Obtained From Chromosome Interaction Data" below).

Hi-C maps are extraordinarily powerful for obtaining insights into the folding of complete genomes. Hi-C data has already led to key discoveries related to the principles and mechanisms of chromosome folding, as summarized in the next section. However, even extremely deeply sequenced Hi-C datasets still do not fully capture the complete quantitative interaction landscape of chromosomes, and are also very costly to generate. Therefore, there continues to be a need for approaches that detect interactions only for targeted regions or loci, such as 4C and 5C. For instance, a complete 5C interaction matrix for a 3 Mb domain can reveal detailed patterns of interactions at 2–4 Kb resolution when sequenced at a depth of about 50–100 million read pairs. To obtain a similar coverage for the same domain by Hi-C one would need many hundreds of billions of read pairs, which is currently not attainable for most laboratories.

For this reason, the field has witnessed the further development of targeted approaches for chromatin interaction analysis for loci of interest by analyzing interactions for only a specific subset of all possible interactions and thereby allowing much more extensive sequencing coverage for the selected set of genomic loci. First, ChIA-PET (26) and HiChIP (27) rely on selective purification of chromatin that is bound by a protein of interest. These methods employ antibodies against proteins of interest, e.g., against RNA polymerase II or modified histone tails to purify only those chromatin fragments bound by those proteins, either directly after chromatin fragmentation (ChIA-PET) or directly after DNA ligation (HiChIP). In this way, one can analyze all genome-wide interactions for each of the active promoters or regulatory elements in the genome in a cost-effective way and at a resolution of single restriction fragments, which is typically not attained with Hi-C. A drawback is that ChIA-PET and HiChIP datasets are difficult to analyze because they are biased by the level of binding of the factor of interest and thus most frequently capture interactions between two sites bound by this factor, and estimating expected background levels for such interactions is complicated. The appropriate normalizing of this bias has not entirely been achieved, but several methods have been developed to start to address this major complication (28, 29).

More recently hybrid capture approaches have been very successfully used to selectively purify ligation products of interest using pools of biotinylated oligo's designed to anneal to loci of interest (30, 31). For instance, a pool of primers to selectively purify all promoter-containing restriction fragments can be used to enrich 3C or Hi-C ligation product libraries for those that are composed of a promoter-containing fragment and a second unknown interacting fragment. This DNA is then directly sequenced. In effect, this produces 4C-like genome-wide interaction profiles for each fragment (e.g., all promoters in the genome) for which capture oligos are designed. Such "Capture C" approaches have proven to be very powerful, allowing very high-resolution interaction mapping (at a single fragment level). Importantly, Capture C datasets do not suffer from the protein binding biases that affect ChIA-PET and HiChIP data, although these datasets also need to be corrected for other biases in interaction detection, like all other 3C-based methods (32, 33).

Finally, there are now several protocols for single-cell analysis of chromosome conformation (34–36). Although interaction matrices derived from single cells are rather sparse, they reveal striking variability in patterns of interactions that point to both dynamic and stochastic processes driving assembly of chromosome structures (8).

There is now a large suite of 3C-based methods available and all have unique strengths and weaknesses, and their own specific applications. These methods have been extensively, and in much more detail than here, described elsewhere (e.g., (37)). The next section discusses some of the key insights into chromosome folding obtained with this suite of technologies.

1.4 INSIGHTS OBTAINED FROM CHROMOSOME INTERACTION DATA

Chromosome conformation capture is now widely used to probe the spatial organization of chromosomes in organisms ranging from bacteria to mammals and plants. Most work has been done on mouse and human chromosomes, and insights from these studies are summarized here. It is important to emphasize that 3C-based data reveals only one aspect of chromosome organization, e.g., it does not report on dynamics directly, and orthogonal data, obtained with (live cell) imaging, are essential for a full understanding of chromosome folding, for interpretation of chromatin interaction data, and for informing model building (see the section on "Polymer Models for Chromosome Folding"). Imaging and 3C-based technologies are complementary and need to be combined for any meaningful generation and testing of models of chromosome conformation and its dynamics.

Intra-chromosomal interactions tend to be much more frequent than inter-chromosomal interactions, even for loci separated by hundreds of Mb (23). This is consistent with the observation by microscopy that chromosomes each occupy their own territory (3). At the periphery of these territories adjacent chromosomes do mingle somewhat (4). Chromosomes themselves are further subdivided into two main types of "compartments": A and B compartments that contain active and open or silent and closed chromatin respectively (20, 23). A and B domains can be up to several Mb in size and occur in alternating fashion along the linear genome. In three-dimensional, space A domains associate with other A domains while B domains cluster with B domains to form spatial A and B compartments. This spatial separation of active and inactive domain corresponds to well-known (electron) microscopy observations of such organization.

At a finer scale, chromosomes form topologically associating domains (TADs) that tend to be up to several hundred Kb (38, 39). The boundaries of many of these domains contain sites bound by the CTCF protein, and these boundaries often loop to each other as evidenced by the appearance of elevated chromatin interactions between these loci (24). Chromatin interactions are mildly enriched within TADs, while interactions between loci located in two adjacent domains are reduced. These domains have been implicated in long-range gene regulation as they correlate with the target range of enhancers (40, 41). Much research effort is now focused on the mechanisms by which these domains form, as also mentioned below.

Finally, specific looping interactions have been detected between promoters and enhancers, between enhancers and other enhancers, between promoters and promoters, and between CTCF-bound sites (30, 31, 42, 43). Interactions between CTCF sites are readily detected in Hi-C datasets (24), but interactions involving promoters and enhancers tend to be much weaker and are more easily detected with any of the targeted 3C-based approaches that allow deeper sequencing of the targeted loci.

Although most cell types analyzed by Hi-C display compartments, TADs, and loops, these can differ in their genomic location and strength. For instance, the locations of A and B compartments is highly cell type–specific, consistent with different parts of the genome being in an active or silent state. Many loops, especially promoter–enhancer interactions, are also cell type–specific while CTCF–CTCF loops and positions vary less (42). These observations strongly indicate that the folding of chromosomes is related and possibly instructive to the regulation of gene expression.

Any of these structural features are dynamic during the cell cycle (Figure 1.2). As cells enter mitosis loops, TADs and A and B compartments all rapidly disappear (44, 45). Hi-C interaction matrices for late prometaphase cells, when chromosomes are fully compacted and condensed, reveal a locus-independent structure: the interaction profile for each locus is very similar to all others and displays a general inverse relationship between the genomic distance between pairs of loci and their interaction frequency. On top of that, there is an elevated interaction frequency for loci separated around 10–12 Mb. Polymer simulations and model building (Figure 1.2) has shown that such interaction frequency matrix is consistent with the chromosome folding as a series of nested loops that rotate around a central helical axis to form the classical rod-shaped mitotic chromosome (45).

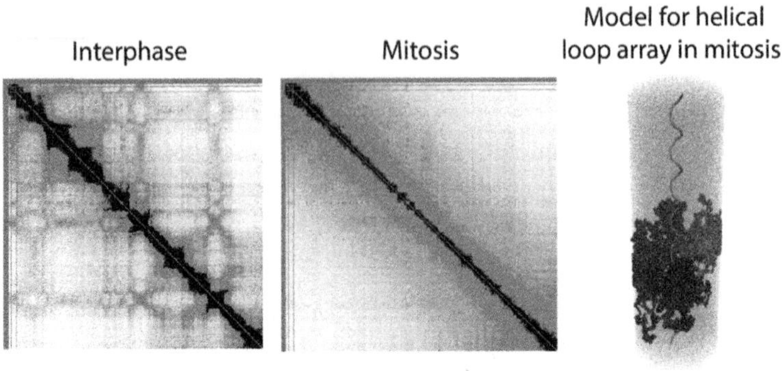

Figure 1.2 Chromatin interaction maps for chromosome 21 in HeLa S3 cells in interphase (left panel) and nocodazole-arrested mitotic cells (from (44)). The interaction maps display plaid patterns that represent A and B compartmentalization (23). In mitosis (middle panel), TADs and compartmentalization are lost and instead a general inverse relationship between interaction frequency and genomic distance is observed, with a slight increase in interactions between loci separated by ~10 Mb (44). Polymer simulations showed that chromatin interaction data are consistent with a structure in which the mitotic chromosome is organized as a helically arranged series of nested loops that emanate from a central spiraling axis. The model (right panel) shows the chromatin in gray, the spiraling axis in red, and a few individual loops rotating around the axis in different colors of blue and green (45).

Further technological innovations and new mechanistic polymer models will undoubtedly reveal additional structural features of chromosomes and their dynamics.

1.5 DYNAMICS AND CELL-TO-CELL VARIATION IN CHROMATIN INTERACTIONS

Many chromatin interaction analyses determine interaction frequencies in large cell populations. The interaction frequency, therefore, can be seen as a proxy for the number of cells in which two loci are in close spatial proximity. This has important consequences for analyzing and interpreting 3C-based data. First, the data typically shows that each locus has many other loci it interacts with, but this does not necessarily indicate that all these interactions occur at the same time in the same cell. In fact, consistent with direct imaging in single cells, and more recent single-cell Hi-C data (34–36), it is clear that chromosome conformation, and hence the matrix of chromatin interactions, is highly variable between cells in the population. For instance, although interaction data from populations of cells shows that all A domains have elevated interaction frequencies with all other A domains, at the single cell level a given domain will interact with only a small set of other domains. Similarly, the frequency of looping interactions, e.g., between two sites bound by CTCF, suggest that even those specific contacts occur in only a fraction of cells at any given moment in time (46).

These observations raise many new questions related to the dynamics of loci in single cells: Are chromosomes highly dynamic in their folding, so that given enough time all interactions detected in the population will at some moment in time occur in each single cell (ergodic), or is the genome sufficiently constrained so that each cell cannot attain all possible conformations in the lifetime of a cell? Knowing the answers to such questions will be critical for a full comprehension and appropriate interpretation of chromatin interaction matrices. Polymer physics and polymer simulations have proven extremely powerful for analysis of chromatin interaction matrices.

1.6 POLYMER MODELS FOR CHROMOSOME FOLDING

Chromosomes are long polymers, and ideas and concepts from polymer physics have proven crucial for interpreting chromatin interaction data obtained with 3C-based assays in at least two major ways. First, the theory of polymer physics makes predictions about the probabilities of interaction between two loci dependent on their position along the length of a polymer. For instance, whether a chromosome is a random coil, or a more orderly packed structure will lead to predictable quantitative differences in how interaction frequency between loci will depend on the genomic distance between them (47). As a result chromatin interaction data will inform on the state of chromatin folding. Further, the presence of loops between specific sites puts additional constraints on the conformation of the entire polymeric chromatin fiber that will have predictable effects on all chromatin interactions along the fiber, including between sites not directly involved in the looping interactions (e.g., (48, 49)).

Second, interpretation of 3C-based data through polymer simulations has led to major insights into the conformation of chromosomes at different cell cycle stages and in single cells. In such simulations, chromosomes are folded according to specific models, allowed to equilibrate and then a 3C or Hi-C experiment is simulated on an ensemble of such conformations to predict the chromatin interaction matrix. Quantitative comparison to experimental chromatin interaction data then allows testing whether or not the folding model accurately predicts experimental data. Such "bottom-up" modeling requires specific ideas about how a chromosome could fold, based on independent data such as observations from imaging or from theoretical considerations. Polymer simulations have been successfully used to test models for domain and loop formation (46, 50), A and B compartmentalization (51, 52), and folding of mitotic chromosomes (Figure 1.2) (44, 45, 53).

Other approaches aim to directly use chromatin interaction frequencies to infer the folding of chromosomes ("top-down" approaches). In such approaches, interaction frequencies can be used to estimate the average distance between pairs of loci, but this will depend on assumptions related to the polymer state of the chromatin (54, 55). This approach will produce a set of very similar structures that represent an average conformation, but these do not fully capture the range of conformation present throughout the cell population. Other approaches build ensembles of conformations that, combined, reproduce the contact frequencies detected by experimental approaches (56). All these approaches add valuable insights into chromosome conformation and its dynamics and have become important analyses for interpreting chromatin interaction datasets. These approaches are discussed in detail throughout this book.

1.7 MECHANISMS OF CHROMOSOME FOLDING AND NUCLEAR ORGANIZATION

During the last several years there has been a major shift from descriptive analysis of chromosome conformation to testing mechanistic models for how chromosomes fold. Two mechanisms for the formation of different aspects of chromosome conformation are now intensely studied. First, it has been proposed that protein complexes such as cohesin and condensin can bind DNA and extrude loops to compact chromatin fibers (46, 50, 57–59). When such complexes bind specific sites and then extrude a loop this will produce characteristic patterns in chromatin interaction matrices that have been detected, e.g., in bacteria (60). A loop extrusion mechanism has been proposed as a means to form dense arrays of loops that define mitotic chromosomes (Figure 1.2) (44, 57, 61–63). Recent polymer simulations show that dynamic formation of such loops indeed reproduces experimental 5C and Hi-C data for mitotic chromosomes. Interestingly, a similar loop extrusion mechanism can lead to the formation of TADs (46, 50). Again, polymer simulations have shown that dynamic loop extrusion, most likely by the cohesin complex, that is blocked at CTCF sites can lead to TADs with enriched interactions within them, sharp boundaries at the sites bound by CTCF and looping interactions between the TAD boundaries. Thus, a single loop extrusion mechanism can lead to mitotic chromosome formation as well as interphase TAD and CTCF loops. Many studies are now focused on understanding the molecular mechanisms of this process and the factors that determine the loading and dissociation of cohesin and condensins.

A second, and distinct mechanism for chromosome folding has been proposed that can explain the formation of A and B compartments, and the general association of segments of chromatin depending on their active or inactive state. A process of phase separation, whereby loci of similar state attract each other through multiple weak interactions, can result in a clustering of chromatin loci with similar states (51, 52). This can explain the formation of A and B compartments. Again, polymer simulations show that such attractions can reproduce key features of chromatin interaction matrices. The molecular mechanism, and molecules involved in such attractions are not well understood, but some insights have recently been obtained. For instance, the HP1 protein that binds methylated histone tails enriched in heterochromatin can mediate clustering of heterochromatic domains (64, 65).

Loop extrusion and phase separation appear to be distinct mechanisms that act independently and sometimes in opposition: Conditions that disrupt loop extrusion and TAD formation do not affect A and B compartmentalization (66). Loop extrusion can sometimes counteract phase separation by actively mixing A and B-type domains (67). Although these two mechanisms have significant explanatory power, it is very likely that there are additional mechanisms at play.

1.8 FUTURE PERSPECTIVE

The field of chromosome conformation analysis is now experiencing a very exciting time of rapid gains in our knowledge of how genomes are organized in space and time and how this relates to regulation of genome function. A combination of high-resolution chromatin interaction maps, super-resolution and live cell imaging, and polymer simulations to test specific models are now leading to mechanistic insights into the mechanisms and dynamics of chromosome folding. Together, these approaches start to unveil how cell-to-cell variation and temporal dynamics in folding is achieved and how this relates to gene regulation, DNA replication, and chromosome segregation.

One key aspect of this field is that it requires a highly interdisciplinary approach, combining imaging, chromatin interaction mapping, and computational and biophysical methods for model building and testing. The recently established 4D Nucleome Network is an example of a highly coordinated effort to bring together scientists from all these disciplines to focus on the chromosome folding problem (68).

Major challenges for the coming years are to identify the molecular machines, and their modes of action, and to identify mechanisms by which chromosome conformation modulates, facilitates, and/or instructs genome regulation including gene expression, DNA replication, and chromosome condensation and segregation.

ACKNOWLEDGMENTS

I thank all my coworkers, collaborators, and colleagues. I apologize for not citing all relevant studies due to space constraints. Here I focus on chromatin interaction mapping approaches, but many other approaches are absolutely crucial for chromosome conformation analysis, including a wide array of imaging methods. Studies in my laboratory are supported by the National Institutes of Health and the Howard Hughes Medical Institute.

REFERENCES

1. Belmont AS, Bruce K. Visualization of G1 chromosomes: A folded, twisted, supercoiled chromonema model of interphase chromatid structure. *J Cell Biol.* 1994;127(2):287–302.
2. Eltsov M, Maclellan KM, Maeshima K, Frangakis AS, Dubochet J. Analysis of cryo-electron microscopy images does not support the existence of 30-nm chromatin fibers in mitotic chromosomes in situ. *Proc Natl Acad Sci U S A.* 2008;105(50):19732–7. Epub 2008/12/10. doi:10.1073/pnas.0810057105. PubMed PMID: 19064912; PMCID: PMC2604964.
3. Cremer T, Cremer C. Chromosome territories, nuclear architecture and gene regulation in mammalian cells. *Nat Rev Genet.* 2001;2(3):292–301.
4. Branco MR, Pombo A. Intermingling of chromosome territories in interphase suggests role in translocations and transcription-dependent associations. *PLoS Biol.* 2006;4(5):e138.

5. Boyle S, Gilchrist S, Bridger JM, Mahy NL, Ellis JA, Bickmore WA. The spatial organization of human chromosomes within the nuclei of normal and emerin-mutant cells. *Hum Mol Genet.* 2001;10(3):211–9.

6. Cremer M, von Hase J, Volm T, Brero A, Kreth G, Walter J, Fischer C, Solovei I, Cremer C, Cremer T. Non-random radial higher-order chromatin arrangements in nuclei of diploid human cells. *Chromosome Res.* 2001;9(7):541–67.

7. Bickmore WA. The spatial organization of the human genome. *Annu Rev Genomics Hum Genet.* 2013;14:1467–84.

8. Gibcus JH, Dekker J. The hierarchy of the 3D genome. *Mol Cell.* 2013;49(5):773–82.

9. Dekker J, Rippe K, Dekker M, Kleckner N. Capturing Chromosome Conformation. *Science.* 2002;295(5558):1306–11.

10. Cullen KE, Kladde MP, Seyfred MA. Interaction between transcription regulatory regions of prolactin chromatin. *Science.* 1993;261(5118):203–6. Epub 1993/07/09. PubMed PMID: 8327891.

11. Ulanovsky L, Bodner M, Trifonov EN, Choder M. Curved DNA: design, synthesis, and circularization. *Proc Natl Acad Sci U S A.* 1986;83(4):862–6. Epub 1986/02/01. PubMed PMID: 3456570; PMCID: PMC322970.

12. Kotlarz D, Fritsch A, Buc H. Variations of intramolecular ligation rates allow the detection of protein-induced bends in DNA. *EMBO J.* 1986;5(4): 799–803. Epub 1986/04/01. PubMed PMID: 3011427; PMCID: PMC1166861.

13. Tolhuis B, Palstra RJ, Splinter E, Grosveld F, de Laat W. Looping and interaction between hypersensitive sites in the active beta-globin locus. *Mol Cell.* 2002;10(6):1453–65.

14. Carter D, Chakalova L, Osborne CS, Dai Y-f, Fraser P. Long-range chromatin regulatory interactions in vivo. *Nat Genet.* 2002;32(4):623–6.

15. Palstra RJ, Tolhuis B, Splinter E, Nijmeijer R, Grosveld F, de Laat W. The beta-globin nuclear compartment in development and erythroid differentiation. *Nat Genet.* 2003;35(2):190–4. PubMed PMID: 14517543.

16. Drissen R, Palstra RJ, Gillemans N, Splinter E, Grosveld F, Philipsen S, de Laat W. The active spatial organization of the beta-globin locus requires the transcription factor EKLF. *Genes Dev.* 2004;18(20):2485–90.

17. Splinter E, Heath H, Kooren J, Palstra RJ, Klous P, Grosveld F, Galjart N, de Laat W. CTCF mediates long-range chromatin looping and local histone modification in the beta-globin locus. *Genes Dev.* 2006;20(17):2349–54.

18. Vernimmen D, De Gobbi M, Sloane-Stanley JA, Wood WG, Higgs DR. Long-range chromosomal interactions regulate the timing of the transition between poised and active gene expression. *EMBO J.* 2007;26(8):2041–51.

19. Vakoc CR, Letting DL, Gheldof N, Sawado T, Bender MA, Groudine M, Weiss MJ, Dekker J, Blobel GA. Proximity among distant regulatory elements at the beta-globin locus requires GATA-1 and FOG-1. *Mol Cell.* 2005;17(3):453–62.

20. Simonis M, Klous P, Splinter E, Moshkin Y, Willemsen R, de Wit E, van Steensel B, de Laat W. Nuclear organization of active and inactive chromatin domains uncovered by chromosome conformation capture-on-chip (4C). *Nat Genet.* 2006;38(11):1348–54.

21. Zhao Z, Tavoosidana G, Sjölinder M, Göndör A, Mariano P, Wang S, Kanduri C, Lezcano M, Sandhu KS, Singh U, Pant V, Tiwari V, Kurukuti S, Ohlsson R. Circular chromosome conformation capture (4C) uncovers extensive networks of epigenetically regulated intra- and interchromo-somal interactions. *Nat Genet.* 2006;38(11):1341–7.

22. Dostie J, Richmond TA, Arnaout RA, Selzer RR, Lee WL, Honan TA, Rubio EH, Krumm A, Lamb J, Nusbaum C, Green RD, Dekker J. Chromosome conformation capture carbon copy (5C): A massively parallel solution for mapping interactions between genomic elements. *Genome Res.* 2006;16(10):1299–309.

23. Lieberman-Aiden E, van Berkum NL, Williams L, Imakaev M, Ragoczy T, Telling A, Amit I, Lajoie BR, Sabo PJ, Dorschner MO, Sandstrom R, Bernstein B, Bender MA, Groudine M, Gnirke A, Stamatoyannopoulos JA, Mirny L, Lander ES, Dekker J. Comprehensive mapping of long-range interactions reveals folding principles of the human genome. *Science.* 2009;326(5950):289–93.

24. Rao SSP, Huntley MH, Durand NC, Stamenova EK, Bochkov ID, Robinson JT, Sanborn AL, Machol I, Omer AD, Lander ES, Lieberman Aiden E. A 3D map of the human genome at kilobase resolution reveals principles of chromatin looping. *Cell.* 2014;159(7):1665–80.

25. Bonev B, Mendelson Cohen N, Szabo Q, Fritsch L, Papadopoulos GL, Lubling Y, Xu X, Lv X, Hugnot JP, Tanay A, Cavalli G. Multiscale 3D genome rewiring during mouse neural development. *Cell.* 2017;171(3):557–72 e24. Epub 2017/10/21. doi:10.1016/j.cell.2017.09.043. PubMed PMID: 29053968; PMCID: PMC5651218.

26. Fullwood MJ, Liu MH, Pan YF, Liu J, Xu H, Mohamed YB, Orlov YL, Velkov S, Ho A, Mei PH, Chew EG, Huang PY, Welboren WJ, Han Y, Ooi HS, Ariyaratne PN, Vega VB, Luo Y, Tan PY, Choy PY, Wansa KD, Zhao B, Lim KS, Leow SC, Yow JS, Joseph R, Li H, Desai KV, Thomsen JS, Lee YK, Karuturi RK, Herve T, Bourque G, Stunnenberg HG, Ruan X, Cacheux-Rataboul V, Sung WK, Liu ET, Wei CL, Cheung E, Ruan Y. An oestrogen-receptor-alpha-bound human chromatin interactome. *Nature.* 2009;462(7269):58–64.

27. Mumbach MR, Rubin AJ, Flynn RA, Dai C, Khavari PA, Greenleaf WJ, Chang HY. HiChIP: Efficient and sensitive analysis of protein-directed genome architecture. *Nat Methods.* 2016;13(11):919–22. Epub 2016/11/01. doi:10.1038/nmeth.3999. PubMed PMID: 27643841; PMCID: PMC5501173.

28. Lareau CA, Aryee MJ. hichipper: a preprocessing pipeline for calling DNA loops from HiChIP data. *Nat Methods.* 2018;15(3):155–6. Epub 2018/03/01. doi:10.1038/nmeth.4583. PubMed PMID: 29489746.

29. Phanstiel DH, Boyle AP, Heidari N, Snyder MP. Mango: A bias-correcting ChIA-PET analysis pipeline. *Bioinformatics*. 2015;31(19):3092–8. Epub 2015/06/03. doi:10.1093/bioinformatics/btv336. PubMed PMID: 26034063; PMCID: PMC4592333.

30. Hughes JR, Roberts N, McGowan S, Hay D, Giannoulatou E, Lynch M, De Gobbi M, Taylor S, Gibbons R, Higgs DR. Analysis of hundreds of cis-regulatory landscapes at high resolution in a single, high-throughput experiment. *Nat Genet*. 2014;46(2):205–12.

31. Schoenfelder S, Furlan-Magaril M, Mifsud B, Tavares-Cadete F, Sugar R, Javierre BM, Nagano T, Katsman Y, Sakthidevi M, Wingett SW, Dimitrova E, Dimond A, Edelman LB, Elderkin S, Tabbada K, Darbo E, Andrews S, Herman B, Higgs A, LeProust E, Osborne CS, Mitchell JA, Luscombe NM, Fraser P. The pluripotent regulatory circuitry connecting promoters to their long-range interacting elements. *Genome Res*. 2015;25(4):582–97.

32. Imakaev M, Fudenberg G, McCord RP, Naumova N, Goloborodko A, Lajoie BR, Dekker J, Mirny LA. Iterative correction of Hi-C data reveals hallmarks of chromosome organization. *Nature Methods*. 2012;9(10):999–1003.

33. Yaffe E, Tanay A. Probabilistic modeling of Hi-C contact maps eliminates systematic biases to characterize global chromosomal architecture. *Nat Genet*. 2011;43(11):1059–65.

34. Nagano T, Lubling Y, Stevens TJ, Schoenfelder S, Yaffe E, Dean W, Laue ED, Tanay A, Fraser P. Single-cell Hi-C reveals cell-to-cell variability in chromosome structure. *Nature*. 2013;502(7469):59–64.

35. Ramani V, Deng X, Qiu R, Gunderson KL, Steemers FJ, Disteche CM, Noble WS, Duan Z, Shendure J. Massively multiplex single-cell Hi-C. *Nat Methods*. 2017;14(3):263–6.

36. Flyamer IM, Gassler J, Imakaev M, Brandao HB, Ulianov SV, Abdennur N, Razin SV, Mirny LA, Tachibana-Konwalski K. Single-nucleus Hi-C reveals unique chromatin reorganization at oocyte-to-zygote transition. *Nature*. 2017;544(7648):110–4. Epub 2017/03/30. doi:10.1038/nature21711. PubMed PMID: 28355183; PMCID: PMC5639698.

37. Denker A, de Laat W. The second decade of 3C technologies: detailed insights into nuclear organization. *Genes Dev*. 2016;30(12):1357–82.

38. Nora EP, Lajoie BR, Schulz EG, Giorgetti L, Okamoto I, Servant N, Piolot T, van Berkum NL, Meisig J, Sedat J, Gribnau J, Barillot E, Blüthgen N, Dekker J, Heard E. Spatial partitioning of the regulatory landscape of the X-inactivation centre. *Nature*. 2012;485(7398):381–5.

39. Dixon JR, Selvaraj S, Yue F, Kim A, Li Y, Shen Y, Hu M, Liu JS, Ren B. Topological domains in mammalian genomes identified by analysis of chromatin interactions. *Nature*. 2012;485(7398):376–80.

40. Symmons O, Uslu VV, Tsujimura T, Ruf S, Nassari S, Schwarzer W, Ettwiller L, Spitz F. Functional and topological characteristics of mammalian regulatory domains. *Genome Res*. 2014;24(3):390–400.

41. Lupiáñez DG, Kraft K, Heinrich V, Krawitz P, Brancati F, Klopocki E, Horn D, Kayserili H, Opitz JM, Laxova R, Santos-Simarro F, Gilbert-Dussardier B, Wittler L, Borschiwer M, Haas SA, Osterwalder M, Franke M, Timmermann B, Hecht J, Spielmann M, Visel A, Mundlos S. Disruptions of topological chromatin domains cause pathogenic rewiring of gene-enhancer interactions. *Cell.* 2015;161(5):1012–25.

42. Sanyal A, Lajoie BR, Jain G, Dekker J. The long-range interaction land-scape of gene promoters. *Nature.* 2012;489(7414):109–13.

43. Li G, Ruan X, Auerbach RK, Sandhu KS, Zheng M, Wang P, Poh HM, Goh Y, Lim J, Zhang J, Sim HS, Peh SQ, Mulawadi FH, Ong CT, Orlov YL, Hong S, Zhang Z, Landt S, Raha D, Euskirchen G, Wei CL, Ge W, Wang H, Davis C, Fisher-Aylor KI, Mortazavi A, Gerstein M, Gingeras T, Wold B, Sun Y, Fullwood MJ, Cheung E, Liu E, Sung WK, Snyder M, Ruan Y. Extensive promoter-centered chromatin interactions provide a topological basis for transcription regulation. *Cell.* 2012;148(1–2): 84–98.

44. Naumova N, Imakaev M, Fudenberg G, Zhan Y, Lajoie BR, Mirny LA, Dekker J. Organization of the mitotic chromosome. *Science.* 2013;342(6161):948–53.

45. Gibcus JH, Samejima K, Goloborodko A, Samejima I, Naumova N, Nuebler J, Kanemaki MT, Xie L, Paulson JR, Earnshaw WC, Mirny LA, Dekker J. A pathway for mitotic chromosome formation. *Science.* 2018;359(6376): ii.

46. Fudenberg G, Imakaev M, Lu C, Goloborodko A, Abdennur N, Mirny LA. Formation of chromosomal domains by loop extrusion. *Cell Rep.* 2016;15(9):2038–49.

47. Fudenberg G, Mirny LA. Higher-order chromatin structure: bridging physics and biology. *Curr Opin Genet Dev.* 2012;22(2):115–24.

48. Rippe K. Making contacts on a nucleic acid polymer. *Trends Biochem Sci.* 2001;26(12):733–40.

49. Doyle B, Fudenberg G, Imakaev M, Mirny LA. Chromatin loops as allo-steric modulators of enhancer-promoter interactions. *PLoS Comput Biol.* 2014;10(10):e1003867.

50. Sanborn AL, Rao SS, Huang SC, Durand NC, Huntley MH, Jewett AI, Bochkov ID, Chinnappan D, Cutkosky A, Li J, Geeting KP, Gnirke A, Melnikov A, McKenna D, Stamenova EK, Lander ES, Aiden EL. Chromatin extrusion explains key features of loop and domain forma-tion in wild-type and engineered genomes. *Proc Natl Acad Sci U S A* 2015;112(47):E6456–65.

51. Jost D, Carrivain P, Cavalli G, Vaillant C. Modeling epigenome folding: Formation and dynamics of topologically associated chromatin domains. *Nucleic Acids Res.* 2014;42(15):9553–61. doi:10.1093/nar/gku698. PubMed PMID: 25092923; PMCID: 4150797.

52. Falk M, Feodorova Y, Naumova N, Imakaev M, Lajoie BR, Leonhardt H, Joffe B, Dekker J, Fudenberg G, Solovei I, Mirny LA. Heterochromatin drives organization of conventional and inverted nuclei. *BioRxiv* 2018; 244038. doi.org/10.1101/244038.

53. Schalbetter SA, Goloborodko A, Fudenberg G, Belton JM, Miles C, Yu M, Dekker J, Mirny L, Baxter J. SMC complexes differentially compact mitotic chromosomes according to genomic context. *Nat Cell Biol.* 2017;19(9):1071–80. Epub 2017/08/22. doi:10.1038/ncb3594. PubMed PMID: 28825700; PMCID: PMC5640152.

54. Baù D, Sanyal A, Lajoie BR, Capriotti E, Byron M, Lawrence JB, Dekker J, Marti-Renom MA. The three-dimensional folding of the alpha-globin gene domain reveals formation of chromatin globules. *Nat Struct Mol Biol.* 2011;18(1):107–14.

55. Marti-Renom MA, Mirny LA. Bridging the resolution gap in structural modeling of 3D genome organization. *PLoS Comput Biol.* 2011;7(7):e1002125. Epub 2011/07/23. PubMed PMID: 21779160; PMCID: 3136432.

56. Giorgetti L, Galupa R, Nora EP, Piolot T, Lam F, Dekker J, Tiana G, Heard E. Predictive polymer modeling reveals coupled fluctuations in chromosome conformation and transcription. *Cell.* 2013;157(4):950–63.

57. Riggs AD. DNA methylation and late replication probably aid cell memory, and type I DNA reeling could aid chromosome folding and enhancer function. *Philos Trans R Soc Lond B Biol Sci.* 1990;326(1235):285–97. Epub 1990/01/30. PubMed PMID: 1968665.

58. Nasmyth K. Disseminating the genome: joining, resolving, and separating sister chromatids during mitosis and meiosis. *Annu Rev Genet.* 2001;35:673–745.

59. Alipour E, Marko JF. Self-organization of domain structures by DNA-loop-extruding enzymes. *Nucleic Acids Res.* 2012;40(22):11202–12.

60. Wang X, Le TBK, Lajoie BR, Dekker J, Laub MT, Rudner DZ. Condensin promotes the juxtaposition of DNA flanking its loading site in Bacillus subtilis. *Genes Dev.* 2015;29(15):1661–75.

61. Nasmyth K. Segregating sister genomes: the molecular biology of chromosome separation. *Science.* 2002;297(5581):559–65.

62. Gibcus JH, Kumiko Samejima K, Goloborodko A, Samejima I, Naumova N, Kanemaki M, Xie L, Paulson JR, Earnshaw WC, Mirny LA, Dekker J. Mitotic chromosomes fold by condensin-dependent helical winding of chromatin loop arrays. *bioRxiv* 2017; 174649. doi: https://doiorg/101101/174649.

63. Goloborodko A, Marko JF, Mirny LA. Mitotic chromosome compaction via active loop extrusion. bioRxiv. 2015; doi: http://dx.doi.org/10.1101/021642.

64. Larson AG, Elnatan D, Keenen MM, Trnka MJ, Johnston JB, Burlingame AL, Agard DA, Redding S, Narlikar GJ. Liquid droplet formation by HP1alpha suggests a role for phase separation in heterochromatin. *Nature.* 2017;547(7662):236–40. Epub 2017/06/22. doi:10.1038/nature22822. PubMed PMID: 28636604; PMCID: PMC5606208.

65. Strom AR, Emelyanov AV, Mir M, Fyodorov DV, Darzacq X, Karpen GH. Phase separation drives heterochromatin domain formation. *Nature.* 2017;547(7662):241–5. Epub 2017/06/22. doi:10.1038/nature22989. PubMed PMID: 28636597.

66. Nora EP, Goloborodko A, Valton AL, Gibcus JH, Uebersohn A, Abdennur N, Dekker J, Mirny LA, Bruneau BG. Targeted degradation of CTCF decouples local insulation of chromosome domains from genomic compartmentalization. *Cell*. 2017;169(5):930–44 e22. Epub 2017/05/20. doi:10.1016/j.cell.2017.05.004. PubMed PMID: 28525758; PMCID: PMC5538188.

67. Schwarzer W, Abdennur N, Goloborodko A, Pekowska A, Fudenberg G, Loe-Mie Y, Fonseca NA, Huber W, Haering CH, Mirny L, Spitz F. Two independent modes of chromatin organization revealed by cohesin removal. *Nature*. 2017;551(7678):51–6. Epub 2017/11/03. doi:10.1038/nature24281. PubMed PMID: 29094699; PMCID: PMC5687303.

68. Dekker J, Belmont AS, Guttman M, Leshyk VO, Lis JT, Lomvardas S, Mirny LA, O'Shea CC, Park PJ, Ren B, Politz JCR, Shendure J, Zhong S, Network DN. The 4D nucleome project. *Nature*. 2017;549(7671):219–26. Epub 2017/09/15. doi:10.1038/nature23884. PubMed PMID: 28905911; PMCID: PMC5617335.

PART 1

First-Principles Models

Modeling the Functional Coupling between 3D Chromatin Organization and Epigenome

CÉDRIC VAILLANT AND DANIEL JOST

2.1 INTRODUCTION

Proper 3D organization and dynamics play essential roles in the operation of many biological processes and are involved in functions as varied as enzymatic activity by well-folded proteins, cell motility generated by architected dynamical cytoskeletons or organ formation by spatially and temporally controlled gene expressions. Although acting on various scales ranging from molecules to organisms, the regulation of such structural and dynamical properties mainly originates from molecular mechanisms. How such microscopic actions are coupled to collectively generate large-scale functions is a long-standing, open, question.

In this context, understanding how the genome self-organizes inside the cell nucleus is one of the major challenges faced in recent years by biology. Thanks to the development of new experimental techniques, especially chromosome conformation capture (Hi-C) technologies [3], supported by parallel confocal and superresolution microscopy studies [4], major progress has been realized in our understanding of the hierarchical chromosome organization: from the local packaging of DNA into a polymer-like chromatin fiber to the large-scale compartmentalization of transcriptionally active or inactive genomic regions (Figure 2.1). Briefly (see Chapter 1 of the present book for a detailed review), a chromosome is locally partitioned into conserved consecutive 200 nm-sized contact domains, the so-called topologically associating domains (TADs), representing the partial folding of kilobasepair (kbp) to megabasepair (Mbp) long genomic regions [5–7]. TADs are defined as highly self-contacting portions of the genome: a sequence inside a TAD has a higher probability to contact sequences inside the same TAD than sequences in neighboring TADs at the same linear distance along the genome, thereby segmenting chromosomes into 3D domains. At the Mbp level, contact maps display a cell-type-specific checkerboard pattern: parts of the genome that share the same transcriptional activity tend to colocalize forming nuclear compartments [1, 7–9], quantifying the older qualitative observations of nuclear organization made by electron microscopy [10]. Inactive regions, the so-called heterochromatin, are preferentially localized at the nuclear periphery while active regions, the so-called euchromatin, occupy more central positions. Reversely, genomic regions that preferentially localize close to the nuclear membrane, the so-called lamina-associated domains (LADs), are mainly heterochromatic [11–13]. At the nuclear level, Hi-C maps [8] confirm that chromosomes occupy distinct spatial territory and do not mix [14, 15]. While the vast majority of Hi-C data is obtained at the population level, very recent single-cell Hi-C experiments [16–19], complemented by superresolution microscopy [7, 20, 21], have highlighted the strong stochasticity of chromosome folding suggesting that chromatin is highly dynamical and plastic along the cell cycle and during differentiation. However, direct *in vivo* characterization of chromatin motion is still challenging [22]. Only a few studies have successfully tracked fluorescently labeled loci during relatively long periods of time up to a few minutes [23–28]. They complete the picture of a fluctuating organization whose dynamics is strongly dependent on transcriptional activity.

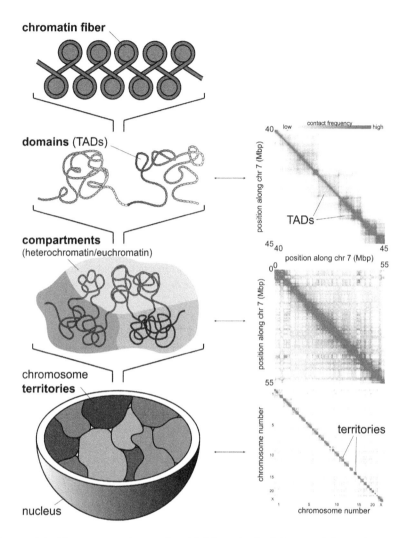

Figure 2.1 Chromosome hierarchical folding. (Left) Scheme of the multi-scale organization of chromosome during interphase. (Right) Hi-C maps for GM12878 cell line at different resolutions [1] plotted using Hi-C JuiceBox [2].

An increasing amount of experimental evidence suggests that genome 3D organization and dynamics adapt to nuclear functions and may play a decisive role in gene regulation and disease [29–31]. Most characterized promoter–enhancer interactions occur within the same TAD [32, 33], suggesting that TADs allow insulation of promoters from enhancers located in neighboring TADs. Disruption of a boundary between two consecutive TADs may cause gene misregulation leading to malformations or cancers [34, 35]. Current experimental knowledge has suggested several molecular mechanisms involved in the local and higher-order organization of the chromosome [36, 37]. Statistical positioning inside the nucleus and formation of active/inactive compartments are putatively driven by chromatin-binding proteins that are known to bind at specific positions along the genome and that have the capacity to self-interact [38–41] or to interact with membrane proteins [42]. TAD formation is partly associated with the translocation along the genome of protein complexes [43–45], cohesin, or condensin rings, that extrude chromatin loops and stop at specific, properly oriented sites where a known transcription factor, the insulator CTCF, binds [1, 46].

However, investigating experimentally how such molecular mechanisms precisely act and cooperate together to control the dynamics and 3D multiscale folding of the genome is very challenging and is limited by the experimental difficulty to capture the dynamical stochastic evolution of chromosomes. In the recent years, to partly circumvent such limitations, physical models have been instrumental in simulating chromosome folding and in testing different molecular mechanisms (see [36, 47–50] for reviews and other chapters in the present book). In this chapter, we review our current efforts to understand the functional coupling between the 3D dynamical organization of chromatin and the 1D segmentation of genome into active and inactive domains using polymer and statistical physics modeling.

2.2 3D CHROMATIN ORGANIZATION AND EPIGENOMICS

All the cells of a multi-cellular organism contain the same genetic information but may have different shapes, physiologies, metabolisms, or functions depending on the cell types, tissues, environments, or differentiation stages. These differences are mainly due to the context-dependent differential regulation of gene expression. Gene expression is regulated at various levels from the binding of transcription factors to the post-translational modifications of the synthesized proteins. Among these different layers of regulation, the modulation of accessibility and specificity of regulators to their cognate DNA sites plays a central role. Locally, the chromatin is characterized by many features like nucleosome positioning, biochemical modifications of DNA, and histones tails or the insertion of histone variants, that contribute significantly to controlling such modulation. In the past decades, advances in sequencing technologies have allowed the detailed characterization of the genomic profiles of various histone modifications or chromatin-binding proteins, shedding light on the association between these so-called epigenomic marks and gene regulation. In many eukaryotes, from yeast to human [51–54], statistical analyses of these patterns along the genome showed that chromosomes are linearly partitioned into 1D cell-type-specific epigenomic domains that extend from few kilobases to megabases and are characterized by the local enrichment of specific epigenomic marks. While based on dozens of profiles, these studies have identified only a small number of main chromatin types for the epigenomic domains (typically four to ten, depending on the resolution): (1) euchromatic states, containing constitutively expressed or activated genes and enhancers; heterochromatic states covering (2) constitutive heterochromatin associated with HP1 proteins and H3K9me3 marks and mainly found in repetitive sequences such as (peri)centromeres, (sub)telomeres, or transposable elements; (3) facultative heterochromatin associated with Polycomb (PcG) complexes and H3K27me3 mark tagging developmentally regulated silent genes; and (4) a less epigenomically defined repressive state, the so-called black or null or quiescent chromatin, that encompasses gene desert, or genes only expressed in few tissues. Typically, in higher eukaryotes, $\sim$20–30% of mappable genomic loci (excluding telomeres and centromeres) correspond to active states, $\sim$5–10% HP1-like states, $\sim$10–20% PcG-like states, and $\sim$40–50% quiescent states, the exact repartition depends on organisms and cell types [55].

From the early studies of nuclear organization made by conventional or electron microscopy [10], it was clear that the active and inactive parts of the genome phase-segregate into (micro) compartments, with heterochromatin localizing mainly at the nuclear periphery and around nucleoli and euchromatin being more internal. Recent developments in Hi-C and superresolution techniques have allowed us to quantify the relation between spatial organization and epigenomics in more detail [1]. At large-scale, for a given cell-type, statistical analyses of specific checkerboard patterns observed in Hi-C maps (Figure 2.1) showed that genomic loci can be clustered into two groups, the so-called A and B compartments [8, 56]: the contact frequency from sequences of the same group (A vs. A or B vs. B) is stronger ($\sim$2 fold) than from sequences in different ones (A vs. B). Genomic regions corresponding to A compartment are gene-rich and are associated with histone marks specific to active genes. In contrast, loci belonging to B compartment harbor a weak gene density and contain more repressed histone modifications. These compartments can be subdivided into subgroups that exhibit peculiar contact patterns and that correspond to different epigenomic states [1]. Reciprocally, epigenomic domains (as defined above) contact domains of the same chromatin type more frequently than domains with different states [57]. Recent single-cell Hi-C experiments [17–19] and high-resolution imaging of multiple probes on the same chromosome [20, 21] have confirmed that loci sharing the same epigenomic content tend to colocalize inside the nucleus. Altogether, these observations demonstrate the large-scale clustering of functionally similar genomic loci.

At the sub-Mbp scale, TADs are also significantly correlated with epigenomic domains [51, 55, 58] (Figure 2.2). In *Drosophila*, positioning of TADs along the genome displays strong similarity with the locations of epigenomic domains [55, 58, 59]: loci within the same TAD tend to have the same chromatin state (Figure 2.2, left), boundaries between TADs are rich in active marks, and the large-scale checkerboard pattern emerges from long-range interactions between TADs of the same chromatin type [56]. Recent superresolution microscopy of individual TADs showed that the epigenomic state also impacts the local 3D chromatin compaction: active TADs being less compact than black/quiescent and PcG-associated TADs [7, 21], confirming the observations that more Hi-C contacts are observed in inactive domains [60]. All this suggests that, in *Drosophila*, TAD formation is strongly associated with epigenomic domains. In mammals, TADs are also significantly associated with the local chromatin state [51, 55] even if the correspondence between TAD and epigenomic segmentations

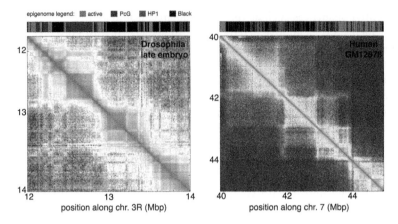

Figure 2.2 Coupling between epigenome and contactome. (Left) Hi-C map of a 2 Mbp-long genomic region of *Drosophila* chromosome arm 3R, obtained from late embryos [56]. On top, we plot the local epigenomic state (see color legend) as obtained by Fillion et al [52] for the embryonic cell line Kc167 (for simplicity we merged the two originally defined active states into one single active state). (Right) Hi-C map for a 5 Mbp-long genomic region of human chromosome 7 obtained for the GM12878 cell line [1]. Epigenomic states were taken from Ho et al [51]. For simplicity, we clustered the 16 originally defined states into the 4 standard chromatin types.

is less clear (Figure 2.2 right). TAD boundaries are mainly characterized by the binding of insulator proteins like CTCF and do not necessarily reflect the frontiers between different epigenomic domains. Recent experimental and modeling works suggest that, in mammals, TADs might emerge from the coupled action of CTCF-cohesin-mediated mechanism (see the presentation of the loop extrusion model in [43, 44] and in Chapter 4 of the present book) and of epigenomically associated mechanism as in *Drosophila* [61, 62].

Altogether, these results highlight the strong interplay between the 1D segmentation of the genome into epigenomic domains, the so-called epigenome, and the 3D compartmentalization of chromosomes into contact domains, the so-called "contactome". This crosstalk is now well documented and has inspired numerous statistical works inferring various 3D organization features like TAD or compartments from epigenomic data [58, 63, 64] or using the 3D contact information to better understand various aspects of gene regulation [65]. However, the *mechanistic* foundations of such coupling are still unclear. In particular, to what extent epigenomically associated mechanisms drive chromosome organization? What is the role of this non-random 3D organization in the establishment and maintenance of stable epigenomic information?

In the next section, we will present and discuss how we addressed the former question using polymer physics in the context of chromatin folding in *Drosophila* (Section 2.3) and how we formalized the latter question with theoretical modeling (Section 2.4).

2.3 EPIGENOME-DRIVEN PHASE SEPARATION OF CHROMATIN

The observed correlations between epigenome and contactome suggest the existence of epigenomic-specific mechanisms playing major roles in chromatin folding. Actually, there is an increasing amount of experimental evidences showing that chromatin-binding proteins associated with specific epigenomic domains possess the molecular capacity to interact or oligomerize, hence promoting directly or effectively physical bridging between genomic loci of the same chromatin type. Indeed, heterochromatin-associated factors like PcG or HP1 display structural domains (respectively, sterile alpha motif (SAM) domains or chromodomains (CD)) that may favor multimerization [39, 38]. In particular, very recent experiments have shown that human and *Drosophila* HP-1 can self-interact, leading eventually to a liquid-like phase separation *in vitro*, in the absence of chromatin, and to the formation of *in vivo* heterochromatic compartments [40, 41]. Similarly, mutualization of transcription machinery resources or DNA looping mediated by promoter–enhancer interactions may also lead to effective attractions between active loci [66–68]. Black/quiescent chromatin is often associated with lamins or is enriched in histone H1 that may also promote binding. In addition, *in vitro* experiments have demonstrated that two nucleosomes may interact directly and that such interactions are sensitive to biochemical modifications of histone tails [69, 70].

All this suggests that the heterochromatin/euchromatin phase separation is driven by specific short-range interactions mediated by epigenomic markers like histone modifications or chromatin-binding proteins.

2.3.1 Block copolymer model

To formalize and test this hypothesis, we developed a general framework by treating chromatin as a block copolymer (Figure 2.3), where each block corresponds to an epigenomic domain and where each monomer interacts preferentially with other monomers of the same chromatin type. While being generic, we focused our approach on chromatin folding in *Drosophila* where the coupling between epigenome and contactome is very strong. Similar approaches have also been applied to mammals by other groups and will be discussed in Section 2.3.6.

More specifically, we modeled chromatin as a semi-flexible, self-avoiding, self-interacting polymer [60, 71–73]. A chain corresponding to a given genomic region is composed of N monomers, each representing n bp. Each bead m is characterized by its epigenomic state $e(m)$. We limit our analysis to the four major classes of chromatin state described above (active, PcG, HP-1, and black/quiescent). A long epigenomic domain will thus be represented by a block of consecutive monomers all sharing the same state. Beads of the same epigenomic state may specifically interact via short-range, transient interactions. The full dynamics of the chain is then governed by two contributions: (i) bending rigidity and excluded volume describing the "null" model of the chain, and (ii) epigenomics-mediated attractive interactions.

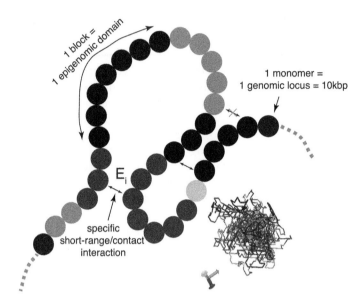

Figure 2.3 Block copolymer model. Each monomer represents a given genomic locus. One block corresponds to one epigenomic domain. Pairwise interactions between monomers depend on the local epigenomic state. (Bottom right corner) Snapshot taken from a kinetic Monte-Carlo simulation of the block copolymer model of *Drosophila* chromosome 3R.

By definition, this model belongs to the generic family of block copolymers. In the past decades, this wide class of models has been extensively studied in physics and chemistry, mainly to characterize the phase diagram of melts of short synthetic chains composed by a few blocks arranged either periodically or randomly [74]. However, the properties of such framework applied to long polymers (the chromosomes) with many blocks of various sizes (the epigenomic domains) are poorly characterized.

2.3.2 Simulation methods

In recent years, we have developed several methods to investigate the behavior of the block copolymer model of chromatin. From the self-consistent Gaussian approximations allowing efficient access to the steady-state behavior of short chains [71, 72] to more detailed numerical simulations of chain dynamics [60, 73]. In this chapter, we will focus on our most recent results using simulations of long chains, recapitulating all our previous findings.

The polymer is modeled as a self-avoiding walk on a Face Centered Cubic (FCC) lattice to allow maximal coordination number (= 12). The energy of a given configuration is given by

$$H = \frac{\kappa}{2} \sum_{m=1}^{N-1} \left(1 - \cos\theta_m\right) + \sum_{l,m} U_{e(l),e(m)} \delta_{l,m}. \tag{2.1}$$

The first contribution accounts for the local stiffness of the chain with κ the bending rigidity and θ_m the angle between bond vectors m and $m+1$. The second contribution accounts for epigenomic-driven interactions with $\delta_{l,m}=1$ if monomers l and m occupy nearest-neighbor (NN) sites on the lattice ($\delta_{l,m}=0$ otherwise), and $U_{e,e'}$ the strength of interaction between a pair of spatially neighboring beads of chromatin states e and e'. For simplicity, we will assume that interactions occur only between monomers of the same chromatin state ($U_{e,e'}=0$ if $e \neq e'$) and that the strength of interaction (that we note E_i) is the same whatever the chromatin state ($U_{e,e} \equiv E_i$ for all e). Confinement and effect of other chains are accounted for by using periodic boundary conditions. The dynamics of the chain follow a kinetic Monte-Carlo (KMC) scheme with local moves developed by Hugouvieux and coworkers [75]. This scheme allows at most two monomers to occupy the same lattice site, but only if they are consecutive along the chain. One Monte Carlo step (MCS) consists of N trial moves where a monomer is randomly chosen and displaced to a nearest-neighbor site on the lattice. Trial moves are accepted according to a Metropolis criterion applied to H and if the chain connectivity is maintained and the self-avoidance criterion is not violated. These simple rules allow efficient simulations of reptation motion in dense – topologically constrained – systems, while still accounting for the main characteristics of polymer dynamics like polymer connectivity, excluded volume, and non-crossability of polymer strands. More details on the lattice model and KMC scheme can be found in [60, 73, 75].

As explained in Chapter 7 of the present book, chromosomes are intrinsically long, topologically constrained – so-called crumpled polymers. These constraints have a strong impact on the dynamics of the chain and lead to peculiar structural and dynamical scalings [50, 76, 77] different from classical Rouse or worm-like chain models [78]. Recently, we derived a coarse-graining strategy [73] that accounts properly for this regime and establishes an intelligible method to fix some model parameters (bending rigidity and number of sites in the simulation box) at a desired resolution. This strategy allows simulation of long chromatin fragments ($N \times n \approx 20$ Mbp) with high numerical efficiency while conserving the structural and dynamical properties of the chain emerging from steric entanglement [73]. In the next section, we will describe the results obtained at a genomic resolution of $n = 10$ kbp and a spatial resolution of ~100 nm (the distance between NN sites on the lattice) which are both typical resolutions achieved in standard Hi-C and microscopy experiments.

For a given set of parameters, the time-unit in our simulations was determined by mapping the predicted time-evolution of the mean-squared displacement (MSD) of individual loci to the typical experimental relation: MSD (in μm^2) $\sim 0.01 t^{0.5}$ (with time t in seconds), observed in higher eukaryotes [23, 26, 27, 28]. For standard parameter values used in the next section, we found that 1 MCS, the temporal resolution of the model, corresponds to ~ 0.01–0.05 sec.

2.3.3 Phase diagram of the model: Towards (micro) phase separation

To illustrate the behavior of the model, we simulated the dynamical folding of a 20 Mbp-long region of *Drosophila* chromosome arm 3R (position 7–27 Mbp) for various values of E_i, the only free parameter of the model. Starting from random, compact, unknotted configurations resembling post-mitotic structures of chromosomes [79], we tracked, for thousands of different trajectories, the dynamical evolution of polymer conformations during 20 hours of "real" time, the typical duration of a cell cycle.

In Figure 2.4A, we plotted the predicted Hi-C maps for a population of unsynchronized cells as in standard Hi-C experiments, i.e., averaged over one cell cycle. At very weak interaction strengths, the polymer behaves as a (nearly) homogeneous chain driven mainly by steric interactions. It has the full characteristics of a crumpled polymer, as explained in detail in Chapter 7 of the present book. As $|E_i|$ is increased, the heteropolymeric nature of the system becomes apparent at the local and large scales. Locally, the contact probability between monomers of the same block increases (Figure 2.4A) and the spatial size of individual epigenomic domains (quantified by the square radius of gyration) decreases (Figure 2.4B, squares), leading to the formation of more or less compact TADs, depending on the strength of E_i and the linear size of the block (longer blocks being more compact at the same interaction strength, data not shown, see [60, 80]). Similarly, on a large-scale, long-range contact between TADs of the same chromatin type are enhanced and TADs of different types phase segregate, leading to a typical checkerboard pattern in predicted Hi-C maps. Structurally, as the strength of interaction augments, monomers of the same epigenomic state aggregate and form larger and more compact distinct 3D domains (Figure 2.4C, D). At high E_i values, this is characteristic of a microphase separation as typically observed in short block copolymer melts [74]. Interestingly, the formation of such large-scale compartmentalization has a strong impact on the local organization. Indeed, the compaction of individual TADs is significantly lower in the presence of long-range contacts than in situations where we only authorize the internal folding of epigenomic domains (circles in Figure 2.4B): in partial or full (micro) phase separation, TADs of the same chromatin type dynamically merge into big 3D clusters allowing conformations of an individual epigenomic block to be more expanded. Such property also explains why, for similar block sizes, the PcG domain in Figure 2.4B (blue squares) is more compact than the active (red) and the black domains: In *Drosophila*, large PcG domains are mainly far from each other along the genome, hence very close to the isolated case, while active and black domains are surrounded by many more domains of the same type.

2.3.4 Comparison to experiments

At each investigated value of E_i, we computed the Pearson correlation between the predicted contact map and the corresponding experimental data obtained by Sexton *et al.* [56] on late *Drosophila* embryos. The correlation was maximal (0.86) for $E_i = -0.1k_B T$. Figure 2.5A illustrates the very good agreement between both maps at the TAD and Mbp levels. For the predicted and experimental maps, we computed the scores on the first principal component of the normalized contact frequency matrix $\bar{C}$ defined as $\bar{C}(l,m) = C(l,m) / P_c(|l-m|)$ with $C(l,m)$ the

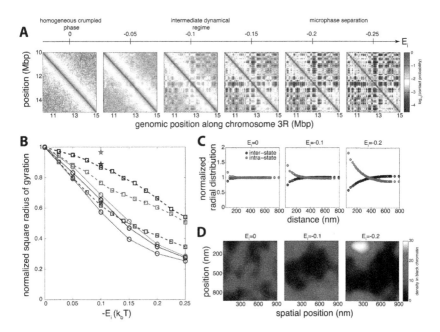

Figure 2.4 Phase diagram of the block copolymer model. (A) Predicted Hi-C maps for a 20 Mbp-long region of *Drosophila* chromosome 3R for increasing strengths of attraction E_i (in k_BT-unit). (B) Evolution of the square radius of gyration (defined as $1/(2N^2)\sum_{l,m}(r_l - r_m)^2$, an estimator of the average square 3D size of a domain) as a function of E_i for 3 large epigenomic domains (red squares: 1 active domain of size 280 kbp; blue squares: 1 PcG of size 330 kbp; black squares: 1 black/quiescent of size 290 kbp). Data were normalized by the corresponding values in the homogeneous case ($E_i = 0$). Circles correspond to situations where we authorized interactions *only* between monomers of the same epigenomic *domain* (no long-range interaction between TADs of the same state). Stars described the case where the specific interaction strength between active monomers was set to zero (PcG, HP1 and black monomers can still interact with monomers of the same type with $E_i = -0.1$). (C) Probability to find a monomer of the same (red circles) or different (black circles) epigenomic state at a given distance from a reference monomer (radial distribution), for three different values of E_i. Data were normalized by the corresponding probability to find a monomer of any state. (D) Typical examples of the volumic density in black monomers in a 2D slice of the simulation box, for three different values of E_i.

contact frequency between loci l and m, and $P_c(s)$ the average contact frequency between two loci separated by a genomic distance s. For one profile, loci with similar scores tend to belong to the same spatial A/B compartment [8, 57]. Both profiles (Figure 2.5A) are strongly correlated (Spearman correlation = 0.74) illustrating how well the checkerboard pattern is reproduced (positions and intensities) by the block copolymer model. Given the simplicity of the model, it is quite remarkable, suggesting that epigenomic-driven forces are the main players of the chromosome folding in *Drosophila*.

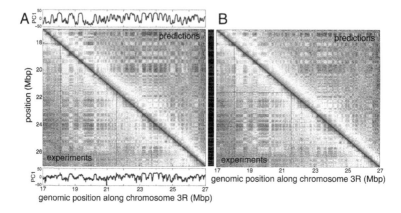

Figure 2.5 Comparison between experimental and predicted data.
(A) (Middle) Predicted ($E_i = -0.1k_bT$, upper triangular part) versus experimental (lower triangular part) Hi-C maps for a 10 Mbp region. Experimental data from [56]. Same color code as in Figure 2.4A. Experimental data divided by a factor 2500 to linearly adjust both scales. (Top, Bottom) A/B compartment analysis (see text) of the predicted (top) and experimental (bottom) Hi-C maps: loci with a negative (resp. positive) scores on the first principal component (PC1) belong to the A (resp. B) compartment. (B) Same as (A) but for the case where the specific interaction strength between active monomers was set to zero (PcG, HP1, and black monomers can still interact with monomers of the same type with $E_i = -0.1$).

Interestingly, experimental data located at an intermediate position in the phase diagram (Figure 2.4A) between the homogeneous – crumpled – phase and the full microphase separation. Interaction strength is weak, TADs are only partially collapsed (Figure 2.4B) and spatial compartments are dynamic and stochastic structures (see below) of typical size ~200–300 nm (Figure 2.4C,D). The model predicted that PcG domains are more compact than black domains, in qualitative agreement with recent measurements in flies of the radius of gyration [7] and of the end-to-end-distances of various epigenomic domains [21]. This means that the observed differences in compaction between PcG and black domains can be explained in a large part by differences in the linear organization of epigenomic blocks along the genome, and not necessarily by differences in interaction strength as stated in [7]. However, as it is, the model failed to predict that active domains are less compact than heterochromatin domains [7]. This discrepancy suggests that interactions between active monomers may be of less use or dispensable in describing chromatin folding in *Drosophila*. Figure 2.5B illustrated indeed that setting the interaction strength between active beads to zero while keeping $E_i = -0.1$ for the others, still allows to globally well describe the Hi-C map (Pearson correlation = 0.86, with a weak loss in phase-segregation) while improving predictions for the compaction of active domains that are now less compact than heterochromatic regions (stars in Figure 2.4B). This suggests that in *Drosophila*, the euchromatin/heterochromatin compartmentalization is mainly driven by the interactions between the dominant black/quiescent – heterochromatic – loci. The formation of the A (euchromatic) compartment is just a by-product of these direct interactions: small active regions are expelled at the periphery of the heterochromatic (micro)compartments leading also to preferential – effective – interactions between active sites.

Looking carefully at the predicted and experimental Hi-C maps, we observed, however, several discrepancies between both maps, suggesting missing ingredients in the model. For example, the model predicted spurious or missing TADs or long-range contacts between TADs. This could be due to an incorrect annotation of the local epigenomic state (we use epigenomic data from an embryonic cell line while Hi-C data were obtained on whole embryos) or the existence of specific interactions driven by other biological processes not accounted in the model, like promoter–enhancer interactions. Refining the model to account more precisely for the local epigenetic content (for example, for the relative levels of histone modifications or chromatin-binding proteins) or differences in the interaction strengths between different states would certainly lead to a better correspondence. We also observed that TADs are more sharply defined in the experiments, particularly in the corners of large TADs. This might be the results of pairing between homologous chromosomes, a phenomenon commonly found in Diptera [81] and not accounted for by the model, or of the presence of extra *cis*-interacting mechanisms, like the recently proposed loop extrusion model in mammals [43, 44] (see Chapter 4 of the present book), that enhance the contact frequencies along the genome.

2.3.5 A dynamical, out-of-equilibrium and stochastic organization

At an interaction strength compatible with biological data ($E_i = -0.1k_bT$), we analyzed the time evolution of chromosome organization. As in [76], we observed that chromatin folding results from the out-of-equilibrium decondensation of the polymeric chain from its initial compact configuration. Figure 2.6A shows $P_c(s)$, the average contact frequency between two loci separated by a genomic distance s, at different time points. Local scales, like the TAD level, reach a (quasi) steady-state within minutes while it takes longer for long-range contacts, ranging from dozens of minutes for Mbp-scale contacts to several hours at the 10 Mbp-scale. These predictions are consistent with experimental observations made on synchronized cells [17, 79] showing that TADs emerge very early in the cell cycle and that the large-scale A/B compartmentalization gradually increases along the cell cycle. Even after a long time (20 hours or more), the model predicts that the system is not at equilibrium, a regime where we should expect that $P_c(s)$ behaves as $\sim s^{-1.5}$ [77]. On contrary, due to strong topological constraints, the chains remain in a "crumpled", unknotted, confined state with $P_c(s) \sim s^{-1.1}$ for $s < 3$ Mbp (crumpling signature [50, 77, 82]) and $P_c(s) \sim s^{-0.5}$ for $s > 3$ Mbp (confinement in chromosome territory [79]), also consistent with Hi-C data [1, 8, 56].

Tracking of the relative distances between pairs of loci revealed that chromatin organization is very stochastic. Figure 2.6B shows three examples of the

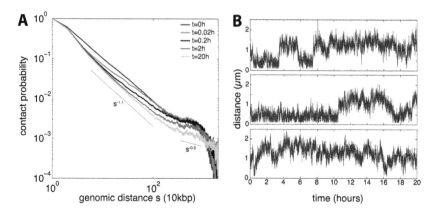

Figure 2.6 Dynamical chromatin folding. (A) Average contact probability between two loci as a function of their relative genomic distance along the genome predicted by the block copolymer model ($E_i = -0.1k_bT$) at different time points during one cell cycle. (B) Examples of the time evolution of the distance between two loci of the same epigenomic state separated by 3 Mbp along the genome.

time evolution along a one-cell cycle of such distance for the same pair of loci separated by 3 Mbp along the genome and having the same epigenomic state. We observed a typical two-state behavior with random transitions between a bound state where both loci remain in contact due to the merging of the TADs they belong to, and an unbound state where both TADs are spatially separated. Analysis of these trajectories for various pairs of loci showed that the transition rate from the unbound to the bound state is a decreasing function of the genomic distance between the two genomic regions, while the transition rate from the bound to the unbound state mainly depends on the respective epigenomic type of each locus, with pairs sharing the same type interacting last. Interestingly, we predicted that a significant proportion (5–15%) of long-range contacts (>1 Mbp) are not established within one cell cycle. This suggests that the genomic distance between regulatory elements, like promoters and enhancers, should not exceed 1 Mbp to ensure that a physical contact between these elements, prerequisite to an activation or repression event, for example, would happen at least once during a one-cell cycle in order to maintain a stable gene expression. It would be interesting to experimentally test such predictions by simultaneously tracking the spatial distance between a promoter and its enhancer and monitoring the current transcriptional activity [83], for various genomic distances between the two elements.

All this suggests that the 3D chromosome organization in higher eukaryotes is out-of-equilibrium, dynamical, and stochastic. This emphasizes the necessity (1) to properly account for the time evolution of such organization in quantitative models of chromosomes, especially for higher eukaryotes where chromosomes are strongly topologically constrained; and (2) to initiate the simulations with proper configurations since the system will keep a partial memory of the large-scale initial structure for a long period of time.

2.3.6 Relation to other approaches

The prediction of long-range interactions is inherent to copolymer models arranged in blocks. Therefore, such models should also be well adapted to describe the active/inactive compartmentalization in mammals. Several approaches have used similar formalisms to model chromatin folding in human or mouse [61, 84–94]. In particular, Wolynes and Onuchic [84, 85], Thirumalai and Hyeon [86, 87], Mirny [88], and Liberman-Aiden [61] have developed block copolymer models, eventually decorated with loop extrusion mechanisms or specific-pairwise interactions between CTCF sites at TAD boundaries. Nicodemi's [89–91] and Marenduzzo's [92–94] groups developed more detailed models accounting for the diffusion and binding of the proteins that mediate epigenomic interactions. Most of these approaches lead to very precise descriptions of chromosome organization and of heterochromatin/euchromatin phase separation in mammals. In many cases, their conclusions were very consistent with ours in *Drosophila*: Interaction strengths between genomic loci are weak leading to a mild (micro) phase separation and to very dynamical and stochastic organization.

This idea that the observed phase separation emerges from heterogeneities in the chromatin primary sequences, in analogy to the well-known physical behavior of synthetic block copolymers [74], is quite general and may arise from other possible mechanisms like active non-equilibrium processes or differences in monomer mobilities [95–97]. At a more phenomenological scale, such compartmentalization may also be interpreted as visco-elastic or liquid phase transitions [94, 98, 99] by using an effective phase-field formalism considering euchromatin and heterochromatin as separated fluids and neglecting the underlying polymeric structure.

2.4 ROLE OF 3D ORGANIZATION IN EPIGENOME STABILITY

As discussed before, the spatial organization of chromatin results in part from the clustering and phase-segregation of epigenomic domains but a still open question is whether this peculiar 3D folding is only a by-product of genome activity or if it is also participating in the regulation of the epigenome assembly and more generally in the regulation of the genome functions.

2.4.1 The "Nano-Reactor" hypothesis

The basic concept behind this structural/functional coupling is the augmentation of the local concentration of regulatory proteins due to spatial co-localization. In bacteria, this "high concentration" paradigm has been evidenced and formalized for many years for the well-known lac operon system [100, 101]:

Molecular crowding and spatial confinement increase the binding affinities of regulators (activators or repressors) to their DNA cognate sequences. This property is enhanced by the presence of a few additional dispersed recruitment sequences (operators) and the ability of the lac repressor to oligomerize, leading to DNA looping. Similarly, in eukaryotes, the nuclear chromatin compartments would correspond to biochemical "nano-reactors" where a small number of regulatory biomolecules are colocalized in space favoring their chemical (co-) activity. At the level of enhancer–promoter genomic modules, the distal action of enhancers is conditioned to their physical proximity with promoters [83]. The presence of different dispersed modules would increase the probability of the first contact between the promoter and one enhancer. The subsequent coalescence of the different modules would then provide both structural stability to the ensemble (i.e., increased duration of gene expression) and robustness and precision through the integration of different signals [102]. Along the same line, Polycomb-mediated repression involves the spatial colocalization of the silencer sequences (the so-called PREs) of several genes. This is mediated by the Polycomb protein complex that forms multi-loop structures, the so-called Polycomb bodies [103–105]. Such clustering operates in cis, i.e., within an epigenomic domain but also in trans between non-consecutive domains along the genome. For example, in Drosophila, strong long-range interactions are observed between the 10-Mbp distant, Polycomb-marked antennapedia (ANT-C) and bithorax (BX-C) domains [56, 106]. Similarly, in the yeast SIR-mediated heterochromatinization system, silencing of subtelomeric genes is associated with the level of SIR-mediated clustering [107]. Such clustering might enhance the local concentration of heterochromatin factors (the SIR proteins) at their telomeric specific recruiting sites and consequently might promote their spreading over the subtelomeric domains.

All this suggests that the spatial confinement of regulatory sequences (enhancers, silencers) may allow sequestering regulatory proteins in the spatial vicinity of the target genomic elements. TADs would correspond to insulated neighborhoods that provide a local, basal level of confinement and of selectivity that is then eventually finely tuned at a lower scale (via promoter–enhancer looping for example) [32, 46, 108, 109]. Similarly, the formation of A/B compartments would reinforce such properties for TADs sharing similar transcriptional activity or an epigenomic state.

2.4.2 Epigenomic 1D–3D positive feedback

In the context of epigenomics, the nano-reactor hypothesis naturally introduces a functional coupling between 3D organization and 1D epigenomic states. Indeed, locally, chromatin states are characterized by specific histone marks that favor the selective binding of regulatory proteins (e.g., PcG for H3K27me3, HP-1 of H3K9me3, or transcription factors for active marks) that can self-associate. Hence, the presence of these marks indirectly promotes 3D clustering and compartmentalization via the mechanisms discussed in 2.3. Moreover, these marks are dynamically deposited and removed by specific enzymatic complexes (e.g., PRC2 or Su(Var)3–0) that physically associate either with the mark they catalyze (e.g., H3K27me3 or H3K9me3) or with the corresponding regulatory proteins (e.g., PRC1 or HP1). This "reader–writer/eraser" property enables the mark and thus the chromatin state to spread, once nucleated, at some specific genomic loci. The crucial point is that spreading might not operate only *in cis*, i.e., unidimensionnally along the genome, but also *in trans* to any chromatin fragments in the spatial vicinity. This would introduce a positive feedback between the epigenomic state dynamics and the compaction of chromatin: Within a given domain the spatial clustering would enhance the "spreading" of the chromatin state over the entire domain (the nano-reactor hypothesis) which in return would enhance compaction (copolymer model).

The ability of enzymes to act *in trans* is clearly a working hypothesis that relies on the assumption that the mechanisms controlling *cis* spreading might also function *in trans*. The molecular processes involved *in trans* (and even *in cis*) spreading of an enzymatic activity to adjacent nucleosomes are still not well understood. Experimental studies on the heterochromatinization in fission yeast have shown that *cis* spreading was not due to allosteric changes of the involved enzymes but more likely to the favorable/stable spatial and orientational arrangement of the enzyme relative to the histone tails of adjacent nucleosomes [110]. Compact chromatin organization induced by architectural proteins such as HP1 or PRC1 might thus reinforce such *cis* activity [111]. Whether or not such a process is restricted to nucleosome *in cis* or can also apply to any spatially proximal nucleosome *in trans*, is unknown. Propagation of silencing *in trans* at the nucleosomal array scale has been evidenced in the Polycomb system [112] but a precise molecular description of this process remains to be elucidated. *In vitro* experiments similar to [110–112] with more extended engineered arrays of nucleosomes will be required for a better understanding of the *cis* versus *trans* spreading mechanisms. At a more coarse-grained scale, some experiments have also pointed out the possible role of *trans*-acting "long-range" spreading in epigenome maintenance as in the heterochromatin domain in yeast [113] or for dosage compensation systems where the propagation of a specific epigenomic signal was associated with the global compaction of sexual chromosomes [109, 114].

2.4.3 The living chromatin model

While theoretical and experimental works on the epigenome assembly based on the "reader-writer/eraser" mechanism have highlighted the role of long-range spreading in the stable formation and maintenance of the epigenomic domain [113, 115–121], all these approaches neglect the effect of the local chromatin state on the spatial folding of the underlying polymer. To formalize and characterize the 1D–3D positive feedback described above, we developed a theoretical framework, the "Living Chromatin" (LC) model, that explicitly couples the spreading of epigenomic marks to the 3D folding of the fiber (Figure 2.7) [72, 122].

This model is a combination of the epigenome regulation model [117, 123] primarily introduced by Dodd *et al.* [115] and the block copolymer model of chromatin [60, 71–73] described in Section 2.3. It belongs to the general class

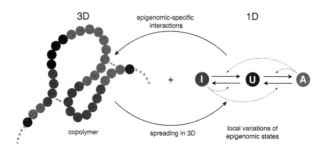

Figure 2.7 The Living chromatin model. The living chromatin model is a combination of the copolymer model where the chromatin organization is driven by epigenomic-specific contact interactions (Right), and of the epigenome regulation model (Left) where the local epigenomic state of each monomer can fluctuate between 3 states: *A*, *U*, and *I*. The inter-conversion (spreading) dynamics between these states depends on the spatial neighborhood of each monomer while the 3D folding depends on the current – primary – epigenomic sequence.

of annealed copolymer models where the physico-chemical state of a monomer can vary according to specific reaction rules [124]. The dynamics of the polymer chain follows the block copolymer model described in Section 2.3.1 with short-range contact interactions between monomers having the same state (only for active (*A*) and inactive (*I*), with no interaction between unmarked (*U*) monomers). For the dynamics of the local epigenomic state, as in [115, 117], we considered a simple case where the state of one monomer can fluctuate only between three variations: an inactive (*I*), an active (*A*), and an intermediate, unmarked (*U*) state. Conversions between *A* and *I* states occur via the first step of mark removal toward the *U* state followed by a step of mark deposition (Figure 2.7 right). Each step can be divided into two contributions: (i) a "noisy" conversion accounting for the leaky activity of modifying enzymes or for nucleosome turnover; and (ii) a recruited conversion, formalizing the "reader–writer/eraser" mechanism, where spreading/erasing of a mark is not restricted to neighboring chromatin elements along the genome but also to any fragments located in the *spatial* neighborhood (Figure 2.7 left). To characterize, in detail, the role of 3D organization in this process, we distinguished between *cis* (only via NN monomers along the chain) and *trans* (3D vicinity) conversions. Physically speaking, the LC model is analogous to a 3-state Ising spin system on a polymer chain with local 3D ferromagnetic coupling: The local epigenomic state stands for spin, random conversions for the temperature and recruited conversions for the coupling.

Practically, we modeled the polymer on a lattice following a KMC scheme slightly different from Section 2.3.2 to account for the dynamics of epigenomic states [122]. One MCS consists of (i) *N* trial monomer state conversions, (ii) *N* / 2 trial binding/unbinding transitions, and (iii) *N* trial monomer moves. In (i) a monomer *m* is randomly picked and a state transition is attempted according to the state-dependent rates:

$$k_{A \to U}(m) = k_{U \to I}(m) = \varepsilon_o + \varepsilon_c \left(\Theta_{e(m-1),I} + \Theta_{e(m+1),I} \right) + \varepsilon_t \sum_{l \neq (m-1:m+1)} \Theta_{e(l),I} \delta_{l,m} \quad (2.2)$$

$$k_{I \to U}(m) = k_{U \to A}(m) = \varepsilon_o + \varepsilon_c \left(\Theta_{e(m-1),A} + \Theta_{e(m+1),A} \right) + \varepsilon_t \sum_{l \neq (m-1:m+1)} \Theta_{e(l),A} \delta_{l,m} \quad (2.3)$$

with $e(l) \in \{A, U, I\}$ the current epigenomic state of monomer l, ε_o the contribution of noisy conversion, ε_c (resp. ε_t) the spreading rate *in cis* (resp. *in trans*), $\Theta_{e(l),X} = 1$ if $e(l) = X$ (0 otherwise), and $\delta_{l,m} = 1$ if monomers l and m occupy NN sites on the lattice (0 otherwise). For simplicity, we assumed that the rates ε_o, ε_c, and ε_t are the same for all the states. In (ii) a monomer m is randomly picked and if its state is either A or I, for every monomer l of the same state occupying a NN site on the lattice and already bound to m, an unbinding event is attempted with a rate k_u. Similarly, for unbound pairs a binding event is realized with a rate k_b. In (iii), a monomer is randomly picked and move to a NN site on the lattice. The move is accepted only if the connexions along the chain and between the bound monomers are maintained. To simplify, we focused our studies on small chains at a steady-state, neglecting crumpling effects described in Section 2.3.

2.4.4 Stability of one epigenomic domain

In a recent study [122], we investigated the behavior of an isolated small chain ($N = 100$) evolving under the LC model as a function of the attraction strength k_b/k_u and of the relative conversion rates $\varepsilon_{c,t}/\varepsilon_o$. Here, we report a similar analysis but for a longer chain ($N \approx 200$) in a semi-dilute environment (10% volumic density), simulated using periodic boundary condition [60]. Following the analogy with an Ising model, we characterized the global epigenomic state S of the system using an effective magnetization:

$$S = \frac{1}{N} \sum_{l=1}^{N} \left(\Theta_{e(l),A} - \Theta_{e(l),I} \right) \quad (2.4)$$

$S \sim +1$ (resp. -1) implies that the full domain is in a coherent A (resp. I) macrostate where most of the monomers have an A (resp. I) state. $S \sim 0$ defines a globally incoherent epigenomic state with a mixture of A, U, and I monomers.

In the absence of *trans* spreading ($\varepsilon_t = 0$), the LC model reduces to a simple system where the epigenomic dynamics is disconnected from the 3D polymeric organization and evolves only under short-range 1D spreading. As expected for a 1D system driven only by NN processes, no phase transition is observed in this case and the distribution of S remains peaked around 0. The existence of stable coherent active (A) or inactive (I) macro-states is unlikely (Figure 2.8I).

In the presence of *trans* spreading, this simple picture is dramatically modified. In Figure 2.8, we plotted the phase diagram of the system as a function of k_b/k_u and $\varepsilon/\varepsilon_o$ where we assumed that $\varepsilon_c = \varepsilon_t \equiv \varepsilon$. At a weak attraction strength ($k_b/k_u \lesssim 0.1$), the polymer has a swollen organization.

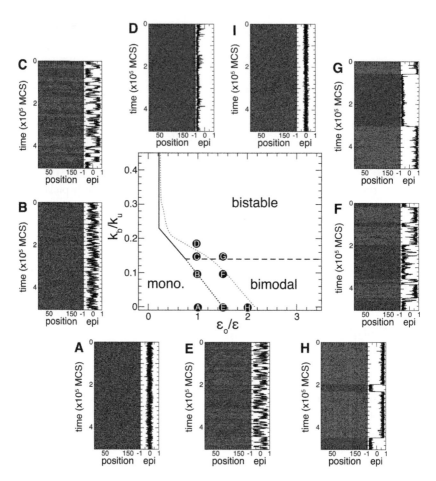

Figure 2.8 Phase diagram of one epigenomic domain ($\varepsilon_c = \varepsilon_t \equiv \varepsilon$, $\varepsilon_o = 0.001$, $k_u = 0.001$). The monostable, bistable and bimodal regions are demarcated by black lines. The corresponding curves for an isolated shorter chain as investigated in [122] is reported for comparison (orange lines). (A-H) Examples of the time evolution of the local epigenomic state (Left: red for A, blue for I and black for U) and of the global epigenomic state S (Right), for various values of $\varepsilon/\varepsilon_o$ and k_b/k_u (noted as black dots in the phase diagram) predicted by the full LC model.(I) Same as (A–H) but when *trans* spreading was neglected ($\varepsilon_c = 0.2, \varepsilon_t = 0$).

While for $\varepsilon/\varepsilon_o \lesssim 1$–1.5 the system remains monostable with a globally incoherent epigenome characterized by short-lived coexisting A and I microdomains (Figure 2.8A,B,E), at high ε the weak *trans* spreading activity due to the presence of (some) random long-range contacts allows the emergence of coherent epigenomic domains (Figure 2.8 H,F). Strictly speaking, this transition from monostability (incoherent state) to bimodality (coherent A and I macro-states) does not reflect a phase transition but rather is a signature of finite size effects. Hence, the stability of a macro-state increases linearly with ε [122].

As k_b/k_u augments, the polymeric system exhibits a collapse transition where the chain passes from a swollen coil to a compact globule [125]. Above this collapse, for ε higher than a critical, k_b/k_u -dependent recruitment strength, we observed a second-order phase transition towards a bistable regime (Figure 2.8C,D,G). In this phase, cooperative effects are dominant and lead to the emergence of super-stable A or I macro-states (stability increases exponentially with ε [122]). This is characteristic of the presence of phase transitions in 1D systems with effective long-range interactions only if the strength of interactions between two mono-mers l and m decreases more slowly than $1/\,|\,m-l\,|^2$ [126], i.e., in our case, only if epigenomic-driven interactions (via k_b/k_u) are strong enough to partially col-lapse the polymer so that the contact probability between two monomers scales are slower than $1/\,|\,m-l\,|^2$.

As already shown in [115, 117], these results confirmed that the emergence and maintenance of stable coherent macro-states require an efficient *trans* spreading activity. Moreover, accounting explicitly for the polymeric structure and for the impact of epigenomic-driven interactions suggested that physical bridging may strongly enhance the stability of coherent epigenomic domains by creating a more compact 3D neighborhood facilitating *trans*-mediated recruited conversions. Comparison with the phase diagram of an isolated chain (orange lines in Figure 2.8) underlines this effect since accounting for an effective con-finement of the chain (via the control of the volumic fraction) reduced the critical value to switch from the monostable, incoherent regime to the bistable/bimodal, coherent one.

2.4.5 Stability of antagonistic epigenomic domains

In the previous section, we discussed how *trans* activity coupled with epigenomic-driven interactions affect the stability of a single epigenomic domain. The next step is to understand how such mechanisms impact the epigenomic stability of a genomic region containing several adjacent antagonistic chromatin states (A and I). As a proof of concept, for parameters leading to bistability, we addressed this issue by following the dynamics of a region initially prepared with one (reported in [122]) or two (reported here in Figure 2.9) I domains directly adjacent to one or two A domains of the same size. In Figure 2.9A–C, we presented some examples for a region initialized with four adjacent epigenomic domains (two active A, two inac-tive I) forming two distinct 3D compartments (one for A, one for I). In particular, for various situations, we quantified the stability of the 1D epigenomic organiza-tion (Figure 2.9E) by measuring the time it takes for the system to switch from this mosaic initial state to a typical steady-state (coherent A or I macro-states or incoherent state depending on the parameters).

In the absence of *trans* spreading activity, each subdomain is very unstable (red dots in Figure 2.9E) and rapidly converges to an incoherent epigenomic organiza-tion. Similarly, accounting for *trans* spreading but neglecting the epigenomic-driven interactions leads to a rapid destabilization of the system (blue dots in Figure 2.9E)

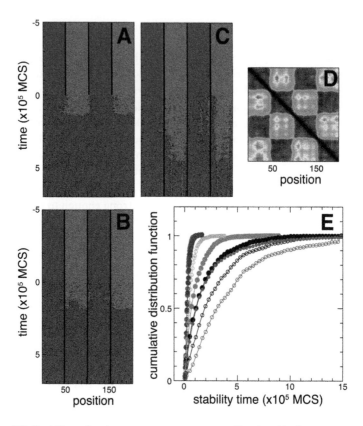

Figure 2.9 Stability of epigenome compartmentalization ($k_b/k_u = 0.28$, $\varepsilon_c = 0.01$, $\varepsilon_t = \varepsilon_o = 0.001$). (A–C) Examples of the time evolution of the local epigenomic state of the genomic region initialized with four adjacent epigenomic domains forming two spatial compartments (Average distance map between any pair of monomers shown in (D)). Initially, the state of each subdomain is forced. At $t > 0$, forcing is switched off (except in C where a weak loading rate of 0.001 is maintained). (E) Cumulative distribution of the stability time τ of the mosaic epigenomic pattern with only *cis*-recruitment (red dots), without epigenomic-driven interactions (blue dots), with weaker interactions ($k_b/k_u = 0.18$, cyan dots), in the absence (orange dots) or the presence (purple dots) of 1D barriers, in the presence of 1D barriers and weak nucleation (purple circles). Black dots and circles correspond to a system with only 2 epigenomic domains, each of size 50 [122].

towards a coherent macro-state. It is only by fully considering the positive feedback between epigenome and polymer dynamics that the four subdomains remain significantly stable (Figure 2.9A), the stronger the interactions the more stable the partition (cyan and orange dots in Figure 2.9E). Indeed, the formation of two distinct, compact spatial compartments for *A* and *I* domains limits the "invasion" *in trans* of one epigenomic domain by the antagonistic state of its neighboring domains. This also leads to strong cooperativity between the subdomains of the same epigenomic state that switch their states always at the same time.

A way to enhance the stability of subdomains is to introduce 1D barriers (Figure 2.9B). By maintaining the monomers at the boundary between two antagonistic subdomains in a neutral U state, we hinder the propagation *in cis* between NN subdomains. Such barriers are biologically relevant with the binding of insulator proteins such as CTCF at TAD boundaries [46] that can physically prevent the action *in cis* of epigenomic enzymes. External "contamination" of one domain by the other can thus only arise from the *trans* spreading activity across the frontier. This leads to a significant stabilization by 2 to 3 fold (purple dots in Figure 2.9E) depending on the size of the barrier [122].

Previously, adjacent antagonistic subdomains were forced to be in one epigenomic state and, at $t > 0$, the system was evolving in the absence of forcing. Here, we asked, in association with 1D barriers, how maintaining a weak permanent forcing of the initial state inside each subdomain influences their stability (Figure 2.9C). This situation mimics the presence of nucleation sites like PREs for H3K27me3/PcG domains. We observed a strong increase of the mean stability time even at low loading rates (purple circles in Figure 2.9E). This is fully consistent with recent experimental studies showing that long-term memory relies on self-propagation (in our case promoted by spatial condensation) and on sequence-specific *cis*-recruitment mechanisms [127–129]. Our results suggest that spatial compaction, by promoting self-propagation *in trans*, might cooperate with *cis*-recruitment to achieve strong stability. This means that a weakening of the recruitment might be compensated by an increase of the compaction. Whether these compensatory mechanisms indeed occur in real systems at both developmental and evolutionary time scales [130] has to be further investigated. Interestingly, compared to the case of a chain with only two adjacent subdomains as studied in [122] (black dots and circles in Figure 2.9E), we observed that stability is enhanced when considering four adjacent domains. This implies that forming a large-scale spatial compartment, like the A/B or heterochromatin/euchromatin compartments, increases the insulation of both antagonistic marks and delays the cooperative switching of subdomains towards a global coherent macro-state.

2.4.6 Towards a quantitative model

The LC model represents a powerful theoretical and numerical formalism to study the dynamical coupling between the 1D epigenomic information along the chain and the 3D chromatin organization: 3D acts on 1D via the *trans* spreading mechanism while the 1D feedbacks the 3D via epigenomically driven contact interactions. This framework is modular and can be easily generalized to any number of epigenomic states and any biochemical reactions or interaction scheme. We showed that an efficient epigenome stability and compartmentalization requires (i) *trans* spreading mechanisms, (ii) eventually 1D barriers and weak permanent nucleation, and (iii) the chain to be collapsed (i.e., around or below the collapse transition). This latter regime is exactly the condition consistent with experimental Hi-C data as we showed in our previous works on chromatin folding for a fixed epigenome (see Section 2.3). However, to be applied to specific *in vivo* situations, the LC model should be extended to consider other biologically relevant ingredients such as titration effects [118, 120], replication and cell cycle duration [115, 123], conversion asymmetries [117], and multicolor epigenome [131]. In order to progress toward a quantitative description of this 1D–3D coupling, a correct parameter inference would require experiments that can record the large-scale dynamics of both the 1D and 3D organization, during the establishment and the maintenance stages, in both wild-type and mutant backgrounds. The corresponding experimental techniques remain to be developed.

Recently, Michieletto *et al.* have also developed a physical model of such 1D–3D coupling of chromatin [132]. In their approach, the dynamics of the epigenome and of the polymer are governed by an identical Hamiltonian, i.e., the spreading of a mark is tightly related to the (pre)existence of chemical bonds with the nearest monomers. This is the main difference with the LC model where spreading *in trans* is not directly coupled to the copolymer dynamics but rather depends only on the presence of monomers in the spatial neighborhood. Compared to their approach, the LC framework is somehow more general since we explicitly treat the local epigenomic dynamics as biochemical reactions and not as a Hamiltonian dynamics. In addition, we decomposed the spreading into two contributions (*cis* and *trans*), that, we think, is crucial to understand clearly the 1D/3D coupling. Our proposed mechanism leads to second-order phase transition while Michieletto *et al.* found a first-order transition within their framework. There is, to date, no experimental evidence for one or the other type of transition. More importantly, the main and similar outcome of these two complementary and pioneering studies is that self-attraction and *trans* spreading activity at the local scale can be translated into a macroscopic coupling between the epigenome and spatial compartmentalization dynamics. As shown in [122], the correlated evolution of the global epigenomic state and of the radius of gyration of the chain at the collapse transition illustrates nicely how the local 1D–3D feedback mechanism induces a large-scale coupling between the epigenome and the spatial chain folding: incoherent epigenomic states tend to be associated with a partial decondensation of the chain while coherent states correspond to more condensed configurations.

2.5 DISCUSSION AND PERSPECTIVES

In this chapter, we discussed how polymer modeling allows us to better understand the coupling between epigenome and 3D chromosome organization.

In a first part, we showed that epigenomically associated mechanisms are the main drivers of chromosome folding: A/B or heterochromatin/euchromatin compartments in *Drosophila* emerge naturally from the mild microphase separation of different chromatin states that lead to a very dynamical and stochastic organization. Our model predicts that active chromatin only weakly interacts with itself. This may reflect a distinct local mode of interaction between chromatin types: active chromatin rather organizes locally via pairwise short-range bridging between discrete specific genomic sites while heterochromatin may interact more continuously via clustering of multiple chromatin loci. This is consistent with more homogeneous internal contact patterns observed for inactive domains and more complex profiles of contact for active domains as observed in human cell lines [67]. Overall, a finer understanding of these different modes of self-association will require a proper inference of the chromatin-state-specific interaction strength. Thanks to higher-resolution Hi-C and epigenomic data, we expect to gain deeper insights into the complexity of the local epigenomic and genomic control of chromatin self-association. Additionally, interactions with nuclear landmarks such as membrane and nuclear pores are known to play a fundamental role in controlling large-scale nuclear organization [114, 133]. Integration of such interactions in our framework would also lead to a more detailed description of chromatin folding.

In the second part, we addressed the role of 3D organization into epigenomic stability and maintenance. Our working hypothesis is that spatial compartmentalization may provide a favorable environment playing a functional role of "nano-reactor" by confining the proper regulators close to the target regions. TADs might have a role in either preventing (by sequestering) or facilitating the long-range communication between distal regulatory genomic elements, thus enhancing the efficiency of gene co-activation or co-repression [134, 135]. In our copolymer framework, we remarked that experimental observations are compatible with a region of the phase diagram that is sensitive to variations in the interaction strength and in the block size. One could hypothesize that by modulating the number of bridging molecules (or their bridging efficiency), cells might finely tune the local condensation and the long-range contacts between epigenomic domains, and thus might regulate gene expression or epigenomics. To test this, we developed an extended copolymer model, the Living Chromatin model that readily couples the local transition between different chromatin states with the spatial organization of the chain. We demonstrated that epigenome plasticity and robustness is ensured when the chain is in a sufficiently collapsed state which is exactly the physiological condition. Building on the classical Waddington picture of epigenomic landscape [136], progression through successive developmental or differentiation stages as well as pathologies may now correspond to different pathways on the folding-epigenome landscape with the enzymatic activity and self-affinity of architectural chromatin-binding proteins as control parameters.

The ultimate goal would be to build a quantitative model that could reproduce both the complex linear epigenomic pattern and the spatial chromatin organization in real systems such as in *Drosophila* and make testable predictions. However, a proper inference of the corresponding parameters would require to account properly for dynamics. Indeed, as discussed above, chromosome folding is out-of-equilibrium, dynamical and stochastic. At the TAD scale (the relevant regulatory scale), the chain rapidly reaches a stationary state. However, at larger scales, convergence towards a metastable state may be slow. The cell cycle duration then may constitute an additional control parameter: The establishment of stable long-range contact might be challenged by cell cycle duration. Efforts toward the development of time-predictive models of the spatial and epigenomic organization are required. We already managed to calibrate the copolymer model from MSD measurement of chromatin loci such that we can have a reliable description of chromosome folding kinetics. However, we still lack a precise time-parameterization of the local epigenome dynamics. Furthermore, we do not consider the out-of-equilibrium effect of replication which is, of course, an important issue to understand epigenetic maintenance. Incorporating all these ingredients into a quantitative, predictive model would represent an intriguing challenging task for the future on both theoretical and experimental sides.

ACKNOWLEDGMENTS

We apologize to colleagues whose work could not be cited due to space constraints. We thank Guido Tiana and Luca Giorgetti for inviting us to write this chapter. We are grateful to Surya Ghosh, Angelo Rosa, Ralf Everaers, Pascal Carrivain, Ivan Junier, Peter Meister, and Giacomo Cavalli for fruitful discussions. We acknowledge our funding agencies: Agence Nationale de la Recherche (ANR-15-CE12-0006 EpiDevoMath), Fondation pour la Recherche Médicale (DEI20151234396), and CNRS. Computing resources were provided by the CIMENT infrastructure at University Grenoble-Alpes (supported by the Rhône-Alpes region, Grant CPER07 13 CIRA) and by the Pôle Scientifique de Modélisation Numérique and Centre Blaise Pascal at ENS Lyon.

REFERENCES

1. Rao, S. S. P., Huntley, M. H., Durand, N. C., Stamenova, E. K., Bochkov, I. D., Robinson, J. T., Sanborn, A. L., Machol, I., Omer, A. D., Lander, E. S., and Aiden, E. L. A 3D map of the human genome at kilobase resolution reveals principles of chromatin looping. *Cell* 159(7), 1665–80 (2014).
2. Durand, N. C., Robinson, J. T., Shamim, M. S., Machol, I., Mesirov, J. P., Lander, E. S., and Lieberman Aiden, E. Juicebox provides a visualization system for Hi-C contact maps with unlimited zoom. *Cell Systems* 3, 99–101 (2016).
3. de Wit, E. and de Laat, W. A decade of 3C technologies: Insights into nuclear organization. *Genes Dev* 26, 11–24 (2012).
4. Cremer, C., Szczurek, A., Schock, F., Gourram, A., and Birk, U. Super-resolution microscopy approaches to nuclear nanostructure imaging. *Methods* 123, 11–32 (2017).
5. Dixon, J. R., Selvaraj, S., Yue, F., Kim, A., Li, Y., Shen, Y., Hu, M., Liu, J. S., and Ren, B. Topological domains in mammalian genomes identified by analysis of chromatin interactions. *Nature* 485(7398), 376–380 (2012).
6. Nora, E. P., Lajoie, B. R., Schulz, E. G., Giorgetti, L., Okamoto, I., Servant, N., Piolot, T., van Berkum, N. L., Meisig, J., Sedat, J., Gribnau, J., Barillot, E., Blüthgen, N., Dekker, J., and Heard, E. Spatial partitioning of the regulatory landscape of the X-inactivation centre. *Nature* 485(7398), 381–385 (2012).

7. Boettiger, A. N., Bintu, B., Moffitt, J. R., Wang, S., Beliveau, B. J., Fudenberg, G., Imakaev, M., Mirny, L. A., Wu, C., and Zhuang, X. Super-resolution imaging reveals distinct chromatin folding for different epigenetic states. *Nature* 529(7586), 418–422 (2016).

8. Lieberman-Aiden, E., van Berkum, N. L., Williams, L., Imakaev, M., Ragoczy, T., Telling, A., Amit, I., Lajoie, B. R., Sabo, P. J., Dorschner, M. O., Sandstrom, R., Bernstein, B., Bender, M. A., Groudine, M., Gnirke, A., Stamatoyannopoulos, J., Mirny, L. A., Lander, E. S., and Dekker, J. Comprehensive mapping of long-range interactions reveals folding principles of the human genome. *Science* 326(5950), 289–293 (2009).

9. Cremer, T., Cremer, M., Hübner, B., Strickfaden, H., Smeets, D., Popken, J., Sterr, M., Markaki, Y., Rippe, K., and Cremer, C. The 4D nucleome: Evidence for a dynamic nuclear landscape based on co-aligned active and inactive nuclear compartments. *FEBS Letters* 589, 2931–2943 (2015).

10. Biggiogera, M., Courtens, J. L., Derenzini, M., Fakan, S., Hernandez-Verdun, D., Risueno, M. C., and Soyer-Gobillard, M. O. Osmium ammine: Review of current applications to visualize DNA in electron microscopy. *Biol. Cell* 87, 121–132 (1996).

11. Mattout, A., Cabianca, D. S., and Gasser, S. M. Chromatin states and nuclear organization in development - a view from the nuclear lamina. *Genome Biology* 16, 174 (2015).

12. Towbin, B. D., Meister, P., and Gasser, S. M. The nuclear envelope—a scaffold for silencing? *Current Opinion in Genetics & Development* 19(2), 180–186 (2009).

13. Meuleman, W., Peric-Hupkes, D., Kind, J., Beaudry, J. B., Pagie, L., Kellis, M., Reinders, M., Wessels, L., and van Steensel, B. Constitutive nuclear lamina-genome interactions are highly conserved and associated with A/T-rich sequence. *Genome Research* 23, 270–280 (2013).

14. Bolzer, A., Kreth, G., Solovei, I., Koehler, D., Saracoglu, K., Fauth, C., Muller, S., Eils, R., Cremer, C., Speicher, M. R., and Cremer, T. Three-dimensional maps of all chromosomes in human male fibroblast nuclei and prometaphase rosettes. *PLoS Biology* 3, e157 (2005).

15. Gerlich, D., Beaudouin, J., Kalbfuss, B., Daigle, N., Eils, R., and Ellenberg, J. Global chromosome positions are transmitted through mitosis in mammalian cells. *Cell* 112(6), 751–764 (2003).

16. Nagano, T., Lubling, Y., Stevens, T. J., Schoenfelder, S., Yaffe, E., Dean, W., Laue, E. D., Tanay, A., and Fraser, P. Single-cell Hi-C reveals cell-to-cell variability in chromosome structure. *Nature* 502(7469), 59–64, Oct (2013).

17. Nagano, T., Lubling, Y., Varnai, C., Dudley, C., Leung, W., Baran, Y., Cohen, N. M., Wingett, S., Fraser, P., and Tanay, A. Cell cycle dynamics of chromosomal organisation at single-cell resolution. *Nature* 547(7661), 61–67 (2017).

18. Stevens, T. J., Lando, D., Basu, S., Atkinson, L. P., Cao, Y., Lee, S. F., Leeb, M., Wohlfahrt, K. J., Boucher, W., and O'Shaughnessy-Kirwan, A. 3D structures of individual mammalian genomes studied by single-cell Hi-C. *Nature* 544, 59–64 (2017).

19. Flyamer, I. M., Gassler, J., Imakaev, M., Brandao, H. B., Ulianov, S. V., Abdennur, N., Razin, S. V., Mirny, L. A., and Tachibana-Konwalski, K. Single-nucleus Hi-C reveals unique chromatin reorganization at oocyte-to-zygote transition. *Nature* 544, 110–114 (2017).

20. Wang, S., Su, J. H., Beliveau, B. J., Bintu, B., Moffitt, J. R., Wu, C., and Zhuang, X. Spatial organization of chromatin domains and compartments in single chromosomes. *Science* 353, 598–602 (2016).

21. Cattoni, D. I., Gizzi, A. M. C., Georgieva, M., Di Stefano, M., Valeri, A., Chamousset, D., Houbron, C., Dejardin, S., Fiche, J. B., Gonzalez, I., Chang, J. M., Sexton, T., Marti-Renom, M. A., Bantignies, F., Cavalli, G., and Nollmann, M. Single-cell absolute contact probability detection reveals chromosomes are organized by multiple low-frequency yet specific interactions. *Nature Communications* 8, 1753 (2017).

22. Bystricky, K. Chromosome dynamics and folding in eukaryotes: Insights from live cell microscopy. *FEBS Letters* 589, 3014–3022 (2015).

23. Hajjoul, H., Mathon, J., Ranchon, H., Goiffon, I., Mozziconacci, J., Albert, B., Carrivain, P., Victor, J. M., Gadal, O., Bystricky, K., and Bancaud, A. High-throughput chromatin motion tracking in living yeast reveals the flexibility of the fiber throughout the genome. *Genome Res* 23(11), 1829–1838 (2013).

24. Germier, T., Kocanova, S., Walther, N., Bancaud, A., Shaban, H. A., Sellou, H., Politi, A. Z., Ellenberg, J., Gallardo, F., and Bystricky, K. Real-time imaging of a single gene reveals transcription-initiated local confinement. *Biophysical Journal* 113, 1383–1394 (2017).

25. Bronstein, I., Israel, Y., Kepten, E., Mai, S., Shav-Tal, Y., Barkai, E., and Garini, Y. Transient anomalous diffusion of telomeres in the nucleus of mammalian cells. *Phys. Rev. Lett.* 103(1), 018102 (2009).

26. Shinkai, S., Nozaki, T., Maeshima, K., and Togashi, Y. Dynamic nucleosome movement provides structural information of topological chromatin domains in living human cells. *PLOS Computational Biology* 12(10), 1–16, 10 (2016).

27. Nozaki, T., Imai, R., Tanbo, M., Nagashima, R., Tamura, S., Tani, T., Joti, Y., Tomita, M., Hibino, K., Kanemaki, M. T., Wendt, K. S., Okada, Y., Nagai, T., and Maeshima, K. Dynamic organization of chromatin domains revealed by super-resolution live-cell imaging. *Molecular Cell* 67, 282–293 (2017).

28. Lucas, J. S., Zhang, Y., Dudko, O. K., and Murre, C. 3d trajectories adopted by coding and regulatory DNA elements: First-passage times for genomic interactions. *Cell* 158, 339–352 (2014).

29. Ruiz-Velasco, M. and Zaugg, J. B. Structure meets function: How chromatin organisation conveys functionality. *Current Opinion in Systems Biology* 1, 129–136 (2017).

30. Kaiser, V. B. and Semple, C. A. When TADS go bad: Chromatin structure and nuclear organisation in human disease. *F1000Research* 6 (2017).

31. Corces, M. R. and Corces, V. G. The three-dimensional cancer genome. *Current Opinion in Genetics & Development* 36, 1–7 (2016).

32. Zhan, Y., Mariani, L., Barozzi, I., Schulz, E., Bluthgen, N., Stadler, M., Tiana, G., and Giorgetti, L. Reciprocal insulation analysis of Hi-C data shows that TADs represent a functionally but not structurally privileged scale in the hierarchical folding of chromosomes. *Genome Research* 27, 479–490 (2017).

33. Quintero-Cadena, P. and Sternberg, P. W. Enhancer sharing promotes neighborhoods of transcriptional regulation across eukaryotes. *G3: Genes, Genomes, Genetics* 6, 4167–4174 (2016).

34. Hnisz, D., Weintraub, A. S., Day, D. S., Valton, A. L., Bak, R. O., Li, C. H., Goldmann, J., Lajoie, B. R., Fan, Z. P., Sigova, A. A., Reddy, J., Borges-Rivera, D., Ihn Lee, T., Jaenisch, R., Porteus, M. H., Dekker, J., and Young, R. A. Activation of proto-oncogenes by disruption of chromosome neighborhoods. *Science* 351, 1454–1458 (2016).

35. Lupiáñez, D. G., Kraft, K., Heinrich, V., Krawitz, P., Brancati, F., Klopocki, E., Horn, D., Kayserili, H., Opitz, J. M., Laxova, R., Santos-Simarro, F., Gilbert-Dussardier, B., Wittler, L., Borschiwer, M., Haas, S. A., Osterwalder, M., Franke, M., Timmermann, B., Hecht, J., Spielmann, M., Visel, A., and Mundlos, S. Disruptions of topological chromatin domains cause pathogenic rewiring of gene-enhancer interactions. *Cell* 161(5), 1012–1025 (2015).

36. Jost, D., Vaillant, C., and Meister, P. Coupling 1D modifications and 3D nuclear organization: Data, models and function. *Curr. Opin. Cell Biol.* 44, 20–27 (2017).

37. Pombo, A. and Dillon, N. Three-dimensional genome architecture: Players and mechanisms. *Nature Reviews Molecular Cell Biology* 16, 245–257 (2015).

38. Canzio, D., Liao, M., Naber, N., Pate, E., Larson, A., Wu, S., Marina, D. B., Garcia, J. F., Madhani, H. D., Cooke, R., Schuck, P., Cheng, Y., and Narlikar, G. J. A conformational switch in HP1 releases auto-inhibition to drive heterochromatin assembly. *Nature* 496(7445), 377–381 (2013).

39. Isono, K., Endo, T. A., Ku, M., Yamada, D., Suzuki, R., Sharif, J., Ishikura, T., Toyoda, T., Bernstein, B. E., and Koseki, H. Sam domain polymerization links subnuclear clustering of PRC1 to gene silencing. *Dev. Cell* 26(6), 565–577 (2013).

40. Larson, A. G., Elnatan, D., Keenen, M. M., Trnka, M. J., Johnston, J. B., Burlingame, A. L., Agard, D. A., Redding, S., and Narlikar, G. J. Liquid droplet formation by HP1α suggests a role for phase separation in heterochromatin. *Nature* 547, 236–240 (2017).

41. Strom, A. R., Emelyanov, A. V., Mir, M., Fyodorov, D. V., Darzacq, X., and Karpen, G. H. Phase separation drives heterochromatin domain formation. *Nature* 547, 241–245 (2017).

42. Ulianov, S., Shevelyov, Y., and Razin, S. Lamina-associated chromatin in the context of the mammalian genome folding. *Biopolymers and Cell* 32, 327–333 (2016).

43. Sanborn, A. L., Rao, S. S. P., Huang, S. C., Durand, N. C., Huntley, M. H., Jewett, A. I., Bochkov, I. D., Chinnappan, D., Cutkosky, A., Li, J., Geeting, K. P., Gnirke, A., Melnikov, A., McKenna, D., Stamenova, E. K., Lander, E.

S., and Aiden, E. L. Chromatin extrusion explains key features of loop and domain formation in wild-type and engineered genomes. *Proc Natl Acad Sci U S A* 112(47), E6456–E6465 (2015).

44. Fudenberg, G., Imakaev, M., Lu, C., Goloborodko, A., Abdennur, N., and Mirny, L. A., Formation of chromosomal domains by loop extrusion. *Cell Rep* 15(9), 2038–2049 (2016).

45. Dekker, J. and Mirny, L. The 3D genome as moderator of chromosomal communication. *Cell* 164, 1110–1121 (2016).

46. Dowen, J. M., Fan, Z. P., Hnisz, D., Ren, G., Abraham, B. J., Zhang, L. N., Weintraub, A. S., Tong Ihn Lee, J. S., and Zhao, K. Control of cell identity genes occurs in insulated neighborhoods in mammalian chromosomes. *Cell* 159, 374–387 (2014).

47. Amitai, A. and Holcman, D. Polymer physics of nuclear organization and function. *Physics Reports* 678, 1–83 (2017).

48. Cortini, R., Barbi, M., Caré, B., Lavelle, C., Lesne, A., Mozziconacci, J., and Victor, J. M. The physics of epigenetics. *Rev. Mod. Phys.* 88, 025002 (2016).

49. Rosa, A. and Zimmer, C. Computational models of large-scale genome architecture. *Int. Rev. Cell Mol Biol.* 307, 275–349 (2014).

50. Halverson, J. D., Smrek, J., Kremer, K., and Grosberg, A. Y. From a melt of rings to chromosome territories: the role of topological constraints in genome folding. *Rep. Prog. Phys.* 77(2), 022601 (2014).

51. Ho, J. W. K., Jung, Y. L., Liu, T., Alver, B. H., Lee, S., Ikegami, K., Sohn, K. A., Minoda, A., Tolstorukov, M. Y., Appert, A., Parker, S. C. J., Gu, T., Kundaje, A., Riddle, N. C., Bishop, E., Egelhofer, T. A., Hu, S. S., Alekseyenko, A. A., Rechtsteiner, A., Asker, D., Belsky, J. A., Bowman, S. K., Chen, Q. B., Chen, R. A. J., Day, D. S., Dong, Y., Dose, A. C., Duan, X., Epstein, C. B., Ercan, S., Feingold, E. A., Ferrari, F., Garrigues, J. M., Gehlenborg, N., Good, P. J., Haseley, P., He, D., Herrmann, M., Hoffman, M. M., Jeffers, T. E., Kharchenko, P. V., Kolasinska-Zwierz, P., Kotwaliwale, C. V., Kumar, N., Langley, S. A., Larschan, E. N., Latorre, I., Libbrecht, M. W., Lin, X., Park, R., Pazin, M. J., Pham, H. N., Plachetka, A., Qin, B., Schwartz, Y. B., Shoresh, N., Stempor, P., Vielle, A., Wang, C., Whittle, C. M., Xue, H., Kingston, R. E., Kim, J. H., Bernstein, B. E., Dernburg, A. F., Pirrotta, V., Kuroda, M. I., Noble, W. S., Tullius, T. D., Kellis, M., MacAlpine, D. M., Strome, S., Elgin, S. C. R., Liu, X. S., Lieb, J. D., Ahringer, J., Karpen, G. H., and Park, P. J. Comparative analysis of metazoan chromatin organization. *Nature* 512(7515), 449–52 (2014).

52. Filion, G. J., van Bemmel, J. G., Braunschweig, U., Talhout, W., Kind, J., Ward, L. D., Brugman, W., de Castro, I. J., Kerkhoven, R. M., and Bussemaker, H. J. Systematic protein location mapping reveals five principal chromatin types in *Drosophila* cells. *Cell* 143(2), 212–224 (2010).

53. Roudier, F., Ahmed, I., Bérard, C., Sarazin, A., Mary-Huard, T., Cortijo, S., Bouyer, D., Caillieux, E., Duvernois-Berthet, E., Al-Shikhley, L., Giraut, L., Després, B., Drevensek, S., Barneche, F., Dèrozier, S., Brunaud, V., Aubourg, S., Schnittger, A., Bowler, C., Martin-Magniette, M. L., Robin,

S., Caboche, M., and Colot, V.. Integrative epigenomic mapping defines four main chromatin states in Arabidopsis. *EMBO J* 30(10), 1928–1938 (2011).

54. Julienne, H., Zoufir, A., Audit, B., and Arneodo, A. Human genome replication proceeds through four chromatin states. *PLoS Com* 9, e1003233 (2013).

55. Haddad, N., Vaillant, C., and Jost, D. IC-finder: Inferring robustly the hierarchical organization of chromatin folding. *Nucleic Acids Res.* 45, e81 (2017).

56. Sexton, T., Yaffe, E., Kenigsberg, E., Bantignies, F., Leblanc, B., Hoichman, M., Parrinello, H., Tanay, A., and Cavalli, G. Three-dimensional folding and functional organization principles of the *Drosophila* genome. *Cell* 148(3), 458–472 (2012).

57. Imakaev, M., Fudenberg, G., McCord, R. P., Naumova, N., Goloborodko, A., Lajoie, B. R., Dekker, J., and Mirny, L. A. Iterative correction of Hi-C data reveals hallmarks of chromosome organization. *Nat. Methods* 9(10), 999–1003 (2012).

58. Rowley, M. J., Nichols, M. H., Lyu, X., Ando-Kuri, M., Rivera, I. S. M., Hermetz, K., Wang, P., Ruan, Y., and Corces, V. G. Evolutionarily conserved principles predict 3D chromatin organization. *Molecular Cell* 67, 837–852 (2017).

59. Ulianov, S. V., Khrameeva, E. E., Gavrilov, A. A., Flyamer, I. M., Kos, P., Mikhaleva, E. A., Penin, A. A., Logacheva, M. D., Imakaev, M. V., Chertovich, A., Gelfand, M. S., Shevelyov, Y. Y., and Razin, S. V. Active chromatin and transcription play a key role in chromosome partitioning into topologically associating domains. *Genome Res.* 26(1), 70–84 (2016).

60. Olarte-Plata, J. D., Haddad, N., Vaillant, C., and Jost, D. The folding landscape of the epigenome. *Phys. Biol.* 13(2), 026001 (2016).

61. Rao, S. S., Huang, S. C., St Hilaire, B. G., Engreitz, J. M., Perez, E. M., Kieffer-Kwon, K. R., Sanborn, A. L., Johnstone, S. E., Bascom, G. D., and Bochkov, I. D. Cohesin loss eliminates all loop domains. *Cell* 171, 305–320 (2017).

62. Schwarzer, W., Abdennur, N., Goloborodko, A., Pekowska, A., Fudenberg, G., Loe-Mie, Y., Fonseca, N. A., Huber, W., Haering, C., Mirny, L., and Spitz, F. Two independent modes of chromosome organization are revealed by cohesin removal. *Nature* 551(7678), 51–56 (2017).

63. Zhu, Y., Chen, Z., Zhang, K., Wang, M., Medovoy, D., Whitaker, J. W., Ding, B., Li, N., Zheng, L., and Wang, W. Constructing 3D interaction maps from 1D epigenomes. *Nat. Commun.* 7, 10812 (2016).

64. Fortin, J. P. and Hansen, K. D. Reconstructing A/B compartments as revealed by Hi-C using long-range correlations in epigenetic data. *Genome Biology* 16, 180 (2015).

65. Pancaldi, V., Carrillo-de Santa-Pau, E., Javierre, B. M., Juan, D., Fraser, P., Spivakov, M., Valencia, A., and Rico, D. Integrating epigenomic data and 3D genomic structure with a new measure of chromatin assortativity. *Genome Biology* 17, 152 (2016).

66. Phillips-Cremins, J. E., Sauria, M. E. G., Sanyal, A., Gerasimova, T. I., Lajoie, B. R., Bell, J. S. K., Ong, C. T., Hookway, T. A., Guo, C., Sun, Y., Bland, M. J., Wagstaff, W., Dalton, S., McDevitt, T. C., Sen, R., Dekker, J., Taylor, J., and Corces, V. G. Architectural protein subclasses shape 3D organization of genomes during lineage commitment. *Cell* 153(6), 1281–1295 (2013).

67. Sofueva, S., Yaffe, E., Chan, W. C., Georgopoulou, D., Vietri Rudan, M., Mira-Bontenbal, H., Pollard, S. M., Schroth, G. P., Tanay, A., and Hadjur, S. Cohesin-mediated interactions organize chromosomal domain architecture. *EMBO J* 32(24), 3119–3129 (2013).

68. Whyte, W. A., Orlando, D. A., Hnisz, D., Abraham, B. J., Lin, C. Y., Kagey, M. H., Rahl, P. B., Lee, T. I., and Young, R. A. Master transcription factors and mediator establish super-enhancers at key cell identity genes. *Cell* 153, 307–319 (2013).

69. Funke, J. J., Ketterer, P., Lieleg, C., Schunter, S., Korber, P., and Dietz, H. Uncovering the forces between nucleosomes using DNA origami. *Science Advances* 2, e1600974 (2016).

70. Bertin, A., Leforestier, A., Durand, D., and Livolant, F. Role of histone tails in the conformation and interactions of nucleosome core particles. *Biochemistry* 43, 4773–4780 (2004).

71. Jost, D., Carrivain, P., Cavalli, G., and Vaillant, C. Modeling epigenome folding: Formation and dynamics of topologically associated chromatin domains. *Nucleic Acids Res.* 42(15), 9553–9561 (2014).

72. Haddad, N., Jost, D., and Vaillant, C. Perspectives: Using polymer modeling to understand the formation and function of nuclear compartments. *Chromosome Research* 25(1), 35 (2017).

73. Ghosh, S. K. and Jost, D. How epigenome drives chromatin folding and dynamics, insights from efficient coarse-grained models of chromosomes. *PLOS Computational Biology* 14, e1006159 (2018).

74. Bates, F. S. and Fredrickson, G. H. Block copolymer thermodynamics: Theory and experiment. *Annual review of physical chemistry* 41, 525–557 (1990).

75. Hugouvieux, V., Axelos, M., and Kolb, M. Amphiphilic multiblock copolymers: From intramolecular pearl necklace to layered structures. *Macromolecules* 42, 392–400 (2009).

76. Rosa, A. and Everaers, R. Structure and dynamics of interphase chromosomes. *PLoS Comput. Biol.* 4(8), e1000153 (2008).

77. Mirny, L. A. The fractal globule as a model of chromatin architecture in the cell. *Chromosome Res.* 19(1), 37–51 (2011).

78. Doi, M. and Edwards, S. F. *The Theory of Polymer Dynamics*. Oxford University Press, New York, (1986).

79. Naumova, N., Imakaev, M., Fudenberg, G., Zhan, Y., Lajoie, B. R., Mirny, L. A., and Dekker, J. Organization of the mitotic chromosome. *Science* 342(6161), 948–953 (2013).

80. Caré, B., Emeriau, P. E., Cortini, R., and Victor, J. M. Chromatin epigenomic domain folding: Size matters. *AIMS Biophysics* 2, 517–530 (2015).

81. Metz, C. W. Chromosome studies on the Diptera. ii. the paired association of chromosomes in the Diptera, and its significance. *Journal of
Experimental Zoology Part A: Ecological Genetics and Physiology* 21,
213–279 (1916).

82. Rosa, A., Becker, N. B., and Everaers, R. Looping probabilities in model
interphase chromosomes. *Biophys. J.* 98, 2410–2419 (2010).

83. Chen, H., Fujioka, M., and Gregor, T. Direct visualization of transcriptional
activation by physical enhancer-promoter proximity. *Nat. Genet.* 50(9),
1296–1303 (2018).

84. Di Pierro, M., Zhang, B., Lieberman-Aiden, E., Wolynes, P. G., and
Onuchic, J. N. Transferable model for chromosome architecture. *Proc.
Natl. Acad. Sci. U S A* 113, 12168–12173 (2016).

85. Di Pierro, M., Cheng, R. R., Aiden, E. L., Wolynes, P. G., and Onuchic,
J. N. De novo prediction of human chromosome structures: Epigenetic
marking patterns encode genome architecture. *Proc. Natl. Acad. Sci. U S A*
114, 12126–12131 (2017).

86. Liu, L., Shi, G., Thirumalai, D., and Hyeon, C. Architecture of human interphase chromosome determines the spatiotemporal dynamics of chromatin loci. *bioRxiv* 1, 223669 (2017).

87. Shi, G., Liu, L., Hyeon, C., and Thirumalai, D. Interphase human chromosome exhibits out of equilibrium glassy dynamics. *Nat. Commun.* 9(1),
3161 (2018).

88. Nuebler, J., Fudenberg, G., Imakaev, M., Abdennur, N., and Mirny, L.
Chromatin organization by an interplay of loop extrusion and compartmental segregation. *Proc. Natl. Acad. Sci. USA* 115, E6697–E6706 (2018).

89. Barbieri, M., Chotalia, M., Fraser, J., Lavitas, L. M., Dostie, J., Pombo, A.,
and Nicodemi, M. Complexity of chromatin folding is captured by the
strings and binders switch model. *Proc. Natl. Acad. Sci. U S A* 109(40),
16173–16178 (2012).

90. Barbieri, M., Xie, S. Q., Triglia, E. T., Chiariello, A. M., Bianco, S., de
Santiago, I., Branco, M. R., Rueda, D., Nicodemi, M., and Pombo, A.
Active and poised promoter states drive folding of the extended HoxB
locus in mouse embryonic stem cells. *Nature Structural & Molecular
Biology* 24, 515–524 (2017).

91. Chiariello, A. M., Annunziatella, C., Bianco, S., Esposito, A., and
Nicodemi, M. Polymer physics of chromosome large-scale 3D organisation. *Sci. Rep.* 6, 29775 (2016).

92. Brackley, C. A., Johnson, J., Kelly, S., Cook, P. R., and Marenduzzo, D.
Simulated binding of transcription factors to active and inactive regions
folds human chromosomes into loops, rosettes and topological domains.
Nucleic Acids Res. 44(8), 3503–3512 (2016).

93. Brackley, C. A., Brown, J. M., Waithe, D., Babbs, C., Davies, J.,
Hughes, J. R., Buckle, V. J., and Marenduzzo, D. Predicting the three-
dimensional folding of cis-regulatory regions in mammalian genomes
using bioinformatic data and polymer models. *Genome Biology* 17, 59
(2016).

94. Brackley, C. A., Liebchen, B., Michieletto, D., Mouvet, F., Cook, P. R., and Marenduzzo, D. Ephemeral protein binding to DNA shapes stable nuclear bodies and chromatin domains. *Biophysical Journal* 112, 1085–1093 (2017).

95. Awazu, A. Nuclear dynamical deformation induced hetero-and euchromatin positioning. *Phys. Rev. E.* 92, 032709 (2015).

96. Smrek, J. and Kremer, K. Small activity differences drive phase separation in active-passive polymer mixtures. *Phys. Rev. Lett.* 118, 098002 (2017).

97. Ganai, N., Sengupta, S., and Menon, G. I. Chromosome positioning from activity-based segregation. *Nucleic Acids Res.* 42(7), 4145–4159 (2014).

98. Iborra, F. J. Can visco-elastic phase separation, macromolecular crowding and colloidal physics explain nuclear organisation? *Theoretical Biology and Medical Modelling* 4, 15 (2007).

99. Lee, S. S., Tashiro, S., Awazu, A., and Kobayashi, R. A new application of the phase-field method for understanding the mechanisms of nuclear architecture reorganization. *Journal of mathematical biology* 74, 333–354 (2017).

100. Oehler, S. and Müller-Hill, B. High local concentration: A fundamental strategy of life. *J. Mol. Biol.* 395, 242–253 (2010).

101. Vilar, J. and Leibler, S. DNA looping and physical constraints on transcription regulation. *J. Mol. Biol.* 331, 981–989 (2003).

102. Spitz, F. Gene regulation at a distance: From remote enhancers to 3D regulatory ensembles. *Semin. Cell Dev. Biol.* 57, 57–67 (2016).

103. Lanzuolo, C., Roure, V., Dekker, J., Bantignies, F., and Orlando, V. Polycomb response elements mediate the formation of chromosome higher-order structures in the bithorax complex. *Nat. Cell Biol.* 9, 1167–1174 (2007).

104. Wani, A. H., Boettiger, A. N., Schorderet, P., Ergun, A., Münger, C., Sadreyev, R. I., Zhuang, X., Kingston, R. E., and Francis, N. J. Chromatin topology is coupled to Polycomb group protein subnuclear organization. *Nat Commun* 7, 10291 (2016).

105. Vieux-Rochas, M., Fabre, P., Leleu, M., Duboule, D., and Noordermeer, D. Clustering of mammalian Hox genes with other H3K27me3 targets within an active nuclear domain. *Proc. Natl. Acad. Sci. U S A* 112, 4672–4677 (2015).

106. Bantignies, F., Roure, V., Comet, I., Leblanc, B., Schuettengruber, B., Bonnet, J., Tixier, V., Mas, A., and Cavalli, G. Polycomb-dependent regulatory contacts between distant Hox loci in *Drosophila*. *Cell* 144(2), 214–226 (2011).

107. Meister, P. and Taddei, A. Building silent compartments at the nuclear periphery: a recurrent theme. *Curr. Opin. Genet. Dev.* 23, 96–103 (2013).

108. Dily, F. L., Baù, D., Pohl, A., Vicent, G. P., Serra, F., Soronellas, D., Castellano, G., Wright, R. H. G., Ballare, C., Filion, G., Marti-Renom, M. A., and Beato, M. Distinct structural transitions of chromatin topological domains correlate with coordinated hormone-induced gene regulation. *Genes Dev* 28(19), 2151–2162 (2014).

109. Giorgetti, L., Galupa, R., Nora, E. P., Piolot, T., Lam, F., Dekker, J., Tiana, G., and Heard, E. Predictive polymer modeling reveals coupled fluctuations in chromosome conformation and transcription. *Cell* 157, 950–963 (2014).

110. Al-Sady, B., Madhani, H., and Narlikar, G. Division of labor between the chromodomains of HP1 and Suv39 methylase enables coordination of heterochromatin spread. *Mol Cell* 51, 80–91 (2013).

111. Yuan, W., Wu, T., Fu, H., Dai, C., Wu, H., Liu, N., Li, X., Xu, M., Zhang, Z., and Niu, T., Han, Z., Chai, J., Zhou, X. J., Gao, S., and Zhu, B. Dense chromatin activates Polycomb repressive complex 2 to regulate H3 lysine 27 methylation. *Science* 337, 971–975 (2012).

112. Lavigne, M., Francis, N., King, I., and Kingston, R. Propagation of silencing; recruitment and repression of naive chromatin in trans by Polycomb repressed chromatin. *Mol. C.* 13, 415–425 (2004).

113. Obersriebnig, M. J., Pallesen, E. M. H., Sneppen, K., Trusina, A., and Thon, G. Nucleation and spreading of a heterochromatic domain in fission yeast. *Nat. Commun.* 7, 11518 (2016).

114. Sharma, R., Jost, D., Kind, J., Gómez-Saldivar, G., van Steensel, B., Askjaer, P., Vaillant, C., and Meister, P. Differential spatial and structural organization of the X chromosome underlies dosage compensation in *C. elegans*. *Genes Dev* 28(23), 2591–2596 (2014).

115. Dodd, I. B., Micheelsen, M. A., Sneppen, K., and Thon, G. Theoretical analysis of epigenetic cell memory by nucleosome modification. *Cell* 129(4), 813–822 (2007).

116. Angel, A., Song, J., Dean, C., and Howard, M. A Polycomb-based switch underlying quantitative epigenetic memory. *Nature* 476, 105–108 (2011).

117. Jost, D. Bifurcation in epigenetics: Implications in development, proliferation, and diseases. *Phys. Rev. E. Stat. Nonlin. Soft Matter Phys.* 89(1), 010701 (2014).

118. Dayarian, A. and Sengupta, A. Titration and hysteresis in epigenetic chromatin silencing. *Phys.* 10, 036005 (2013).

119. Erdel, F. and Greene, E. C. Generalized nucleation and looping model for epigenetic memory of histone modifications. *Proceedings of the National Academy of Sciences* 113, E4180–E4189 (2016).

120. Sneppen, K. and Dodd, I. Cooperative stabilization of the SIR complex provides robust epigenetic memory in a model of SIR silencing in Saccharomyces cerevisiae. *Epigenetics* 10(4), 293–302 (2015).

121. Berry, S., Dean, C., and Howard, M. Slow chromatin dynamics allow Polycomb target genes to filter fluctuations in transcription factor activity. *Cell Systems* 4, 1–13 (2017).

122. Jost, D. and Vaillant, C. Epigenomics in 3D: Importance of long-range spreading and specific interactions in epigenomic maintenance. *Nucleic Acids Res.* 46, 2252–2264 (2018).

123. Zerihun, M. B., Vaillant, C., and Jost, D. Effect of replication on epigenetic memory and consequences on gene transcription. *Phys. Biol.* 12(2), 026007 (2015).

124. Grosberg, A. Y. Disordered polymers. *Physics-Uspekhi* 40, 125–158 (1997).

125. Grosberg, A. Y. and Khokhlov, A. R. *Statistical Physics of Macromolecules.* AIP Press, (1994).

126. Mukamel, D., Ruffo, S., and Schreiber, N. Breaking of ergodicity and long relaxation times in systems with long-range interactions. *Phys. Rev. Lett.* 95, 240604 (2005).

127. Laprell, F., Finkll, K., and J., M. Propagation of Polycomb-repressed chromatin requires sequence-specific recruitment to DNA. *Science* 356, 85–88 (2017).

128. Wang, X. and Moazed, D. DNA sequence-dependent epigenetic inheritance of gene silencing and histone H3K9 methylation. *Science* 356, 88–91 (2017).

129. Coleman, R. and Struhl, G. Causal role for inheritance of H3K27me3 in maintaining the OFF state of a *Drosophila* HOX gene. *Science* 356, eaai8236 (2017).

130. Schuettengruber, B., Elkayam, N. O., Sexton, T., Entrevan, M., Stern, S., Thomas, A., Yaffe, E., Parrinello, H., Tanay, A., and Cavalli, G. Cooperativity, specificity, and evolutionary stability of Polycomb targeting in *Drosophila*. *Cell Rep.* 9, 219–233 (2014).

131. Sneppen, K. and Mitarai, N. Multistability with a metastable mixed state. *Phys. Rev. Lett.* 109, 100602 (2012).

132. Michieletto, D., Orlandini, E., and Marenduzzo, D. Polymer model with epigenetic recoloring reveals a pathway for the de novo establishment and 3D organization of chromatin domains. *Phys. Rev. X.* 6, 041047 (2016).

133. Jerabek, H. and Heermann, D. W. How chromatin looping and nuclear envelope attachment affect genome organization in eukaryotic cell nuclei. *Int. Rev. Cell Mol. Biol.* 307, 351–381 (2014).

134. Sexton, T. and Cavalli, G. The role of chromosome domains in shaping the functional genome. *Cell* 160(6), 1049–1059 (2015).

135. Tolhuis, B., Blom, M., Kerkhoven, R. M., Pagie, L., Teunissen, H., Nieuwland, M., Simonis, M., de Laat, W., van Lohuizen, M., and van Steensel, B. Interactions among Polycomb domains are guided by chromosome architecture. *PLoS Genet.* 7(3), e1001343 (2011).

136. Waddington, C. H. *The strategy of the genes*, volume 20. Routledge, (2014).

3

The Strings and Binders Switch Model of Chromatin

SIMONA BIANCO, ANDREA M. CHIARIELLO, CARLO ANNUNZIATELLA, ANDREA ESPOSITO, LUCA FIORILLO, AND MARIO NICODEMI

3.1 INTRODUCTION

Innovative technologies such as Hi-C (Lieberman-Aiden, 2009) or the more recent GAM (Beagrie et al., 2017) have revolutionized the field of chromatin architecture as they allow us to measure the frequency of physical contacts between DNA sites genome-wide. The resulting contact maps return a picture of the regulatory network of contacts across genomic regions, including for instance the physical loops between enhancers and promoters established by transcription factors (TFs) and other molecules. Such contacts have complex patterns in the cell nuclei of higher organisms, showing that chromatin has a formidable, non-random 3D structure (Bickmore and van Steensel, 2013, Tanay and Cavalli, 2013, Dekker and Mirny, 2016).

Chromosomes are folded into an arrangement of megabase-sized domains, known as Topological Associating Domains (TADs) (Nora et al., 2012, Dixon et al., 2012), which are enriched in internal interactions. TADs have, in turn, domain-specific, inner patterns (Phillips-Cremins et al., 2013) and, additionally, they interact with each other forming higher-order structures, named meta-TADs (Fraser et al., 2015). Meta-TADs encompass the so-called 'A/B compartments' of open and closed chromatin (Lieberman-Aiden et al., 2009), and extend up to chromosomal scales. A number of molecular factors involved in chromatin folding have been discovered, such as CTCF/Cohesin (Sanborn et al., 2015), active and poised polymerases (Barbieri et al., 2017), PRC1 (Kundu et al., 2017), MLL3/4 (Yan Biorxiv, 2017) and many more.

To make sense of the complexity of the experimental contact patterns and to understand the molecular mechanisms shaping chromosome architecture, concepts from polymer physics have been introduced (see, e.g., a review in Nicodemi and Pombo 2014, other contributions in this volume and original papers such as Kreth et al., 2004, Nicodemi and Prisco, 2009, Bohn and Heermann, 2010, Barbieri et al., 2012, Brackley et al., 2013, Giorgetti et al., 2014, Jost et al., 2014, Sanborn et al., 2015, Brackley et al., 2016, Fudenberg et al., 2016, Chiariello et al., 2016, Di Stefano et al., 2016, Brackley et al., 2017). In particular, in this review, we focus on the Strings and Binders Switch model, which was specifically proposed to explain, in a quantitative, predictive framework, the effects on folding of the interactions between chromosomes and their cognate binding molecules (Nicodemi and Prisco, 2009, Barbieri et al., 2012).

3.2. THE BASIC FEATURES OF THE STRINGS AND BINDERS SWITCH (SBS) MODEL

3.2.1 The strings and binders switch model

In the string & binders switch (SBS) model (Nicodemi and Prisco, 2009) a chromatin filament (a "string") is described as a self-avoiding walk (SAW) polymer chain of beads (Figure 3.1a). The chain includes specific beads that act as binding sites for diffusing molecules (the "binders"). The concentration, c_m, of the binders and the scale of their binding energy, E_{int}, to the chain are the key parameters of the model. The SBS model has been shown to well recapitulate Hi-C, GAM, and FISH data across chromosomal scales and cell types, and to explain the 3D structure of a variety of genomic loci with high accuracy (Barbieri et al., 2012, Fraser et al., 2015, Barbieri et al., 2017).

3.2.2 The phase diagram of the SBS homopolymer

As a first simple application, we discuss the thermodynamic phase diagram of the SBS homopolymer model, i.e., a chain where all beads are equal and can interact

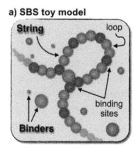

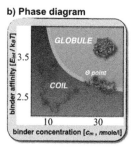

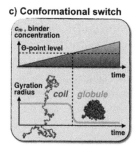

Figure 3.1. The Strings & Binders Switch (SBS) model of chromatin. a. Our SBS model is a polymer chain with binding sites for diffusing binding molecules; molecule concentration is c_m, and affinity is E_{int}. Adapted from (Bianco et al., 2017). The binders can bridge and loop the polymer, hence changing its architecture. b. Stable architectural classes of the system correspond to its emergent thermodynamic phases. The toy model of panel a), in particular, has a coil–globule phase transition, where its conformations switch from open, randomly folded states to more compact, globular ones. Adapted from (Chiariello et al., 2016). c. Conformational changes (monitored by the values of the polymer gyration radius) can be sharply controlled (switch-like) by, e.g., gene up/down-regulation (acting on c_m) or chemical modifications (acting on E_{int}) with no need of parameter fine-tuning.

with one type of binders. The thermodynamic state of the SBS homopolymer at equilibrium only depends on the interaction energy E_{int} and concentration c_m of the binders. Importantly, the scaling properties of polymer physics dictate that there is a one-to-one correspondence between the thermodynamic states of the system and its conformational folding classes. The phase diagram includes three main thermodynamic phases (Figure 3.1b, adapted from (Chiariello et al., 2016)). The system is in an open, coil state for relatively small values of c_m and E_{int}, because the binders form only a few unstable loops. The conformations of the system belong to the self-avoiding-walk conformational class, i.e., are those corresponding to an open, randomly folded chain. At higher values of c_m and E_{int}, the polymer undergoes a coil–globule phase transition and becomes folded in a collapsed, disordered lump. The conformations in the globular state are compact, and have different critical exponents with respect to the SAW state. Interestingly, at even higher values of c_{ms} and E_{int}, the binders attached to the polymer undergo an order–disorder transition and an ordered structure is established, albeit the binders have no direct interactions with each other apart from hard-core repulsion. In the ordered state the polymer is compact, arranged by the binders in a crystal-like structure (Chiariello et al., 2016).

3.2.3 A switch-like control of folding

The coil–globule transition is typical of interacting polymers and does not depend on the finer details of the system (de Gennes, 1979). Hence, the sharp phase transitions between the open and compact states envisaged within the SBS model can also explain how conformational changes in chromatin topology can be regulated in a switch-like way by robust, yet simple biological mechanisms, such as up- or down-regulation of the concentration of the polymer-binding molecules or by epigenetic modifications affecting the binding sites along the polymer (Figure 3.1c). The former mechanism increases or decreases c_m, while the latter can change E_{int}, as a way to cross the sharp transition threshold between the different phases, thus providing a switch between chromatin states. These are robust mechanisms, requiring no fine-tuning of the binder concentration or affinity. Additionally, the order of magnitude of the binder concentration, c_m, and energy, E_{int}, where the transitions are predicted, fall well within biological expectations for known transcription factors in the cell nucleus (Barbieri et al., 2012).

3.2.4 Critical exponents of the contact probability

The average pairwise contact probability, $P(s)$, of bead pairs at a given contour distance, s, characterizes the folding state of the polymer. In the coil phase, $P(s)$ has an asymptotic power-law decay with s, $P(s) \sim s^{-\alpha}$. The exponent, α, is named the critical exponent of the contact probability. In the coil state, it is equal to $\alpha \sim 2.1$, which belongs to the SAW universality class, which is well known in polymer physics. The exponent is $\alpha \sim 1.5$ at the coil–globule transition (named the polymer Θ-point (de Gennes, 1979)). The behavior of $P(s)$ in the globular state is different in the disordered and in the ordered phases. In the latter, for instance, a power–law decay with $\alpha \sim 1.0$ is found (Chiariello et al., 2016).

3.3. A MODEL OF CHROMATIN FOLDING

3.3.1 The mixture model of chromatin

Since microscopy and Hi-C experiments have shown that DNA is typically organized in eu- and hetero-chromatic domains, we hypothesized that a single chromosome is likely to be a mixture of differently folded regions (Figure 3.2a). This view has been named the *mixture model* of chromatin organization (Barbieri et al., 2012, Chiariello et al., 2016). The conformational states of the different regions of a chromosome must belong, at least at a first approximation, to the folding classes predicted by polymer physics. As the coil–globule transition is universal across interacting polymers, the phase diagram of the SBS homopolymer discussed in the previous section can provide a guideline of the main folding classes of the different regions along a chromosome.

The *mixture model* of chromatin can be tested against experimental data. For instance, the average pairwise contact probability, $P(s)$, between loci at a fixed genomic distance, s, derived from Hi-C data can be compared against the $P(s)$ predicted by the model. In a simple coarse-grained approach, where chromatin at large scales is folded in the states of a homopolymer, the predicted $P(s)$ is just a linear combination of the contact probability of the main conformational classes described in the SBS phase diagram above, each weighted by the fraction

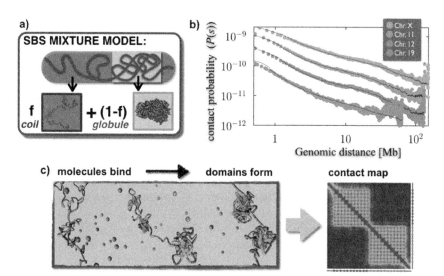

Figure 3.2. The SBS mixture model of chromatin. a. In the mixture model, chromosomes are composed of a mixture of different regions folded into the different thermodynamic states predicted by polymer physics. Adapted from (Chiariello et al., 2016). b. The model explains well genome-wide average contact Hi-C data (IMR90 cells from Dixon et al., 2012) from the sub-Mb to chromosomal scales. Adapted from (Chiariello et al., 2016). c. Patterns in the contact matrix, such as TADs, can be explained by the presence of different combinations of bridging factors along the genomic sequence. Adapted from (Barbieri et al., 2012).

of chromatin in the corresponding class. Considering the simplicity of the model, it is interesting that it can explain experimental data about $P(s)$ from the sub Mb up to chromosomal scales, over three orders of magnitude in genomic distances, as shown in Figure 3.2b (data from (Dixon et al., 2012), figure adapted from (Chiariello et al., 2016)). The fit based on the *mixture model* also gives an insight into the mixture compositions of each chromosome. Interestingly, chromosome X has the largest fraction of regions in the closed state (above 60%); gene-rich chromosomes, such as chromosome 17 or 19, are composed of much more open mixtures (>60%), as expected by biological considerations (Chiariello et al., 2016).

Note that, in this view, the exponent derived from a power-law fit of the $P(s)$ extracted from experimental data is an effective exponent, depending on the mixture composition. The critical exponents are only those corresponding to the system pure states described previously.

3.3.2 Pattern formation (TADs) in the SBS block copolymer model

The SBS model also gives insights on how patterns, such as TADs or metaTADs, can be formed in chromatin architecture (Figure 3.2c). Without entering the details of polymer physics, consider for instance the toy case of a block copolymer, i.e., a chain with two types of beads (red/green) arranged in two blocks, each interacting with a different type of binder (red/green, Figure 3.2c). As the two blocks do not interact with each other, they fold independently in one of the states envisaged within the SBS homopolymer model. For instance, in case the blocks fold into their globular states, the contact matrix of the block copolymer is marked by two distinct domains, visually similar TADs (Figure 3.2c).

3.4 THE SBS MODEL OF THE *SOX9* LOCUS IN MESC

A more realistic version of the SBS polymer model, including different types of binding sites (represented by different colors in the following figures), can explain Hi-C pairwise contact data of real loci with good accuracy. This has been shown in a variety of cases, such as the *Bmp7*, the 7q11.23, or the *Xist* loci in mouse (Chiariello et al., 2016, 2017). For illustration, here we review the case of the *Sox9* locus. This is a fundamental region in tissue development and a number of mutations there, including human structural variants, have been associated with limb malformations and congenital diseases. The SBS model of *Sox9* in mouse embryonic stem cells (mESC) (Figure 3.3) has been shown to reproduce the patterns of Hi-C data with 95% accuracy (Chiariello et al., 2016). A 3D snapshot of the locus model is shown in Figure 3.3b.

In this model, the binding domains (colors) are the physical determinants of chromatin folding patterns. Their position along the sequence is shown in Figure 3.3c (Chiariello et al., 2016). The location of the main binding domains roughly coincides with the known TADs (Dixon et al., 2012) of the locus (Figure 3.3c). However, relevant overlaps exist between the binding domains, which originate the complex sub-TAD structures visible in Hi-C data, as well as the interactions across TAD borders responsible for the formations of higher-order chromatin domains, such as metaTADs (Chiariello et al., 2016). As Hi-C data can be comparatively well explained by the SBS model, the latter offers a view of the mechanisms underlying the regulation of chromosome architecture. From the SBS model we can also derive a single-allele picture of how the locus is folded in 3D and dynamically breaths. A snapshot of the *Sox9* 3D conformation in mESC is shown in Figure 3.3b. From the 3D structure it has also been discovered that abundant many-body contacts are present, beyond pair-wise interactions, such as triplets and multi-way contacts (Chiariello et al., 2016), as experimentally confirmed by technologies such as GAM (Beagrie et al., 2017).

3.4.1 Molecular nature of the binding domains

Insights into the molecular nature of model-predicted binding domains (colors) and their cognate binding factors can be gained by cross-referencing the information on their genomic location with available epigenomic datasets (e.g., ENCODE, 2012). Figure 3.3c shows the correlations between binding domains and chromatin features in the *Sox9* region in mESC (Bianco et al., 2017). Such analysis shows that each color correlates with a combination of different

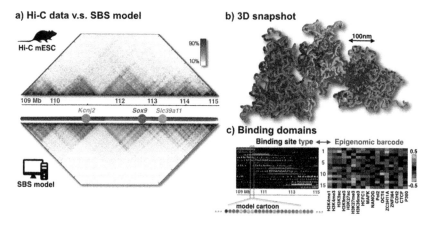

Figure 3.3. The SBS model of the *Sox9* locus in mESC. a. The SBS model can explain the folding of the *Sox9* locus with good accuracy. Hi-C (Dixon et al., 2012) v.s. model contact data have a Pearson correlation of r = 0.95. Adapted from (Chiariello et al., 2016). b. From the SBS model, single-molecule 3D conformations of chromatin can be derived. A 3D snapshot of the *Sox9* locus model is shown here. The color scheme is given in panel a). Adapted from (Chiariello et al., 2016). c. Combinations of factors, also including CTCF, correlating with the model-binding domains (colors) shaping the locus 3D architecture. Adapted from (Chiariello et al., 2016) and (Bianco et al., 2017).

factors, rather than a single mark. Some domains correlate with active marks and Pol-II, others with repressive marks. CTCF also correlates with some of the domains, but not with all. This picture is consistent with recent discoveries that a number of factors play a role in chromatin folding, ranging from architectural TFs (e.g., CTCF/Cohesin, see Sanborn et al., 2015) as well as PRC1 (Kundu et al., 2017), MLL3/4 (Yan Biorxiv, 2017), Active/Poised Pol-II (Barbieri et al., 2017), and more. These results support the view that a combinatorial code of a variety of molecular factors produces the complex architecture visible from Hi-C data.

3.5. PREDICTING THE EFFECTS OF MUTATIONS ON GENOME 3D ARCHITECTURE

The SBS polymer model of a specific locus can be employed to make predictions on the impact of genomic mutations on the 3D structure of the locus and, hence, on the network of contacts between genes and their regulators. Specifically, the variant of interest can be implemented into the polymer model of the locus and its new 3D conformation derived under only the laws of physics. The resulting contact map provides detailed information on the 3D effects of the variant.

This type of approach has been implemented in several loci, including the *Xist* (Chiariello et al., 2016) and the *Epha4* locus (Bianco et al., 2018). The successful comparison of model predictions against Hi-C data in cells bearing the same mutations provides a stringent test of the model, as no free-fitting parameters are adjusted.

3.6 CONCLUSIONS

Important advancements have occurred in recent years in the development of polymer models of chromatin. The *Fractal Globule* model (Lieberman-Aiden et al., 2009, Rosa and Everaers, 2008), an early hypothesis on chromosome folding, has turned out to be only a qualitative, pictorial description of the structure of chromosomes as, e.g., its predicted contact matrix have no patterns (neither loops, TADs, sub-TADs, metaTADs, A/B compartments, etc.) and hence cannot explain the folding of real loci.

The *SBS model* has proposed a description of chromosome folding based on basic concepts of polymer thermodynamics, where chromatin 3D structures are physically produced by local micro-phase separation of cognate binding sites along the sequence (Nicodemi and Prisco, 2009), as supported by recent experiments (Hnisz et al., 2017). The *SBS model* has been employed to dissect folding at several loci, including *Xist, Sox9, Bmp7*, and the *HoxB* regions in mESC (Scialdone, 2011, Chiariello et al., 2016, Annunziatella, 2016, Barbieri, 2017). SBS variants, informed by protein binding site data from Chip-Seq or DHS maps, have been also used to explain contact patterns across genomic regions, ranging from murine erythroblasts (Brackley et al., 2016), to *Drosophila* (Jost et al., 2014) and budding yeast (Cheng et al., 2015).

The SBS model and its variants focus on the role of thermodynamic states of the system. However, additional mechanisms, including off-equilibrium processes, could have a role in shaping chromatin 3D architecture. An important recent example is the *Loop Extrusion* (LE) model (Sanborn et al., 2015, Fudenberg et al., 2016), described in Chapter 4 in this volume, which describes the case where an active motor (possibly the cohesin complex) binds to DNA and actively extrudes chromatin loops up to the moment when it halts at oppositely oriented CTCF sites. It is still unclear whether an active extrusion mechanism is indeed systematically required as a diffusive slip-link loop to form mechanisms that could explain the (Brackley et al., 2017). Anyway, the active *LE* or its passive, slip-link version appear to well describe the folding of loci where CTCF is a main driving force. However, in a number of cases those models have been unable to explain Hi-C contact data. For instance, the 3D structure of the globin loci has been shown to be preserved in knockout experiments of CTCF and other single TFs (Brackley et al., 2016), while its folding can be described with good accuracy by the thermodynamic mechanisms envisaged by the SBS model. Another example is a recent study of the murine *HoxB* locus that illustrated how promoter interactions, based on active and poised Poll-II, direct higher-order chromatin folding in embryonic stem cells by a mechanism well described by the SBS model and not consistent with CTCF-based loop extrusion (Barbieri, 2017).

Interestingly, the basic concepts of polymer physics envisaged by the SBS can explain chromatin 3D structures from the sub-Mbp to chromosomal scales, across cell types and chromosomes. Approaches have been developed to identify the SBS model best describing the folding of any locus of interest, based only on architectural data without requiring previous knowledge of DNA-binding proteins (Chiariello et al., 2016, Bianco et al., 2018). That approach is interesting because the determinants of folding and their mechanisms of actions can be identified with no a priori assumptions beyond that chromosome conformations reflect polymer physics.

The general impression emerging from recent studies is that the folding of real chromatin is likely to result from a combination of different polymer physics mechanisms, also including thermodynamic processes, as envisaged by the SBS model, or off-equilibrium processes, as envisaged by the Loop Extrusion or Slip-Link models.

Progress in the development of principled approaches to dissect chromosome architecture can strongly improve our understanding of the fundamental molecular mechanisms shaping chromatin architecture and their profound influences on how genes are regulated in healthy tissues and in diseases. Polymer physics can, thus, help identify new diagnostic and treatment tools for diseases linked to chromatin mis-folding, such as congenital disorders and cancer (Ong and Corces, 2014, Lupiáñez et al., 2015, Valton and Dekker, 2016).

ACKNOWLEDGMENTS

Work supported by the NIH ID 1U54DK107977-01, CINECA ISCRA ID HP10CYFPS5 and HP10CRTY8P, and Einstein BIH Fellowship grants to MN, computer resources from INFN, CINECA, and *Scope/ReCAS* at the University of Naples.

REFERENCES

Annunziatella C., Chiariello A.M., Bianco S., Nicodemi M. 2016. Polymer models of the hierarchical folding of the Hox-B chromosomal locus. *Phys. Rev. E.* 94: 042402.

Barbieri M., Chotalia M., Fraser J. et al. 2012. Complexity of chromatin folding is captured by the strings and binders switch model. *Proc. Natl. Acad. Sci. USA* 109: 16173–16178.

Barbieri M., Xie S.Q., Torlai Triglia E. et al. 2017. Active and poised promoter states drive folding of the extended HoxB locus in mouse embryonic stem cells. *Nature Struct. Mol. Bio,* 24: 515.

Beagrie R.A., Scialdone A., Schueler M. et al. 2017. Complex multi-enhancer contacts captured by Genome Architecture Mapping (GAM), a novel ligation-free approach. *Nature* 543: 519.

Bianco S., Chiariello A. M., Annunziatella C., Esposito A., Nicodemi, M. 2017. Predicting chromatin architecture from models of polymer physics. *Chromosome Res.* 25: 25–34.

Bianco S., Lupiáñez D.G., Chiariello A.M. et al. 2018. Polymer physics predicts the effects of structural variants on chromatin architecture. *Nat. Gen.* 50: 662–667.

Bickmore W.A., van Steensel B. 2013. Genome architecture: Domain organization of interphase chromosomes. *Cell* 152: 1270–1284.

Bohn M., Heermann D.W. 2010. Diffusion-driven looping provides a consistent framework for chromatin organisation. *PLoS ONE* 5:e12218.

Brackley C.A., Taylor S., Papantonis A., Cook P.R., Marenduzzo D. 2013. Nonspecific bridging-induced attraction drives clustering of DNA-binding proteins and genome organisation. *Proc. Natl. Acad. Sci. USA* 110, E3605-11.

Brackley C.A., Johnson J., Kelly S., Cook P.R., Marenduzzo D. 2016. Simulated binding of transcription factors to active and inactive regions folds human chromosomes into loops, rosettes and topological domains. *Nucl. Acids Res.* 44: 3503–3512.

Brackley C.A., Johnson J., Michieletto D. et al. 2017. Non-equilibrium chromosome looping via molecular slip-links. *Phys. Rev. Lett.* 119: 138101.

Cheng T.M.K., Heeger S., Chaleil R.A.G. et al. 2015. A simple biophysical model emulates budding yeast chromosome condensation. *eLife* 4: e05565.

Chiariello A.M., Annunziatella C., Bianco S., Esposito A., Nicodemi M. 2016. Polymer physics of chromosome large-scale 3d organisation. *Scientific Reports* 6: 29775.

Chiariello A.M., Esposito A., Annunziatella C. et al. 2017. A polymer physics investigation of the architecture of the murine orthologue of the 7q11.23 human locus. *Front. Neurosci.* 11: 559.

de Gennes, P. G. 1979. *Scaling Concepts in Polymer Physics.* Cornell Univ. Press, Ithaca, NY.

Dekker J., Mirny L. 2016. 3D genome as moderator of chromosomal communication. *Cell* 164: 1110–1121.

Di Stefano M., Paulsen J., Lien T.G., Hovig E., Micheletti C. 2016 Hi-C-constrained physical models of human chromosomes recover functionally-related properties of genome organization. *Scientific Reports* 6: 35985.

Dixon J.R., Selvaraj S., Yue F., et al. 2012. Topological domains in mammalian genomes identified by analysis of chromatin interactions. *Nature* 485: 376–380.

Fraser J., Ferrai C., Chiariello A.M. et al. 2015. Hierarchical folding and reorganisation of chromosomes are linked to transcriptional changes during cellular differentiation. *Mol. Syst. Biol.* 11: 852.

Fudenberg G., Imakaev M., Lu C. et al. 2016. Formation of chromosomal domains by loop extrusion. *Cell Reports* 15: 1–12.

Giorgetti L., Galupa R., Nora E.P. et al. 2014. Predictive polymer modeling reveals coupled fluctuations in chromosome conformation and transcription. *Cell* 157: 950–963.

Hnisz, D., Shrinivas, K., Young, R.A., Chakraborty, A.K., and Sharp, P.A. 2017. A phase separation model for transcriptional control. *Cell* 169: 13–23.

Jost D., Carrivain P., Cavalli G., Vaillant C. 2014. Modeling epigenome folding: Formation and dynamics of topologically associated chromatin domains. *Nucleic Acids Res.* 42: 9553–9561.

Kreth G., Finsterle J., von Hase J., Cremer M., Cremer C. 2004. Radial arrangement of chromosome territories in human cell nuclei: a computer model approach based on gene density indicates a probabilistic global positioning code. *Biophys. J.* 86: 2803–2812.

Kundu S., Ji F., Sunwoo H. et al. 2017. Polycomb repressive complex 1 generates discrete compacted domains that change during differentiation. *Molecular Cell* 65: 432.

Lieberman-Aiden E., van Berkum N.L., Williams L. et al. 2009. Comprehensive mapping of long-range interactions reveals folding principles of the human genome. *Science* 326: 289–293.

Lupiáñez D.G., Kraft K., Heinrich V. et al. 2015. Disruptions of topological chromatin domains cause pathogenic rewiring of gene-enhancer interactions. *Cell* 161: 1.

Nicodemi M., Pombo A. 2014. Models of chromosome structure. *Curr. Opin. Cell Biol.* 28C: 90–95.

Nicodemi M., Prisco A. 2009 Thermodynamic pathways to genome spatial organization in the cell nucleus. *Biophys. J.* 96: 2168–2177.

Nora E.P., Lajoie B.R., Schulz E.G. et al. 2012. Spatial partitioning of the regulatory landscape of the X-inactivation centre. *Nature* 485: 381–385.

Ong C.T., Corces V.G. 2014. CTCF: An architectural protein bridging genome topology and function. *Nat. Rev. Genet.* 15: 234–246.

Phillips-Cremins J.E., Sauria M.E., Sanyal A. et al. 2013. Architectural protein subclasses shape 3D organization of genomes during lineage commitment. *Cell* 153: 1281–1295.

Rosa A., Everaers R. 2008. Structure and dynamics of interphase chromosomes. *PLoS Comput. Biol.* 4: e1000153.

Sanborn A.L., Rao S.S.P., Huang S.C et al. 2015. Chromatin extrusion explains key features of loop and domain formation in wild-type and engineered genomes. *Proc. Natl. Acad. Sci. USA* 112: E6456–E6465.

Scialdone A., Cataudella I., Barbieri M., Prisco A., Nicodemi M. 2011. Conformation regulation of the X chromosome inactivation center: a model. *PLoS Comput. Biol.* 7: e1002229.

Tanay A., Cavalli G. 2013. Chromosomal domains: Epigenetic contexts and functional implications of genomic compartmentalization. *Curr. Opin. Genet. Dev.* 23: 197–203.

The ENCODE Project Consortium. 2012. An integrated encyclopedia of DNA elements in the human genome. *Nature* 489: 57–74.

Valton, A.L., Dekker J. 2016. TAD disruption as oncogenic driver. *Curr. Opin. Genet. Dev.* 36: 34–40.

Yan J., Chen S.A., Local A., et al. 2017. Histone H3 Lysine 4 methyltransferases MLL3 and MLL4 Modulate Long-range Chromatin Interactions at Enhancers. *Biorxiv* 110239: 1–61. dx.doi.org/10.1101/110239.

4

Loop Extrusion: A Universal Mechanism of Chromosome Organization

LEONID A. MIRNY AND ANTON GOLOBORODKO

4.1 INTRODUCTION

Challenges and goals of biophysical modeling of chromosome organization

One of the main challenges in characterizing chromosome organization is its almost paradoxical duality. On the one hand, conformations of the chromatin chain are highly variable from cell to cell, and changing in time, thus forming a diverse ensemble of structures. On the other, Hi-C data from recent years have revealed several distinct and locus-specific patterns of interaction prominent in this ensemble, though present in a small fraction of cells (Cattoni et al., 2017; Anon, n.d.). These observations are in stark contrast to the structures of proteins where domains are either folded into distinct conformations with little or very localized structural variation, or remain largely unfolded and disordered (i.e. natively unfolded proteins). Together, these observations suggest that chromosome organization can be described by a conformational ensemble of polymers, where sufficiently weak *interactions* and *active processes* maintain transient, and likely functional, elements of organization (loops, domains, contacts, etc.) within polymers that are disordered on other length scales (e.g., within loops and between loops).

Efforts to reconstruct such ensembles have been recently reviewed and classified (Imakaev et al., 2015; Marti-Renom & Mirny, 2011). Our team has been developing different *de novo* approaches where we model genome organization based on prior assumptions and mechanisms, and then select specific models and their parameters based on agreement with experimental Hi-C and microscopy data. Such models can include phenomenological physical models (e.g., coarse-grained models in Gibcus et al., 2018) or polymer simulations of various levels of details (e.g., Fudenberg et al., 2016; Goloborodko et al. 2016; Falk et al., 2018; Nuebler et al., 2017; Tjong et al., 2012; Rosa & Everaers, 2008; Jerabek & Heermann, 2014; Rosa et al., 2010; see details in Section 4.3). These models typically have three goals: The first goal is to determine the general class of polymer architectures that are consistent with experimental observations (e.g., the non-equilibrium fractal globule (Lieberman-Aiden et al., 2009; Rosa et al., 2010), the bottlebrush of consecutive loops (Naumova et al., 2013) or random-walk-like organization in budding yeast (Tjong et al., 2012)). The second goal is to characterize these architectures more quantitatively, estimating their specific parameters (e.g., loop sizes (Fudenberg et al., 2016), chromosome geometry (Tjong et al., 2012), properties of the chromatin fiber (Arbona et al., 2017) or relative strength of interactions e.g., (Falk et al., 2018)). The third goal is to reveal mechanisms and processes that underlie the formation of such architectures and shape chromosome organization (e.g., microphase separation in (Falk et al., 2018), condensation by "binders-and-switchers" (Barbieri et al., 2012), or loop extrusion with barriers in (Fudenberg et al., 2016)).

Models of conformational ensembles allow inference of structural elements not otherwise visible in experimental data

These goals are better achieved by modeling approaches that generate ensembles of polymer conformations, either *de novo* (Tjong et al., 2012) or by inference from Hi-C data (Serra et al., 2017), rather than by models of individual conformations (Lesne et al., 2014; Mercy et al., 2017). Statistical properties of ensembles can then be compared to features of experimental data. For example, the curves of the contact probability $P(s)$ with genomic distance s can be directly compared between models and Hi-C data (Rosa et al., 2010; Lieberman-Aiden et al., 2009; Fudenberg et al., 2016; Gibcus et al., 2018; Naumova et al., 2013); distance between genomic elements can be compared between models and microscopy data (Therizols et al., 2010; Tjong et al., 2012; Tark-Dame et al., 2014; Rosa & Everaers, 2008). Similarly, other macroscopic features such as shapes and ordered banding of prophase chromosomes can be compared to simulations as well (Zhang & Heermann, 2011; Goloborodko, et al. 2016). Sweeping parameter values or other forms of inference (Arbona et al., 2017) helps to find specific values of sizes, shapes, and characteristics of these structures.

Importantly, modeling allows us to identify the elements of structural organization and infer their quantitative characteristics that are not otherwise visible in Hi-C data or microscopy (e.g., the sizes or arrangements of loops in mitotic chromosomes (Naumova et al., 2013; Gibcus et al., 2018), or the spiraling of bacterial chromosomes (Umbarger et al., 2011)). While such structural models provide insights about elements of organization, they are not always based on biologically and physically realistic assumptions (e.g., assuming loops are pinned in space or unrealistically long-range or position-dependent potentials of interactions) and hence do not necessarily provide information about microscopic processes that generate such structures or shape conformational ensembles. Achieving the third goal of suggesting microscopic molecular mechanisms that shape macroscopic chromosome organization is a major challenge for modeling.

From equilibrium to non-equilibrium to active models of chromosomes

Chromatin is a polymer and hence its organization can be described by statistical polymer physics. The first studies describing conformations of long DNA molecules by the model of a freely jointed chain (i.e., a random walk) date back to the 1960s and 70s (Cohen and Eisenberg, 1966; Schellman, 1974), while one of the first quantitative comparisons of interphase chromosomal organization to the random walk polymer model was done in van den Engh et al., (1992). Since then, numerous studies have described chromosomes with equilibrium polymer models (Dekker et al., 2002; Hahnfeldt, 1993) models with fixed loops (Bohn et al., 2007; Sachs et al., 1995; Münkel & Langowski, 1998) or more complicated rosettes of loops (Münkel et al., 1999), but were limited by the imaging data available for a handful of loci (Mateos-Langerak et al., 2009; Yokota et al., 1995). Models of loops used in (Münkel et al., 1999; Bohn et al., 2007) assumed fixed, predefined positions of loops, and are drastically different from the dynamically extruded loops discussed in Section 4.3. While models with fixed loops can serve as adequate structural models, the model of loop extrusion actually provides a *mechanism* for how such loops can be formed.

One of the first non-equilibrium polymer models of chromosomes was considered in Rosa & Everaers (2008), who argued that topological constraints, i.e., the inability of chains to pass through each other, should slow the mixing of chromosomes, preventing their equilibration between cell divisions. This study further demonstrated that such non-equilibrium polymer organization can lead to observed chromosomal territories. Another non-equilibrium model was proposed in the first Hi-C study (Lieberman-Aiden et al., 2009) and suggested that topologically constrained compact states, conjectured earlier (Yu. Grosberg et al., 1988) and known as the crumpled (fractal) globules are consistent with the contact probability scaling *P(s)* observed in Hi-C.

As the resolution of Hi-C data increased, new patterns of chromosome organization because apparent, most prominently Topologically Associating Domains (TADs). Explaining mechanisms of TAD formation using equilibrium or nonequilibrium models (Barbieri et al., 2012; Jost et al., 2014) was fraught with serious difficulties: First, TADs form only local enrichments of intra-TAD contacts without showing the long-range patterns that emerged in many models (Jost et al., 2014) based on preferential interactions. Second, models based on preferential interactions failed to reproduce experiments which showed that neighboring TADs merge when a small CTCF-containing boundary element between them has been deleted. More recent attempts to adopt such models to reproduce TADs required a model with as many as 21 types of monomers (Bianco et al., 2018) interacting with each other in different ways. Surprisingly, the authors found that the behavior of the models does not change when simulated CTCF boundaries are removed.

A model that was able to reproduce both TADs (Fudenberg et al., 2016) and changes in chromosome organization in various mutants (Fudenberg et al., 2018) is based on the concept of loop extrusion with barriers, which was originally proposed to explain mitotic compaction (Nasmyth, 2001; Alipour & Marko, 2012), and similar to earlier proposals of loop reeling (Riggs, 1990). Proposed purely on theoretical grounds, the mechanism of loop extrusion remained hypothetical until recently (Ganji et al., 2018; Fudenberg et al., 2018).

Loop extrusion – a new paradigm of active chromosome folding by motor proteins.

In the process of loop extrusion, a *Loop Extruding Factor* (LEF) associates with the DNA or the chromatin fiber and forms a progressively expanding loop (Figure 4.1). This process is driven by energy consumption (*active process*) and hence does work to change chromosome conformation. A LEF can be thought of as two connected molecular motors that progressively move along chromatin in opposite directions. It is not known whether both motors move (two-sided extrusion (Fudenberg et al., 2016; Goloborodko, Imakaev, et al. 2016)), or only one moves while the other is anchored (one-sided extrusion (Ganji et al., 2018), whether they move at the same speed, or at the same time (see (Fudenberg et al., 2018) for discussion).

SMC proteins are likely candidates for loop extruding factors. SMC proteins are present in all forms of life.

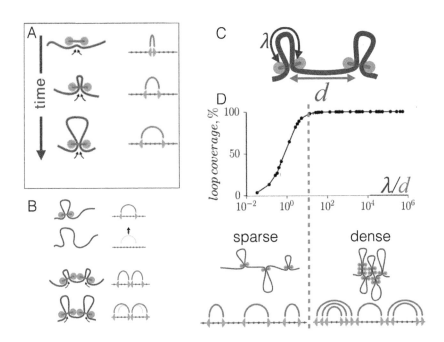

Figure 4.1 The process of loop extrusion (A); rules of interactions between LEFs (B) as implemented in (Fudenberg et al., 2016; Fudenberg et al., 2018). (C) Two length-scale parameters that determine dynamics of the system (see text); (D) Coverage by the loops and two steady-state regimes of compaction (Goloborodko, Marko, et al. 2016).

Emerging experimental evidence (see Section 4.2) point at SMC (Structural Maintenance of Chromosomes) complexes as potential LEFs. Strikingly, SMC proteins are present in all forms of life, including bacteria, archaea, and eukarya (Haering & Gruber, 2016a; Haering & Gruber, 2016b). SMC proteins play several key roles in chromosome organization. In well-characterized bacteria with single circular chromosomes, loop extrusion by SMCs begins at a specific location close to the origin of replication; and extrusion along the two chromosomal arms leads to their juxtaposition (Wang et al., 2017; Badrinarayanan et al., 2015). In mammals, the canonical role of the SMC protein complex *cohesin* is to co-align sister chromatids, while the role of SMC complexes of *condensin* is to compact mitotic chromosomes. As discussed here, cohesin also plays a crucial role in the formation of chromosomal domains (Fudenberg et al., 2016; Sanborn et al., 2015; Sofueva et al., 2013). On the contrary, in budding yeast, it is *cohesin* that compacts chromosomal arms during mitosis (Schalbetter et al., 2017), while in *S. pombe*, cohesin appears to establish interphase domains (Mizuguchi et al., 2014). In summary, SMCs play crucial roles in mitotic compaction, alignment of sister chromatids, meiotic chromosome organization, dosage compensation, and in the formation of interphase domains. Loop extrusion can be central for many of the functions of SMCs.

In Section 4.2 we review recent theoretical and experimental studies that suggest how loop extrusion can organize and reorganize chromosomes.

4.2 PHYSICS OF LOOP EXTRUSION AND CHROMOSOME ORGANIZATION

Extrusion by multiple LEFs can lead to the formation of reinforced loops.

Recent studies (Alipour & Marko, 2012; Goloborodko, Marko, et al. 2016) focused on the behavior of a dynamically growing system of loops, while setting aside the effects of these loops on 3D polymer conformation. These studies considered dynamics in the system of multiple LEFs that extrude loops and dissociate and associate back to the polymer in a random location, including other loops. LEFs, however, cannot pass by each other, i.e., loops cannot cross each other, consistent with topological entrapment by SMCs (Figure 4.1). What kind of loop structures can emerge in this system: branching tree-like structures or a regular array of loops? Does the system reach a steady state with a finite loop size or do loops grow till they reach the size of a whole polymer?

Since a LEF can associate within already existing loops, nested loops can be formed (Figure 4.1). The emergence of nested loops creates an interesting hierarchy of larger "parental" loops, and smaller "child" loops. If the growth of a parental loop is blocked, a child loop can grow until it reaches the size of the parental loop, creating a "reinforced" loop with multiple LEFs at its base. As individual LEFs dissociate, surplus LEFs can drop in to replace them, like the planks of the ship of Theseus, enabling reinforced loops to remain intact. If several LEFs associate within the same parent at about the same time, they can grow until they split the parental loop. These processes are central to dynamics in the dense regime discussed below.

Loop extrusion is characterized by two length scales, achieves a steady state, and shows two dynamical regimes: a sparse regime and a dense regime.

Simulations and analytical treatment (see (Goloborodko, Marko, et al. 2016) and below) shows that the dynamics of the system can be fully described by two length scales: (Figure 4.1)

λ – the processivity of a LEF, i.e., the average size of a loop extruded over LEF's residence time on DNA if the extrusion process were unobstructed. If the mean residence time is τ and the velocity of the loop extrusion is υ, then $\lambda = \upsilon\tau$.

d – the average separation between LEFs along the polymer. For N, LEFs associate with a polymer of length L, $d = L/N$.

Two regimes of the system naturally emerge in the two limiting cases: the sparse regime when $d \gg \lambda$, and the dense regime when $d \ll \lambda$. In the sparse regime, growing loops are separated by large distances, hence they do not interact often, and the system rapidly reaches steady state with average loop size $l = \lambda$. Although little linear compaction is achieved by such sparse loops, this regime does facilitate interactions between neighboring (cis) regions of the genome and may serve as a framework for chromosome organization during interphase (see Section 4.2.2).

In the dense regime ($d \ll \lambda$), neighboring loops constantly collide and block each other as they expand, and individual loops get reinforced by multiple LEFs. The system reaches a steady state because of a balance between two processes: loop death (when all LEFs at a base of a loop dissociate) and loop division (when a loop gets split into two child loops). This results in a steady state where the average loop size is defined by λ and d as $l \approx d \, log(\lambda/d) < \lambda$, and where each loop is reinforced by multiple LEFs ($\approx log(\lambda/d)$). No tree-like branching structures are formed because LEFs rapidly reach loop bases, causing them to reinforce existing loops. Interestingly, this dense steady state is rapidly reached, requiring just a few LEF turnover times. This regime also leads to efficient chromosome compaction as 100% of the fiber is extruded into loops and any spontaneously emerging gaps are rapidly extruded by nearby LEFs (Goloborodko et al. 2016; Goloborodko et al. 2016).

This simple model provides a framework for understanding the effects of loop extrusion with the interphase corresponding to the sparse regime (provided additional extrusion barriers are established), and the metaphase corresponding to the dense regime (accompanied by poor solvent conditions). While several biological details may be missing from this physical model, it successfully reproduces several vital biological phenomena that could not be explained by other processes (see the following sections).

4.2.1 Loop extrusion during mitosis

Loop extrusion in the dense regime ($d \ll \lambda$) can reproduce compaction and segregation of mitotic chromosomes.

Two phenomena are central to chromosome reorganization during cell division: (i) compaction of individual chromosomes and (ii) their concomitant segregation from one another.

Several aspects of these processes are remarkable from the physics point of view. First, chromosomes are compacted (~500-fold in length, vertebrates) forming elongated rather than globular structures. This is in striking contrast to compaction by a poor solvent or by crosslinking agents (Barbieri et al., 2012) as these would necessarily minimize the surface area and hence lead to a spherical collapsed state (Grosberg, 1994). Second, each chromosome in this compacted state maintains its linear order, i.e., the order of genomic elements in the elongated chromosome resembles their order along the genome. Third, the compaction machinery is able to distinguish sister chromatids (identical copies of the same chromosome) and compact them separately, i.e., form interactions (crosslinks) exclusively within individual chromatids rather than between them. Fourth, the process of compaction coincides with the segregation of sister chromatids. Fifth, originally intertwined sister chromatids become topologically disentangled, which is surprising given the general tendency of polymers to become more intertwined as they are concentrated (Marko, 2011). Finally, and most strikingly, all these processes are accomplished by the collective action of protein complexes each being a thousand times smaller than compacted chromosomes. These complexes cannot distinguish chromosomes from each other (due to their similar DNA sequence) and are unable to communicate with each other (due to the screening of long-range interactions in physiological solvents). Chromosome reorganization for cell division constitutes a remarkable multiscale self-organization phenomenon where hundreds of thousands of nanometer-sized molecules act on meter-long polymers, assembling them into regular and organized micron-sized mitotic chromosomes (Figure 4.2).

These remarkable processes take place due to activities of two protein enzymes: topoisomerase II and condensin (Hirano & Mitchison, 1994; Wood & Earnshaw, 1990; Hirano, 1995), with topoisomerase II playing an essential but largely supportive function (Hirano & Mitchison, 1993). Strikingly, 3D simulations of loop extrusion in the dense regime ($\lambda/d > 20$) (Goloborodko, Imakaev, et al. 2016) show that myriads of LEFs, representing *condensins*, can accomplish all these tasks by producing compacted and elongated chromatids, that maintain linear order in the compact state, and that faithfully segregate from each other (Figure 4.2). The activity of topoisomerase II was essential for complete segregation and was modeled by weak topological constraints allowing chains to pass through each other. Details of these simulations are presented in Section 4.3.

Loop extrusion can accomplish this because it generates an array of consecutive loops without gaps in individual chromosomes, as has been proposed in Nasmyth (2001). Classical microscopy studies and recent Hi-C experiments show that mitotic chromosomes are indeed formed by arrays of consecutive loops (Paulson & Laemmli, 1977; Earnshaw & Laemmli, 1983; Naumova et al., 2013). Formation of such arrays naturally results in lengthwise chromosome compaction while maintaining genomic order. Steric repulsion between chromosomes shaped as bottlebrushes of loops further leads to segregation of chromosomes (Goloborodko, Imakaev, et al. 2016).

Simulations also show that loop extrusion naturally leads to the formation of the central condensin-rich scaffold, that has been observed by microscopy in

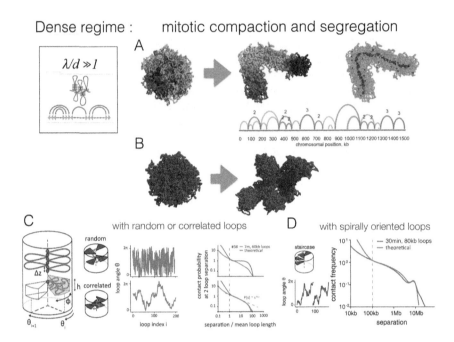

Figure 4.2 The dense regime as a model of mitotic chromosomes compactions. (A) Snapshots of simulations from (Goloborodko, Imakaev, et al. 2016) illustrating that the loop extrusion leads to the formation of the prophase-line chromosome that has (i) organized linear structures (reflected by colors); (ii) elongated morphology of prophase chromosome; and (iii) central condensin-rich scaffold. The loop diagram shows structures of emerging loops with nested and reinforced loops (shown by the number above arches). (B) Loop extrusion leads to segregation of two identical chromatids. (C and D) Analytically solvable models of mitotic chromosomes with ordered loop arrays and they comparison with time-resolved Hi-C from (Gibcus et al., 2018).

mitotic chromosomes (Saitoh et al., 1994; Maeshima & Laemmli, 2003). As loops are extruded, condensins accumulate next to each other, forming a continuous condensin scaffold, which is in turn surrounded by extruded loops that run away from the scaffold to maximize their conformational entropy. Thus, loop extrusion (in the presence of simulated topo II) can compact polymers into structures that reproduce several hallmarks of mitotic chromosomes (Goloborodko et al. 2016).

Time-resolved Hi-C data for mitotic compaction allow testing various polymer models of loop organization

Most recently, Hi-C for several time points through prophase and prometaphase of highly synchronized cells have been produced (Gibcus et al., 2018). These data allowed testing models of mitotic chromosome compaction and organization. As the first step in this direction, models for individual time-points sampled by Hi-C have been developed (Gibcus et al., 2018). While models are consistent with the dense arrays of loops organized around the central scaffold, as produced by loop extrusion, they suggest additional features of chromosome organization that may not emerge from loop extrusion alone.

While loop extrusion in the dense regime leads to the formation of a bottle-brush of loops, relative orientations of loops that emanate from the scaffold are not specified by this model (Figure 4.2). Cylindrical symmetry of mitotic chromosomes and radial organization of loops in the bottlebrush, however, allowed the development of analytically solvable polymer models for which contact probability *P(s)* curves can be computed (Gibcus et al., 2018). In these models (Figure 4.2), each loop is assumed to be a Gaussian polymer blob of a particular height and angular width. Loops are located consecutively along the chromosomal axis (z axis). The individual models differ in the rules governing the relative orientations of the loops: random independent orientations, random but correlated orientations of neighboring loops (angular random walk), and orientations that follow a spiral path (modeled by Ornstein-Uhlenbeck angular random walk with an angular drift) (Gibcus et al., 2018). For each model, one can analytically compute the contact probability curve *P(s)* as a function of parameters of the model, such as the linear density of loops, their high and angular width, and model-specific parameters (correlations between neighboring loops, pitch of the spiral etc). Despite the existence of several free parameters, models yield different *P(s)* curves, some of which don't agree with the *P(s)* from Hi-C data, while others do. Different time points of mitotic compaction can be reproduced by different models. Models found analytically can then be implemented as polymer simulations and investigated further.

Prophase and prometaphase chromosomes are formed by loop arrays with nested loops and spiral scaffolds

Polymer models that can reproduce Hi-C reveal two novel aspects of chromosome organization. First, while prophase chromosomes (first 5–10 minutes of compaction) are formed by dense arrays of loops where consecutive loops are pointing in correlated directions, chromosomes at later stages of compaction (30–60 min) are formed by loops that emanate from the scaffold following a spiral path (Figure 4.2). Spiral organization in animal mitotic chromosomes has been observed for decades (Jane, 1934; Ohnuki, 1965; Manton, 1950) but these observations were largely attributed to artifacts of chromosome isolation. The reporting of spirals in Hi-C data suggests that spiral scaffolds may indeed be present in native chromosomes. Second, while loop sizes grow from 40 Kb to 80 Kb in the early stages of compaction, data for later stages suggest that very high linear densities of loops can be achieved through nested loops, i.e., larger loops that are split into smaller ones.

Although dense arrays of loops emanating from the central scaffold are consistent with the basic loop-extrusion model, novel aspects of chromosome organization such as nested loops and spiral scaffolds discovered in this study go far beyond simple loop extrusion, requiring significant extensions of the model to include two types of condensins and interactions between condensins.

Several important biophysical questions about mitotic compaction remain to be answered.

First, the formation of nested loops can in principle be accomplished by two types of loop extruders with different processsvities and/or timing of action. This agrees with two types of condensins, with condensin II that starts earlier in prophase and condensin I that gains access to chromatin later (after the nuclear envelope breakdown). The activity of the two types of loop extruders however constitutes a new class of models that need to be studied and can have rich behaviors.

Second, the formation of co-oriented loops at early stages of compaction and spiral formation at later stages need to be explained. One possibility is that specific protein–protein interactions between condensins at the bases of neighboring loops lead to their coorientation and to further ordering into a spiral. Alternatively, polymer mechanisms can lead to coorientation and spiraling of the scaffold as the linear density of loops increases (Maritan et al., 2000).

Third, models of mitotic chromosomes discussed above require compaction of extruded loops by some external mechanism, akin to poor solvent conditions. A mechanism that leads to a transition from good solvent conditions during interphase to poor solvent remains to be found. Poor solvent conditions during prophase and prometaphase, however, may lead to sticking of chromosomes and chromatids to each other. Preventing such aggregation of chromosomes under poor solvent conditions may require some additional surfactant that can coat chromosomes in a similar way to the Ki67 (Booth et al., 2016; Cuylen et al., 2016) or BAF (Samwer et al., 2017).

Fourth, the recent observation of single-sided loop extrusion (Ganji et al., 2018) would require revisiting these models, aiming to understand whether single-sided extrusion can achieve required levels of compaction and other elements of organization. It is possible, however, that the single-sided extrusion that was observed for condensin from *S.cerevisiae* is a specific process for this class of organisms where chromosomes are short, requiring little compactions (Lazar-Stefanita et al., 2017) and where cohesin, rather than condensin, plays a major role in mitotic compaction (Schalbetter et al., 2017).

4.2.2 Loop extrusion during interphase

The concept of loop extrusion developed for mitotic compaction have been extended to model interphase domains (TADs).

Loop extrusion in the interphase has been recently thoroughly reviewed (Dekker & Mirny, 2016; Fudenberg et al., 2018), so here we will summarize the main results and challenges. Topologically Associated Domains (TADs) are a hallmark of interphase chromosomes (Dixon et al., 2012; Nora et al., 2012). Each domain constitutes a consecutive region of the genome enriched in interactions with itself and partially insulated (~2-fold) from neighboring regions. Two features of domains are central to their function. First, the insulation of domains from their neighbors is established by domain boundaries that constitute DNA sequences occupied by specific proteins (CTCF, YY1, and others); when such sequences or proteins are removed, domains merge with their neighbors (Sanborn et al., 2015; Nora et al., 2017; Wutz et al., 2017a; Guo et al.,

2015; Narendra et al., 2015). Second, domains establish regulatory "neighborhoods", i.e., regulatory DNA sequences (e.g., enhancers) located within a domain largely act on genes within the same domain (Lupiáñez et al., 2015; Symmons et al., 2014).

These aspects of domain function are remarkable from the physics point of view:

How can the binding of ~5 nm-size proteins to DNA provide insulation or control interactions between genomic regions located ~1000 nm away along the genome and ~100 nm away in space? How can two contacting genomic regions (e.g., a regulatory region and a promoter) recognize that they belong to the same domain, rather than to two neighboring domains? These examples of *action across scales* are reminiscent of the phenomena of self-organization and segregation of mitotic chromosomes by much smaller proteins.

Loop extrusion in the sparse regime ($\lambda/d \sim 1$) and with extrusion barriers can explain the formation and function of interphase domains

To model interphase domains one needs to add extrusion barriers (Figure 4.3), which can block propagation of LEFs once they meet the barrier (Alipour & Marko, 2012; Fudenberg et al., 2016; Sanborn et al., 2015). Polymer simulations of the loop extrusion with barriers can reproduce TADs as seen in Hi-C, both qualitatively and quantitatively (Fudenberg et al., 2018; Fudenberg et al., 2016). Interestingly, the best agreement with Hi-C data is achieved for the sparse regime ($d \geq \lambda$), closer to its upper limit of $\lambda/d \sim 1$, when about ~50% of chromatin is extruded into loops and hence chromosomes are not compacted. This result from simulations agrees with WaplKO experiments where a ~2-fold decrease in d and a ~10-fold increase in λ has led to the formation of extended and compacted "vermicelli" chromosomes (Wutz et al., 2017b; Tedeschi et al., 2013; Haarhuis et al., 2017; Gassler et al., 2017), characteristic of the dense regime (likely $\lambda/d \sim 10$).

Importantly, loop extrusion explains seemingly paradoxical *action across scales*, allowing small proteins to facilitate and control interactions at much larger scales (Dekker & Mirny, 2016): Loop extrusion leads to the formation of additional transient contacts at and near active LEFs. By stopping a LEF (likely, *cohesin*) at a domain border a small CTCF protein prevents the formation of extruded loops across domain borders. While an extrusion barrier cannot control spatial contacts that happen across domain borders, as evident from single-nucleus Hi-C (Flyamer et al., 2017), it suppresses additional inter-domains contact facilitated by loop extrusion.

If such extrusion-mediated interactions are central for gene regulation, loop extrusion with barriers at domain borders would also facilitate such interactions within domains and suppress them between domains.

This model can also take into account the directionality of CTCF boundaries (Vietri Rudan et al., 2015), partial permeability of boundaries due to rapid CTCF exchange (Hansen et al., 2017), boundaries of different strength, and other phenomena (see (Fudenberg et al., 2018) for review). A model with semipermeable and directional boundaries can reproduce not only the formation of domains, but also enrichment of contacts between domain boundaries ("corner peaks"), scanning of domain interiors by boundary elements ("flames" first reported in

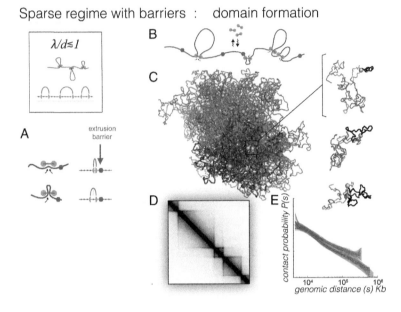

Sparse regime with barriers : domain formation

Figure 4.3 The sparse regime of loop extrusion as a model for interphase chromosome organization. (A) Extrusion barriers block propagation of LEFs (cohesin) in one direction while allowing progressive loop extrusion at the other end of the LEF. (B) A diagram of loop extrusion in the sparse regime with barriers. (C) Simulations of 30,000 beads representing 15Mb of chromatin in periodic boundary conditions, with the loop extruding factors (shown in yellow). The inset shows three snapshots of a loop extruded by a LEF. (D and E) Results of the simulations: (D) The average contact map clearly shows insulated domains, with "flames" and "corner peaks" similar to such Hi-C features. (E) The scaling of the contact probability *P(s)* with genomic distance s for 50 best models (lines) within (green) and between (magenta) domains, as compared to experimental within and between domain *P(s)* for 300–500 Kb domains from Hi-C data (Fudenberg et al., 2016).

Fudenberg et al., 2016), and the emergence of more complex patterns of contact peaks in wild-type and mutant cells.

Loop extrusion forms domains by creating dynamic intra-domain contacts rather than by forming stable CTCF-CTCF loops.

It is a common misconception that central to the formation of a domain is the establishment of a stable loop between CTCFs at its boundaries. Several studies (Doyle et al., 2014; Benedetti et al., 2014; Fudenberg et al., 2016), however, have demonstrated that a stable loop between two boundaries can neither lead to the enrichment of interactions inside a domain, nor insulate neighboring domains from each other. Although CTCF-occupied boundaries show enrichment of contacts between them, according to microscopy such contacts are present in only ~5–10% of cells (Cattoni et al., 2017).

Loop extrusion with barriers, on the contrary, increases the frequency of intra-domain contacts by creating transient contacts at and around extruding cohesin. Since cohesins are expected to extrude at high speed (~20–40 Kb/min (Fudenberg et al., 2018)) and exchange with the nucleoplasm every 5–20 minutes (Gerlich et al., 2006; Wutz et al., 2017b), these induced contacts are highly dynamic and variable from cell to cell. However, these dynamic contacts are central to establishing domains and can be insulated by loop extrusion barriers.

Models of passive barriers cannot establish reliable insulation.

Other models of domain formation suggested that the unique physical properties of boundary elements can make them insulate neighboring domains (Dixon et al., 2016; Stadler et al., 2017). Dixon et al. suggested that more rigid boundary elements can insulate interactions between neighboring domains made by more flexible chromatin. Stadler et al suggest that less compacted boundary regions can stretch in space, thus increasing the distance between domains and achieving insulations. Polymer simulations of boundaries formed by long rigid or longer, less compacted stretches of chromatin, however, show that such types of boundary cannot achieve robust insulation ((Fudenberg et al., 2016) Fig S5 there). While regions proximal to stiff or extended boundaries are insulated, regions further away do not feel the presence of the border and contact across boundaries as frequently as within a domain. From the physics point of view it is clear that the conformational ensemble of long and spatially confined polymer corresponding to two consecutive domains (e.g., ~200 Kb ~400 persistence lengths) cannot be altered by the increased length or rigidity of ~1–5 persistence length fragments.

Loop extrusion can interfere with spatial segregation of active and inactive chromatin

Recent experiments that targeted the loading of cohesin on DNA (Schwarzer et al., 2017) have demonstrated not only that loop extrusion by cohesin is central to the formation of domains, but also that another layer of chromatin organization, namely A/B *compartments* are formed by a cohesin-independent mechanism. Spatial segregation of active (eu-) and inactive (hetero-) chromatin in metazoan nuclei has been known for decades (Wrinch, 1934; Frolova, 1938), and is visible in Hi-C maps as a checkerboard pattern reflecting preferential interactions of active, A compartment regions, with other active regions, and inactive, B compartment regions, with other inactive ones (Lieberman-Aiden et al., 2009). This segregation is likely driven by interactions between different types of chromatin, with inactive–inactive interactions being the strongest (Falk et al., 2018) and possibly mediated by HP1 or other similar proteins (Strom et al., 2017; Larson et al., 2017). Several studies suggested that this interaction-driven segregation constitutes a microphase separation of block copolymers, a well-known model in polymer physics (Jost et al., 2014; Nuebler et al., 2017).

Interestingly, the phase separation of active and inactive chromatin is diminished by cohesin-mediated loop extrusion. Experiments where chromatin-associated cohesin was significantly depleted revealed stronger and finer compartmentalization (Schwarzer et al., 2017). Simulations of the (mirco)phase separation of chromatin subject to additional loop extrusion indicate that the phase separation can be diminished by the process of active extrusion (Nuebler et al., 2017).

Several sources of evidence support the presence of loop extrusion and pose new intriguing questions

Proposed purely on theoretical grounds, to explain compaction and segregation in mitosis (Nasmyth, 2001; Alipour & Marko, 2012) or domain formation (Fudenberg et al., 2016) and cis interactions (Riggs, 1990) during interphase, the process of loop extrusion is now supported by three sources of evidence (see (Fudenberg et al., 2018) for review).

The first source of evidence constitutes cohesin depletion and enrichment experiments, which show not only the loss of domain upon depletion of cohesin, but also the loss of local compaction, consistent with the loop extrusion by cohesin. While CTCF depletion also leads to the loss of domains, peaks, and flames, it has no effect on local compaction, consistent with its primarily instructive function in establishing domain boundaries (Nora et al., 2017). Enrichment of cohesin by Wapl deletion leads not only to stronger domains and their associated "peaks" and "flames", but also to overcompaction of chromosomes as evidenced in Hi-C and by microscopy (Tedeschi et al., 2013; Wutz et al., 2017b; Gassler et al., 2017; Haarhuis et al., 2017), and consistent with the increased loop extrusion activity.

The second source of evidence comes from the recovery of chromosomal features upon reactivation of loop extrusion in mammalian cells (Rao et al., 2017; Wutz et al., 2017b) or in bacteria where sites of SMC loading were relocated (Wang et al., 2017; Tran et al., 2017). Time-lapse Hi-C in these bacterial studies also provided direct measurements of the SMC speed of extrusion as 25–50 Kb/min.

The third source of evidence comes from *in vitro* single-molecule experiments on isolated SMCs. These experiments clearly demonstrated that SMCs are ATP-dependent motors, which can either translocate along DNA at a speed of ~4 Kb/min (Terakawa et al., 2017) or extrude DNA loops at 40–80 Kb/min (Ganji et al., 2018), while consuming ~2 ATP molecules per sec (Eeftens et al., 2017). Importantly, these and a recent ATP and transcription inhibition study (Vian et al., 2018) show that SMCs have ATP-dependent loop extruding activity without requiring the assistance of other known DNA motors (e.g., polymerases or helicases).

New questions and challenges for understanding loop extrusion and beyond

The first challenge is the detection of extruded loops and the process of extrusion *in vivo*. These loops are highly dynamic (~5–20 minutes) and vary from cell to cell. While they can be captured by single-cell Hi-C or imaging, they are not distinct from any other contacts. Since these loops are formed at different places in different cells they are not immediately visible or distinct in population Hi-C or multiplexed imaging (Anon n.d.) either. Detecting extruded loops would require the development of new techniques that can unambiguously establish their presence from static or dynamic data.

The second challenge is to understand the roles that loop extrusion can play in a variety of processes that involve genomic DNA. The critical role of loop extrusion in the formation of interphase domains suggests a role in regulating gene expression by modulating enhancer–promoter interactions. Although perturbation of cohesin/CTCF system has led to changes in gene expression, these changes were localized and modest. Direct evidence of expression modulation by loop extrusion are yet to be found.

The discovery of this novel motor-driven extrusion is similar to the discovery of cytoplasmic motor proteins in the 1980s, which, as we now know, play crucial roles in a variety of cellular processes from cell and tissue motility and to cell division and signaling. Similarly, one may expect loop extrusion to play important roles not only in mitosis and regulation of gene expression (Merkenschlager & Odom, 2013) but also in meiosis, allele-specific expression (Savova et al., 2013; Horta et al., 2018; Guo et al., 2012) and dosage compensation (Bonora et al., 2018; Crane et al., 2015), replication, splicing (Kim et al., 2018), DNA repair (Vian et al., 2018), switching of differentiation programs, and mutagenesis and recombination (Jain et al., 2018).

The third challenge is to understand the molecular mechanism behind the regulation of loop extrusion and its interplay with other cellular processes. The models described previously assumed uniform loading and unloading of cohesin and condensins, exponential residence times and a constant extrusion speed, and a lack of specific interactions with other DNA-bound proteins other than occlusion by boundary elements. In reality, certain loci and processes may recruit additional cohesins, stop, move, pause or evict them. Boundary elements can also be modified, stabilized or evicted by protein or DNA modifications. The slowdown of SMC-mediated loop extrusion by transcription, evident in mammals (Wutz et al., 2017b) and bacteria (Tran et al., 2017; Wang et al., 2017), can also play an important role. Another complex interplay can arise between loop extruding SMCs and cohesins connecting sister chromatids (Stanyte et al., 2018); and possibly similar interplay between static and dynamic SMCs can occur in the meiotic synaptonemal complex. As discussed above, loop extrusion can interfere with other mechanisms of chromosome folding; understanding the interplay of extrusion with functional RNAs, polycomb, homolog pairing, and other players and processes of chromosome organization can bring many new biological insights.

Understanding the evolution of the universal process that is present in all forms of life, the origin of new SMCs, the origin and evolution of boundary proteins, and parallel evolution innovations in body plan and development in metazoans can shed light on their roles of SMCs and loop extrusion in development.

Loop extrusion also creates new avenues of research in polymer physics. As an active process acting on polymers it can maintain the polymer system of chromosomes away from equilibrium, leading to new phenomena of self-organization typical for active systems (e.g., Grosberg & Joanny, 2015).

Finally, understanding molecular mechanisms by which ATP hydrolysis is translated into the motor activity of loop extrusion is an emerging biophysical challenge (Eeftens et al., 2017; Kschonsak et al., 2017; Marko et al., 2018). Characterizing and understanding the molecular mechanism of force generation in cytoplasmic motor proteins took decades and required the concerted efforts of cell biologists, biochemists, single-molecule biophysics, and theoreticians. We hope that in the following years we will gain an equally deep understanding of the universal process of loop extrusion and its role in genome and cell function.

4.3 ELEMENTS OF POLYMER SIMULATIONS

Setting up in silico simulations.

In coarse-grained polymer simulations of chromosomes, chromatin is modeled as a chain of particles connected by elastic bonds (Figure 4.4). Depending on the goal of the simulation and available computational resources, each particle may represent as much as a few kb of chromatin (Nuebler et al., 2017; Falk et al., 2018) and as little a single nucleosome (~200 bp) (Gibcus et al., 2018). These particles are subject to various forces, which model the key aspects of the physics of the chromatin fiber and the nuclear environment. Among the typical forces used in simulations are:

1. interparticle repulsion. The most basic and essential force, which is modeled on the fact that chromatin is a self-avoiding polymer. Often the repulsion energy for completely overlapping particles is capped at an intermediate value (~3–5 kT), which enables chain passing and allows the polymer to achieve topological equilibration. From the biological standpoint, this trick models the action of the nuclear enzyme topoisomerase II, which passes two adjacent strands of DNA through each other;
2. bending stiffness. This force is used to impose the directional persistence of DNA and chromatin fiber; importantly, the exact parameters of the chromatin stiffness are currently disputed (Maeshima et al., 2010; Arbona et al., 2017) and likely vary genome-wide and between the experimental conditions;

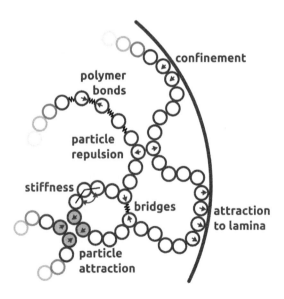

Figure 4.4 The typical system of force used in polymer simulations of chromosomes. Black spheres show particles making up the chromatin fiber, red arrows indicate forces.

3. particle-to-particle attraction. This force models the "sticky" behavior of chromatin fiber, which, under certain conditions, leads to condensation of chromatin into a dense phase. Such forces have been successfully used to explain compartmentalization of active and inactive chromatin inside the nucleus (Nuebler et al., 2017) and formation of inverted nuclei (Falk et al., 2018);

4. confinement. This constraint accounts for the fact that chromatin occupies a major fraction of the nuclear volume. Depending on the type of the simulated system, different types of confinement are used: in whole genome simulations, spherical confinement represents the entire nuclear envelope (Schalbetter et al., 2017; Tjong et al., 2012); in simulations of smaller genomic regions, periodic boundary conditions constrain the chromatin density while avoiding edge effects (Fudenberg et al., 2016); finally, models of vertebrate mitotic chromosomes use cylindrical confinement to impose the characteristic geometry of condensed chromosomes during cell division (Naumova et al., 2013; Gibcus et al., 2018);

5. attraction to the lamina of specific genomic loci. This force is used to impose peri-nuclear localization of inactive chromatin, or telomeres or centromeres in certain types of nuclei (Tjong et al., 2012; Falk et al., 2018);

6. bridges, i.e., extra bonds connecting non-adjacent loci of the polymer. Such forces can be used to model the formation of loops by SMC complexes condensins and cohesins, and to represent molecular links between sister chromosomes at centromeres (Goloborodko et al., 2016).

In a typical polymer simulation, the dynamics of chromatin is modeled using the Langevin equation of motion. This equation represents a modification of Newton's equation of motion, describing the action of forces on individual particles, and additionally accounts for the thermal motion of simulated particles caused by random collisions with the solvent particles. This equation can be solved numerically using several existing computational packages, including OpenMM (Eastman et al., 2013), LAMMPS (Plimpton, 1995) and HOOMD-blue (Anderson et al., 2008).

In most cases, the goal of a polymer simulation is to sample the statistics of chromatin conformations in the thermal equilibrium, which can be achieved by running the simulation for a sufficiently long time. In equilibrium, the statistics of chromatin conformations depends neither on the initial conditions, nor on the exact dynamics of the system. This property simplifies equilibrium polymer simulations in three major ways. First, the statistics of equilibrium conformations does not depend on the initial conformation, which is essential for simulations of a real biological system, where most of the time we do not know the initial arrangement of the chromatin. Second, the statistics of the equilibrium conformations does not depend on the parameters describing the dynamics of the system, e.g., the mass of the chromatin, the density, and viscosity of the solvent, the frequency of collisions with solvent particles, etc. This means that deviations of these parameters from their real values, which often are unknown, do not affect the result of the simulations. Third, the freedom to choose the dynamic parameters of the system means that we can use "unrealistic" values of these parameters to speed up the convergence to the equilibrium by 4 or even 5 orders of magnitude.

Simulating loop extrusion.

This simulation framework can be easily extended to simulate the process of loop extrusion on chromatin using two-stage simulations. In the first stage, we model the motion of loop extruding factors (LEFs) along the chromatin chain, ignoring the 3D conformation of chromatin. At this stage, we describe the chromatin as a 1-dimensional lattice of sites, with individual LEFs occupying some of these sites. We then iteratively update the positions of LEFs over time, accounting for their motor-like motion along the chromatin chain, collisions with each other and other obstacles (e.g., chromatin-bound proteins) as well as the stochastic unbinding and rebinding of LEFs to chromatin. As a result, we obtain the dynamics of LEF positions in time.

In the second stage, we model how the LEF-bridged loops affect chromatin conformation. At this stage, we model chromosomes using the polymer simulations, as described above, and additionally impose bridging interactions between pairs of loci connected by LEFs, either in a quasi-static or in a dynamic fashion. In quasi-static simulations, we pick a single random snapshot of LEF positions, as produced in the first stage of the simulations, and generate an equilibrium ensemble of chromatin conformations for these LEF positions. We then repeat such simulations for a large number of LEF snapshots and as a result obtain the steady-state ensemble of conformations of a chromosomal region undergoing active loop extrusion. In dynamic simulations, we repeatedly update the positions of LEFs within the same polymer simulation to reflect the LEF dynamics, simulated at the first stage. As a result, dynamic loop extrusion simulations explicitly model how chromatin conformation changes in time due to the active loop extrusion.

The key differences between these two styles of simulations are in the underlying assumptions. The quasi-static simulations assume that LEFs extrude loops sufficiently slowly so that the chromatin fiber has enough time to equilibrate between the consecutive steps of LEFs. The dynamic simulations explicitly model the motion of chromatin between consecutive steps of LEFs dynamics, and as a result, explicitly depend on the speed of extrusion as well as the dynamics of chromatin motion (Fudenberg et al., 2016; Nuebler et al., 2017). Each approach has their strengths and weaknesses – the quasi-static approach is robust to the variations in the dynamic parameters, while the dynamic approach is more generic and thus potentially more accurate. More research on the dynamics of chromatin motion and loop extrusion is needed before we can definitively decide which of the two approaches is superior.

Open questions and challenges in polymer simulations of chromosomes

Despite significant progress in coarse-grained polymer simulations of chromosomes, there remain a number of open questions and challenges:

1. *The effect of loop extrusion on the global dynamics of chromatin.* Until very recently, most studies of chromatin dynamics modeled chromosomes as passively diffusing polymer fibers. However, the discovery of loop extrusion

revealed that at the microscopic level chromosomes are constantly "stirred" by motor proteins that actively consume and dissipate energy. How this active motion affects the polymer dynamics on different length and time scales (Grosberg & Joanny, 2015; Prost et al., 2015; Smrek & Kremer, 2017), how it impacts other aspects of chromosome organization (Nuebler et al., 2017), and whether such systems can still be accurately approximated as high-temperature passive systems remains to be studied.

2. *Position-dependent variation of chromatin dynamics.* Recent experimental studies revealed a number of molecular mechanisms that affect the conformation of specific loci, e.g., compartmentalization of active and inactive chromatin, attachment of specific loci to the lamina as well as a sophisticated extrusion landscape, with some loci serving as extrusion boundaries and others possibly serving as loading sites for LEFs. Importantly, all of these mechanisms may affect the dynamics in a locus-specific manner. This has neither been explored theoretically, nor taken into account in analyses of the experimental data. Emerging live-cell imaging data (Gu et al., 2018; Germier et al., 2017; Bronshtein et al., 2016) will provide a source of information for such studies.

3. *The effect of the hydrodynamic interactions on the chromatin motion.* Extensive theoretical and experimental studies of polymer dynamics have shown that the correlated motion of solvent particles can increase the polymer mobility over a broad range of timescales (Rubinstein & Colby, 2003). The importance of such hydrodynamic interactions for chromatin dynamics remains poorly understood and not properly captured by current simulation techniques.

4. *Large-scale simulations of genomes.* Simulating relatively large genomic systems (e.g., the entire mammalian genome or even a single chromosome) for physiologically relevant time scales (hours to days) currently requires heavy coarse-graining of chromosome geometry (by using a large amount of DNA per particle) and dynamics (by reducing viscosity many orders of magnitude below its true value). The effect of such coarse-graining on the accuracy of the simulations remains to be studied. Performing large-scale computations without such approximations is a major technical challenge that would require breakthroughs, both in hardware and in software.

5. *Accurate dynamic chromatin simulations of loop extrusion.* Due to the scarcity of experimental data on loop extrusion, the first generation of loop extrusion simulations had to make a number of assumptions regarding the dynamics of polymer motion and LEFs action. As a result, such simulations currently have a limited quantitative accuracy in reproducing the experimental contact maps. To improve the predictive power of such simulations, we need to fully understand the molecular mechanisms of loop extrusion as well as accurately measure the basic parameters of this process *in vivo*, such as the speed of loop extrusion and the number of active LEFs. On the theoretical side, accurate prediction of contact maps needs careful investigation of the interplay between the diffusive chromatin motion and active extrusion.

6. *The influence of chromatin conformation on LEF dynamics.* The recent experimental study of loop extrusion by yeast condensins (Ganji et al., 2018)

revealed that these LEFs have a low stalling force *in vitro* and cannot extrude loops on DNA under tension. On the experimental side, it remains to be shown if this stalling effect occurs *in vivo* and with other types of LEFs. On the theoretical side, it remains completely unknown how stalling affects the extrusion dynamics and thus chromosome conformation; moreover, incorporating stalling into the standard simulation framework is not straightforward and may require a novel type of simulations.

ACKNOWLEDGMENTS

We are grateful to all members of the Mirny lab for many productive discussions, and specifically to Nezar Abdennur for providing feedback on the manuscript. This work was supported by MIT-France Seed Funds, NIH (GM114190) and NSF, Physics of Living Systems (15049420) grants and by the Center for 3D Structure and Physics of the Genome of NIH 4DN Consortium (DK107980).

REFERENCES

Alipour, E. & Marko, J.F., 2012. Self-organization of domain structures by DNA-loop-extruding enzymes. *Nucleic Acids Research*, 40(22), pp.11202–11212.

Anderson, J.A., Lorenz, C.D. & Travesset, A., 2008. General purpose molecular dynamics simulations fully implemented on graphics processing units. *Journal of Computational Physics*, 227(10), pp.5342–5359.

Anon, *Heterogeneity and Intrinsic Variation in Spatial Genome Organization.* Paperpile. Available at: https://paperpile.com/shared/oZB7zg [Accessed June 2, 2018].

Arbona, J.M. et al., 2017. Inferring the physical properties of yeast chromatin through Bayesian analysis of whole nucleus simulations, *Genome Biology*, 18(1), p. 81.

Badrinarayanan, A., Le, T.B.K. & Laub, M.T., 2015. Bacterial chromosome organization and segregation. *Annual Review of Cell and Developmental Biology*, 31, pp.171–199.

Barbieri, M. et al., 2012. Complexity of chromatin folding is captured by the strings and binders switch model. *Proceedings of the National Academy of Sciences*, 109(40), pp.16173–16178.

Benedetti, F. et al., 2014. Models that include supercoiling of topological domains reproduce several known features of interphase chromosomes. *Nucleic Acids Research*, 42(5), pp.2848–2855.

Bianco, S. et al., 2018. Polymer physics predicts the effects of structural variants on chromatin architecture. *Nature Genetics*, 50(5), pp.662–667.

Bohn, M., Heermann, D.W. & van Driel, R., 2007. Random loop model for long polymers. *Physical Review. E, Statistical, Nonlinear, and Soft Matter Physics*, 76(5 Pt 1), p.051805.

Bonora, G. et al., 2018. Orientation-dependent Dxz4 contacts shape the 3D structure of the inactive X chromosome. *Nature Communications*, 9(1), p.1445.

Booth, D.G. et al., 2016. 3D-CLEM reveals that a major portion of mitotic chromosomes Is not chromatin. *Molecular Cell*, 64(4), pp.790–802.

Bronshtein, I. et al., 2016. Exploring chromatin organization mechanisms through its dynamic properties. *Nucleus*, 7(1), pp.27–33.

Cattoni, D.I. et al., 2017. Single-cell absolute contact probability detection reveals chromosomes are organized by multiple low-frequency yet specific interactions. *Nature Communications*, 8(1), p.1753.

Cohen, G. & Eisenberg, H., 1966. Conformation studies on the sodium and cesium salts of calf thymus deoxyribonucleic acid (DNA). *Biopolymers*, 4(4), pp.429–440.

Crane, E. et al., 2015. Condensin-driven remodelling of X chromosome topology during dosage compensation. *Nature*, 523(7559), pp.240–244.

Cuylen, S. et al., 2016. Ki-67 acts as a biological surfactant to disperse mitotic chromosomes. *Nature*, 535(7611), pp.308–312.

Dekker, J. et al., 2002. Capturing Chromosome Conformation. *Science*, 295, p.1306.

Dekker, J. & Mirny, L., 2016. The 3D genome as moderator of chromosomal communication. *Cell*, 164(6), pp.1110–1121.

Dixon, J.R. et al., 2012. Topological domains in mammalian genomes identified by analysis of chromatin interactions. *Nature*, 485(7398), pp.376–380.

Dixon, J.R., Gorkin, D.U. & Ren, B., 2016. Chromatin domains: The unit of chromosome organization. *Molecular Cell*, 62(5), pp.668–680.

Doyle, B. et al., 2014. Chromatin loops as allosteric modulators of enhancer-promoter interactions. *PLoS Computational Biology*, 10(10), p.e1003867.

Earnshaw, W.C. & Laemmli, U.K., 1983. Architecture of metaphase chromosomes and chromosome scaffolds. *The Journal of Cell Biology*, 96(1), pp.84–93.

Eastman, P. et al., 2013. OpenMM 4: A reusable, extensible, hardware independent library for high performance molecular simulation. *Journal of Chemical Theory and Computation*, 9(1), pp.461–469.

Eeftens, J.M. et al., 2017. Real-time detection of condensin-driven DNA compaction reveals a multistep binding mechanism. *The EMBO Journal*, p.e201797596.

van den Engh, G., Sachs, R. & Trask, B.J., 1992. Estimating genomic distance from DNA sequence location in cell nuclei by a random walk model. *Science*, 257(5075), pp.1410–1412.

Falk, M. et al., 2018. Heterochromatin drives organization of conventional and inverted nuclei. *Biorxiv*, p.244038. Available at: https://www.biorxiv.org/content/early/2018/01/09/244038.abstract [Accessed January 25, 2018].

Flyamer, I.M. et al., 2017. Single-nucleus Hi-C reveals unique chromatin reorganization at oocyte-to-zygote transition. *Nature*, 544(7648), pp.110–114.

Frolova, S., 1938. development of the inert regions of the salivary gland chromosomes of Drosophila. *Nature*, 142, pp.357–358.

Fudenberg, G. et al., 2018. Emerging evidence of chromosome folding by loop extrusion. *Cold Spring Harbor Symposia on Quantitative Biology*. Available at: http://dx.doi.org/10.1101/sqb.2017.82.034710.

Fudenberg, G. et al., 2016. Formation of chromosomal domains by loop extrusion. *Cell Reports*, 15(9), pp.2038–2049.

Ganji, M. et al., 2018. Real-time imaging of DNA loop extrusion by condensin. *Science*, 360(6384), pp.102–105.

Gassler, J. et al., 2017. A mechanism of cohesin-dependent loop extrusion organizes zygotic genome architecture. *The EMBO Journal*, 36(24), pp.3600–3618.

Gerlich, D. et al., 2006. Live-cell imaging reveals a stable cohesin-chromatin interaction after but not before DNA replication. *Current Biology: CB*, 16(15), pp.1571–1578.

Germier, T. et al., 2017. Real-time chromatin dynamics at the single gene level during transcription activation. *Biorxiv*. Available at: https://www.biorxiv.org/content/early/2017/02/23/111179.abstract.

Gibcus, J.H. et al., 2018. A pathway for mitotic chromosome formation. *Science*. Available at: http://dx.doi.org/10.1126/science.aao6135.

Goloborodko, A. et al., 2016. Compaction and segregation of sister chromatids via active loop extrusion. *Elife*, 5, pp. 1–16. e14864. Available at: http://dx.doi.org/10.7554/eLife.14864.

Goloborodko, A., Marko, J.F. & Mirny, L.A., 2016. Chromosome compaction by active loop extrusion. *Biophysical Journal*, 110(10), pp.2162–2168.

Grosberg, A.Y. & Khokhlov, A.R 1994. *Statistical Physics of Macromolecules*, AIP Press, New York.

Grosberg, A.Y. & Joanny, J.F., 2015. Nonequilibrium statistical mechanics of mixtures of particles in contact with different thermostats. *Physical Review. E, Statistical, Nonlinear, and Soft Matter Physics*, 92(3), p.032118.

Gu, B. et al., 2018. Transcription-coupled changes in nuclear mobility of mammalian cis-regulatory elements. *Science*, 359(6379), pp.1050–1055.

Guo, Y. et al., 2015. CRISPR inversion of CTCF sites alters genome topology and enhancer/promoter function. *Cell*, 162(4), pp.900–910.

Guo, Y. et al., 2012. CTCF/cohesin-mediated DNA looping is required for protocadherin α promoter choice. *Proceedings of the National Academy of Sciences of the United States of America*, 109(51), pp.21081–21086.

Haarhuis, J.H.I. et al., 2017. The cohesin release factor WAPL restricts chromatin loop extension. *Cell*, 169(4), pp.693–707.e14.

Haering, C.H. & Gruber, S., 2016a. SnapShot: SMC protein complexes part I. *Cell*, 164(1–2), pp.326–326.e1.

Haering, C.H. & Gruber, S., 2016b. SnapShot: SMC protein complexes part II. *Cell*, 164(4), p.818.e1.

Hahnfeldt, P. et al., 1993. Polymer models for interphase chromosomes. *Proceedings of the National Academy of Sciences of the United States of America*, 90, p.7854.

Hansen, A.S. et al., 2017. CTCF and cohesin regulate chromatin loop stability with distinct dynamics. *Elife*, 6, e25776. Available at: https://elifesciences.org/articles/25776.

Hirano, T., 1995. Biochemical and genetic dissection of mitotic chromosome condensation. *Trends in Biochemical Sciences*, 20(9), pp.357–361.

Hirano, T. & Mitchison, T.J., 1994. A heterodimeric coiled-coil protein required for mitotic chromosome condensation in vitro. *Cell*, 79(3), pp.449–458.

Hirano, T. & Mitchison, T.J., 1993. Topoisomerase II does not play a scaffolding role in the organization of mitotic chromosomes assembled in Xenopus egg extracts. *The Journal of Cell Biology*, 120(3), pp.601–612.

Horta, A. et al., 2018. Cell type-specific interchromosomal interactions as a mechanism for transcriptional diversity. *Biorxiv*, 287532. Available at: https://www.biorxiv.org/content/early/2018/03/23/287532.abstract.

Imakaev, M.V., Fudenberg, G. & Mirny, L.A., 2015. Modeling chromosomes: Beyond pretty pictures. *FEBS Letters*, 589(20 Pt A), pp.3031–3036.

Jain, S. et al., 2018. CTCF-binding elements mediate accessibility of RAG substrates during chromatin scanning. *Cell*, 74(1), pp.102–116. Available at: http://dx.doi.org/10.1016/j.cell.2018.04.035.

Jane, F.W., 1934. Memoirs: The structure of the somatic chromosomes of Alstroemeria and Bomarea. *Journal of Cell Science*, s2–77(305), pp.49–75.

Jerabek, H. & Heermann, D.W., 2014. How chromatin looping and nuclear envelope attachment affect genome organization in eukaryotic cell nuclei. *International Review of Cell and Molecular Biology*, 307, pp.351–381.

Jost, D. et al., 2014. Modeling epigenome folding: formation and dynamics of topologically associated chromatin domains. *Nucleic Acids Research*, 42(15), pp.9553–9561.

Kim, J.S. et al., 2018. Cohesin interacts with a panoply of splicing factors required for cell cycle progression and genomic organization. *Biorxiv*, p.325209. Available at: https://www.biorxiv.org/content/early/2018/05/17/3 25209.abstract [Accessed June 4, 2018].

Kschonsak, M. et al., 2017. Structural basis for a safety-belt mechanism that anchors condensin to chromosomes. *Cell*, 171(3), pp.588–600.e24.

Larson, A.G. et al., 2017. Liquid droplet formation by HP1α suggests a role for phase separation in heterochromatin. *Nature*, 547(7662), pp.236–240.

Lazar-Stefanita, L. et al., 2017. Cohesins and condensins orchestrate the 4D dynamics of yeast chromosomes during the cell cycle. *The EMBO Journal*, 36(18), pp.2684–2697.

Lesne, A. et al., 2014. 3D genome reconstruction from chromosomal contacts. *Nature Methods*, 11(11), pp.1141–1143.

Lieberman-Aiden, E. et al., 2009. Comprehensive mapping of long-range interactions reveals folding principles of the human genome. *Science*, 326, pp.1–7.

Lupiáñez, D.G. et al., 2015. Disruptions of topological chromatin domains cause pathogenic rewiring of gene-enhancer interactions. *Cell*, 161(5), pp.1012–1025.

Maeshima, K., Hihara, S. & Eltsov, M., 2010. Chromatin structure: does the 30-nm fibre exist in vivo? *Current Opinion in Cell Biology*, 22(3), pp.291–297.

Maeshima, K. & Laemmli, U.K., 2003. A two-step scaffolding model for mitotic chromosome assembly. *Developmental Cell*, 4(4), pp.467–480.

Manton, I., 1950. The spiral structure of chromosomes. *Biological Reviews of the Cambridge Philosophical Society*, 25(4), pp.486–508.

Maritan, A. et al., 2000. Optimal shapes of compact strings. *Nature*, 406(6793), pp.287–290.

Marko, J.F. et al., 2018. DNA-segment-capture model for loop extrusion by structural maintenance of chromosome (SMC) protein complexes. *Biorxiv*, p.325373. Available at: https://www.biorxiv.org/content/early/2018/05/17/3 25373 [Accessed May 29, 2018].

Marko, J.F., 2011. Scaling of linking and writhing numbers for spherically confined and topologically equilibrated flexible polymers. *Journal of Statistical Physics*, 142(6), pp.1353–1370.

Marti-Renom, M.A. & Mirny, L.A., 2011. Bridging the resolution gap in structural modeling of 3D genome organization. *PLoS Computational Biology*, 7(7), p.e1002125.

Mateos-Langerak, J. et al., 2009. Spatially confined folding of chromatin in the interphase nucleus. *Proceedings of The National Academy of Sciences of the United States of America*, 106(10), pp.3812–3817.

Mercy, G. et al., 2017. 3D organization of synthetic and scrambled chromosomes. *Science*, 355(6329). Available at: http://dx.doi.org/10.1126/science. aaf4597.

Merkenschlager, M. & Odom, D.T., 2013. CTCF and cohesin: linking gene regulatory elements with their targets. *Cell*, 152(6), pp.1285–1297.

Mizuguchi, T. et al., 2014. Cohesin-dependent globules and heterochromatin shape 3D genome architecture in S. pombe. *Nature*, 516(7531), pp.432–435.

Münkel, C. et al., 1999. Compartmentalization of interphase chromosomes observed in simulation and experiment. *Journal of Molecular Biology*, 285(3), pp.1053–1065.

Münkel, C. & Langowski, J., 1998. Chromosome structure predicted by a polymer model. *Physical Review E*, 57(5), pp.5888–5896.

Narendra, V. et al., 2015. CTCF establishes discrete functional chromatin domains at the Hox clusters during differentiation. *Science*, 347(6225), pp.1017–1021.

Nasmyth, K., 2001. Disseminating the genome: Joining, resolving, and separating sister chromatids during mitosis and meiosis. *Annual Review of Genetics*, 35, p.673.

Naumova, N. et al., 2013. Organization of the mitotic chromosome. *Science*, 342(6161), pp.948–953.

Nora, E.P. et al., 2012. Spatial partitioning of the regulatory landscape of the X-inactivation centre. *Nature*, 485(7398), pp.381–385.

Nora, E.P. et al., 2017. Targeted degradation of CTCF decouples local insulation of chromosome domains from genomic compartmentalization. *Cell*, 169(5), pp.930–944.e22.

Nuebler, J. et al., 2017. Chromatin organization by an interplay of loop extrusion and compartmental segregation. *Proceedings of the National Academy of Sciences*, 115(29), pp.E6697-E6706.

Ohnuki, Y., 1965. Demonstration of the spiral structure of human chromosomes. *Nature*, 208(5013), pp.916–917.

Paulson, J.R. & Laemmli, U.K., 1977. The structure of histone-depleted metaphase chromosomes. *Cell*, 12(3), pp.817–828.

Plimpton, S., 1995. Fast parallel algorithms for short-range molecular dynamics. *Journal of Computational Physics*, 117(1), pp.1–19.

Prost, J., Jülicher, F. & Joanny, J.F., 2015. Active gel physics. *Nature Physics*, 11, p.111.

Rao, S.S.P. et al., 2017. Cohesin loss eliminates all loop domains. *Cell*, 171(2), pp.305–320.e24.

Riggs, A.D., 1990. DNA methylation and late replication probably aid cell memory, and type 1 DNA reeling could aid chromosome folding and enhancer function. *Philosophical Transactions of the Royal Society of London. Series B, Biological Sciences*, 326(1235), pp.285–297.

Rosa, A., Becker, N.B. & Everaers, R., 2010. Looping probabilities in model interphase chromosomes. *Biophysical Journal*, 98(11), pp.2410–2419.

Rosa, A. & Everaers, R., 2008. Structure and dynamics of interphase chromosomes. *PLoS Computational Biology*, 4(8), p.e1000153.

Rubinstein, M. & Colby, R., 2003. *Polymer Physics*, Oxford University Press.

Sachs, R.K. et al., 1995. A random-walk/giant-loop model for interphase chromosomes. *Proceedings of the National Academy of Sciences of the United States of America*, 92(7), pp.2710–2714.

Saitoh, N. et al., 1994. ScII: An abundant chromosome scaffold protein is a member of a family of putative ATPases with an unusual predicted tertiary structure. *The Journal of Cell Biology*, 127(2), pp.303–318.

Samwer, M. et al., 2017. DNA Cross-Bridging Shapes a Single Nucleus from a Set of Mitotic Chromosomes. *Cell*, 170(5), pp.956–972.e23.

Sanborn, A.L. et al., 2015. Chromatin extrusion explains key features of loop and domain formation in wild-type and engineered genomes. *Proceedings of the National Academy of Sciences of the United States of America*, 112(47), pp.E6456–E6465.

Savova, V., Vigneau, S. & Gimelbrant, A.A., 2013. Autosomal monoallelic expression: Genetics of epigenetic diversity? *Current Opinion in Genetics & Development*, 23(6), pp.642–648.

Schalbetter, S.A. et al., 2017. SMC complexes differentially compact mitotic chromosomes according to genomic context. *Nature cell biology*, 19(9), pp.1071–1080.

Schellman, J.A., 1974. Flexibility of DNA. *Biopolymers*, 13(1), pp.217–226.

Schwarzer, W. et al., 2017. Two independent modes of chromatin organization revealed by cohesin removal. *Nature*, 551(7678), pp.51–56.

Serra, F. et al., 2017. Automatic analysis and 3D-modelling of Hi-C data using TADbit reveals structural features of the fly chromatin colors. *PLoS Computational Biology*, 13(7), p.e1005665.

Smrek, J. & Kremer, K., 2017. Small activity differences drive phase separation in active-passive polymer mixtures. *Physical Review Letters*, 118(9), p.098002.

Sofueva, S. et al., 2013. Cohesin-mediated interactions organize chromosomal domain architecture. *The EMBO Journal*, 32(24), pp.3119–3129.

Stadler, M.R., Haines, J.E. & Eisen, M.B., 2017. Convergence of topological domain boundaries, insulators, and polytene interbands revealed by high-resolution mapping of chromatin contacts in the early Drosophila melanogaster embryo. *Elife*, 6. Available at: http://dx.doi.org/10.7554/eLife.29550.

Stanyte, R. et al., 2018. Dynamics of sister chromatid resolution during cell cycle progression. *The Journal of Cell Biology*. Available at: http://dx.doi.org/10.1083/jcb.201801157.

Strom, A.R. et al., 2017. Phase separation drives heterochromatin domain formation. *Nature*, 547(7662), pp.241–245.

Symmons, O. et al., 2014. Functional and topological characteristics of mammalian regulatory domains. *Genome Research*, 24(3), pp.390–400.

Tark-Dame, M. et al., 2014. Depletion of the chromatin looping proteins CTCF and cohesin causes chromatin compaction: Insight into chromatin folding by polymer modelling. *PLoS Computational Biology*, 10(10), p.e1003877.

Tedeschi, A. et al., 2013. Wapl is an essential regulator of chromatin structure and chromosome segregation. *Nature*, 501(7468), pp.564–568.

Terakawa, T. et al., 2017. The condensin complex is a mechanochemical motor that translocates along DNA. *Science*, 11. Available at: http://www.sciencemag.org/lookup/doi/10.1126/science.aan6516.

Therizols, P. et al., 2010. Chromosome arm length and nuclear constraints determine the dynamic relationship of yeast subtelomeres. *Proceedings of the National Academy of Sciences of the United States of America*, 107(5), pp.2025–2030.

Tjong, H. et al., 2012. Physical tethering and volume exclusion determine higher-order genome organization in budding yeast. *Genome Research*, 22(7), pp.1295–1305.

Tran, N.T., Laub, M.T. & Le, T.B.K., 2017. SMC progressively aligns chromosomal arms in Caulobacter crescentus but is antagonized by convergent transcription. *Cell Reports*, 20(9), pp.2057–2071.

Umbarger, M.A. et al., 2011. The three-dimensional architecture of a bacterial genome and its alteration by genetic perturbation. *Molecular Cell*, 44(2), pp.252–264.

Vian, L. et al., 2018. The energetics and physiological impact of cohesin extrusion. *Cell*, 173(5), pp.1165–1178.e20.

Vietri Rudan, M. et al., 2015. Comparative Hi-C reveals that CTCF underlies evolution of chromosomal domain architecture. *Cell Reports*, 10(8), pp.1297–1309.

Wang, X. et al., 2017. Bacillus subtilis SMC complexes juxtapose chromosome arms as they travel from origin to terminus. *Science*, 355(6324), pp.524–527.

Wood, E.R. & Earnshaw, W.C., 1990. Mitotic chromatin condensation in vitro using somatic cell extracts and nuclei with variable levels of endogenous topoisomerase II. *The Journal of Cell Biology*, 111(6 Pt 2), pp.2839–2850.

Wrinch, D.M., 1934. Chromosome behaviour in terms of protein pattern. *Nature*, 134, p.978.

Wutz, G., et al., 2017a. CTCF, WAPL and PDS5 proteins control the formation of TADs and loops by cohesin. *Biorxiv.* Available at: http://dx.doi.org/10.1101/177444.

Wutz, G., et al., 2017b. Topologically associating domains and chromatin loops depend on cohesin and are regulated by CTCF, WAPL, and PDS5 proteins. *The EMBO Journal*, 36(24), pp.3573–3599.

Yokota, H. et al., 1995. Evidence for the organization of chromatin in megabase pair-sized loops arranged along a random walk path in the human G0/G1 interphase nucleus. *The Journal of Cell Biology*, 130(6), pp.1239–1249.

Yu. Grosberg, A., Nechaev, S.K. & Shakhnovich, E.I., 1988. The role of topological constraints in the kinetics of collapse of macromolecules. *Journal De Physique*, 49(12), pp.2095–2100.

Zhang, Y. & Heermann, D.W., 2011. Loops determine the mechanical properties of mitotic chromosomes. *PloS One*, 6(12), p.e29225.

5

Predictive Models for 3D Chromosome Organization: The Transcription Factor and Diffusive Loop Extrusion Models

C. A. BRACKLEY, M. C. PEREIRA,
J. JOHNSON, D. MICHIELETTO, AND D. MARENDUZZO

5.1 HI-C EXPERIMENTS: COMPARTMENTS, DOMAINS AND LOOPS

The three-dimensional spatial organization of chromosomes *in vivo* is currently a topic of very intense research in the biological and biophysical communities, among both experimentalists and modelers. This is because it is intimately coupled to gene regulation and expression (Yu and Ren, 2017). Additionally, chromatin structure has been shown to change substantially during development as cells differentiate (Dixon et al. 2012), in senescence and aging (Chandra et al. 2015; Criscione et al.

2016; Zirkel et al. 2017), and in disease (Lupianez et al. 2016). An important experimental technique which has allowed dramatic progress in the field in the last few years is "Hi-C" – a high-throughput and genome-wide version of "chromosome conformation capture", which uses restriction enzymes, crosslinking, and ligation with biotin-labeling, followed by high-throughput sequencing, to build interaction or "contact" maps showing which chromosome loci are spatially proximate in 3D (Dixon et al. 2012; Lieberman-Aiden et al. 2009; Rao et al. 2014; Sati and Cavalli, 2017). These contact maps have revealed a number of key principles underlying chromosomal organization. First, at large scales (>1–10 Mbp), transcriptionally active regions interact more with other active regions than inactive regions, and similarly inactive regions more often interact with other inactive regions. This is consistent with microscopy which shows a spatial segregation, or phase separation, of euchromatin and heterochromatin. Thus, the term "compartments" is used to describe the differently interacting regions determined by Hi-C: Active regions are referred to as the A compartment, and the inactive regions as the B compartment (Lieberman-Aiden et al. 2009). Higher resolution contact maps showed that at shorter length scales, each chromosome is partitioned into distinct "topologically associating domains" (TADs): There are higher than average levels of interaction within TADs, but reduced interaction between TADs. Interestingly TADs are present during interphase, but not during mitosis when transcription ceases (Naumova et al. 2013). Hi-C data at different resolutions tend to reveal domains of different sizes: In the first data sets they had a typical size of 1 Mbp (Dixon et al. 2012), whereas more recent higher-resolution studies uncovered smaller TADs with sizes in the 100 kbp range (Rao et al. 2014). This domain organization is evolutionarily conserved, as they have been found in budding yeast (Hsieh et al. 2015) and *Caulobacter crescentus* (where they usually are called "chromosomal interaction domains" or CIDs (Le et al. 2013)). Bacterial CIDs were found to be separated by strong promoters, and are eliminated by inhibiting transcription.

Within metazoans, some TADs also seem to be determined by chromatin state, as active and inactive regions typically form separate domains (Dixon et al. 2012; Lieberman-Aiden et al. 2009; Rao et al. 2014; Sexton et al. 2012), with the CCCTC-binding transcription factor (CTCF) and active transcription units (binding sites for RNA polymerase II) being enriched at domain boundaries (Dixon et al. 2012; Rao et al. 2014). Many of the interactions identified by Hi-C (and more focused methods such as 4C (van de Werken et al. 2012), Capture C (Hughes et al. 2014) or capture-Hi-C (Mifsud et al. 2015)) are between enhancers and promoters, consistent with a looping mechanism for enhancer activity where *cis*-regulatory elements come into physical contact. Loops between the binding sites of CTCF are also often observed; these have attracted particular attention because of some puzzling features they display. As the CTCF binding sequence is non-palindromic, it can be assigned a directionality on the genome, and it was found that almost all CTCF-mediated loops form between pairs of sites which are in a convergent arrangement (Rao et al. 2014), and there are far fewer loops between divergent or parallel pairs of sites. This bias is surprising because the 3D organization of a chromosome loop with divergent or convergent CTCF sites is the same: The observation is difficult to explain with conventional polymer physics models. CTCF-mediated loops are widely thought to be associated with the SMC complex cohesin, so named because of its role in sister chromatid cohesion. Cohesin is a ring-like complex which binds DNA and chromatin by topologically embracing it (Murayama and Uhlmann, 2015); how it can hold together two distant chromatin loci as a loop is not well understood: two possibilities are that it binds as single ring embracing two chromatin fibers (Nasmyth, 2001), or as a complex of two rings arranged as a molecular "hand-cuff", where each ring encloses a single DNA or chromatin fiber.

Hi-C data is normally obtained using populations of many cells, but a number of recent studies have applied the technique to single cells. Although the resolution is still limited, and the results are consistent with the domain organization at the megabase scale, these experiments show that there is significant cell-to-cell variation (Nagano et al. 2013), and TADs and compartments vary in strength through the cell cycle (Nagano et al. 2017).

Despite this wealth of new experimental data on the organization of chromosomes, we are to some extent still in the dark as to the biophysical and molecular mechanisms which drive structure formation microscopically. Two main classes of models are currently popular in the field, and these have been studied extensively using computer simulations based on polymer physics principles. The first is the "transcription factor" (TF) model, also known as the strings-and-binders model (Barbieri et al. 2012; Brackley et al. 2016b, 2013; Chiariello et al. 2016). This is based on the idea that multivalent transcription factor complexes (which can bind chromatin at more than one point to form molecular bridges) are the main genome organizers; this affects a scenario where transcription orchestrates chromosome organization. The second is the "loop extrusion" (LE) model (Fudenberg et al. 2016), which views cohesin and CTCF as the master organizers of the genome, and postulates that cohesin is a powerful molecular motor which can extrude chromatin loops of hundreds of kilo-bases. We have recently proposed an alternative "diffusive loop extrusion" variation of the model, which dispenses for the need to assume a motor activity for cohesin.

In this chapter we review the basic principles behind the TF and diffusive LE models, before discussing their predictions and consequences; we also consider whether both mechanisms may, in fact, be at work, playing complementary roles in 3D genome organization. We note that these are bottom-up, or mechanistic simulation approaches, where the models are based on simple biophysical principles. Another common approach is to start from Hi-C data and use sophisticated fitting procedures to generate the most likely structures which reproduce the data (Giorgetti et al. 2014; Tiana et al. 2016) – those "inverse modeling" techniques, which we do not discuss here, are reviewed in Serra et al. (2015).

5.2 THE TRANSCRIPTION FACTOR MODEL: THE BRIDGING-INDUCED ATTRACTION, PROTEIN CLUSTERS AND NUCLEAR BODIES

The simplest version of the TF model is described schematically in Figure 5.1a: a chromatin fiber (represented by a flexible bead-and-spring chain) interacts non-specifically with multivalent spheres. The latter represents transcription

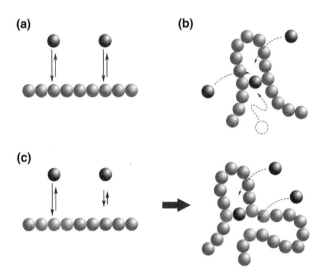

Figure 5.1 Schematic representation of the TF model and bridging-induced attraction. (a,b) A chromatin fiber is modeled as a bead-and-spring polymer, with spherical monomers (blue beads). Proteins (modeled as spherical red beads) bind to the chromatin fiber non-specifically. As proteins are multivalent, upon binding they create bridges which increase local chromatin density (b). Due to the density increase, more proteins can bind (b), creating a feedback loop. This effect has been called 'bridging-induced attraction'. (c,d) Here proteins (red beads) interact with chromatin both non-specifically (low-affinity binding, to blue beads) and specifically (high-affinity binding, to pink beads). Again, the concentration of binding sites increases upon bridge formation, ultimately triggering the formation of clusters of proteins via bridging-induced attraction.

factors or other complexes that can bind to two or more sites on the fiber simultaneously, forming "molecular bridges" that stabilize loops. In this version of the model, the TFs bind to the chromatin fiber via a non-specific attractive interaction. If the interaction strength is large enough, then the bound proteins spontaneously form clusters, a phenomenon first discussed in (Brackley et al. 2013). The general principle underlying this clustering – which occurs even in the absence of attractive DNA–DNA or protein–protein interactions – has been called "bridging-induced attraction", as it requires multivalent binding, or bridging.

Bridging-induced attraction arises through a simple positive feedback loop, as follows (Figure 5.1b). First, proteins bind to the chromatin; as they are multivalent, they can form molecular bridges between different chromatin segments – this increases the local chromatin concentration. In turn, this facilitates further binding of proteins from the soluble pool. The cycle repeats, and a cluster of chromatin-binding proteins forms. The effect is very general, as it applies to any multivalent chromatin-binding proteins, and is appealing as a possible mechanism for the formation of structures such as nuclear bodies (Brackley et al. 2017b), which are essentially clusters of specialized proteins found in eukaryotic nuclei. Examples include polycomb, Cajal and promyelocytic bodies, nucleoli and transcription factories (Mao et al. 2011).

For the simple case described in Figures 5.1a and 5.1b, where TFs can bind to any point along the chromatin, the bridging-induced attraction generates protein clusters which only superficially resemble nuclear bodies. In the simulations, after the proteins are introduced to the system, they very quickly form clusters; these clusters continue to grow in size and merge until ultimately there is only one single large cluster in the steady state (Johnson et al. 2015). Nuclear bodies on the other hand, only grow up to a finite size. To correctly capture this behavior the model must be refined.

An important feature which needs to be added to the model is that, while most transcription factors *do* interact non-specifically with DNA and chromatin (e.g., via electrostatic interactions), they also interact with the genome specifically, via strong affinity to cognate binding sites with well-defined sequences. A more detailed TF model includes stronger specific binding (of, e.g., red proteins to pink chromatin beads in Figure 5.1c). With these interaction rules, protein clusters still form via the bridging-induced attraction mechanism, but no longer grow indefinitely, and reach a self-limiting size – more like nuclear bodies. Why is cluster merging, or "coarsening", arrested in the presence of specific interaction? The answer to this question is that clusters now involve several high-affinity sites joined by chromatin loops, and coarsening therefore involves the creation of networks of more and more loops (which resemble a rosette). While the number of chromatin–protein interactions which stabilize these structures increases approximately linearly with cluster volume (or the number of high-affinity chromatin beads in the cluster), the entropic cost associated with the formation of a rosette grows super-linearly with the number of loops. Cluster growth stops when the entropic cost outweighs the enthalpic gain from chromatin–protein interactions.

A further fundamental aspect of nuclear bodies which is not captured by the protein clusters generated by the TF model as described above, is that they are highly dynamic. Microscopy experiments show that nuclear bodies recover their fluorescence quickly after photobleaching (typically over minutes); this means that the constituents of the bodies are rapidly exchanging with proteins in the soluble pool. The simulated protein clusters, on the other hand, are static, as their formation requires strong chromatin–protein interactions. How can proteins bind strongly yet turn over rapidly? Once again, the TF model can be refined to explain this (Brackley et al. 2017b). Figure 5.2 shows schematically a model where the proteins switch back and forth between an "on" (chromatin-binding) state and an "off" (non-binding) state; this mimics, for example, post-translational modifications, such as phosphorylation, which are known to change DNA-binding affinities. In simulations with switching proteins clusters still form through the bridging-induced attraction, but they are now both stable and dynamic (Figure 5.2). Importantly, this is only possible because switching drives the system away from thermodynamic equilibrium – proteins unbind when they switch off, independent of their interactions with the chromatin and the 3D chromatin

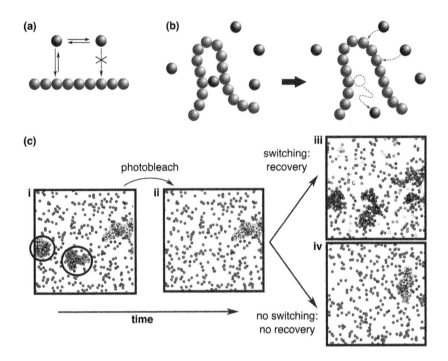

Figure 5.2 Protein switching and recycling nuclear bodies. (a,b) Schematic of a TF model where proteins constantly switch from an "on" state (chromatin-binding, red beads) to an "off" state (non-binding, grey beads). Switching may model, for instance, post-translational protein modification. (c) Snapshots taken during an *in silico* FRAP experiment (only proteins – and not chromatin beads – are shown for clarity). These results were reported in Brackley et al. (2017b). (i) At the beginning, we consider $N=2000$ non-switching proteins, half of which are able to bind the chromatin fiber (both specifically and non-specifically, see Brackley et al. (2017b for details)). The snapshot shown here is obtained after clusters have formed, and each of the 5 clusters is shown in different colors (unbound proteins are colored grey). (ii) Photo-bleaching is simulated by making bound proteins in the highlighted circular areas invisible (although they are still included in the simulation). Hereafter, colors are not changed, hence they can be used to visualize cluster dynamics. (iii) If proteins can switch, new proteins replace their "bleached" counterparts and clusters reappear. (iv) If proteins cannot switch, the proteins which make them up do not recycle, hence clusters do not recover. Figure adapted from Brackley et al. (2017b), with permission.

conformation. Simulated FRAP shows that clusters recover after a time close to the inverse of the switching rate, and that they retain the memory of their shape for longer time scales, just as nuclear bodies do. This model is also very general in that the switching could model different post-translational modifications, active protein degradation and *de novo* replacement, programmed unbinding of RNA polymerase at transcription termination, or any other active (free energy generating) unbinding process.

5.3 THE TRANSCRIPTION FACTOR MODEL: THE BRIDGING-INDUCED ATTRACTION DRIVES CHROMOSOME CONFORMATION

In the TF model, clustering is accompanied by the formation of chromatin "domains", in which intra-domain contacts are enriched over inter-domain contacts. These arise because once a cluster forms, any chromatin segment within the cluster is highly likely to come into close proximity with other segments within the cluster. In order for the model to give specific predictions of real chromosome interactions, it can be extended to include the positions of specific TF binding sites, based on available data (importantly, this is a truly predictive, fitting-free approach which does not require Hi-C data as an input). One could imagine a complex simulation with many species of multivalent transcription factors, using binding data for each – but actually a much simpler model can capture the compartmentalization and TAD formation observed in Hi-C.

In a version of the TF model proposed in (Brackley et al. 2016b), a whole chromosome (chromosome 19 in human lymphoblastoid cells) was simulated (Figure 5.3), with a resolution where each chromatin bead represented 3 kbp. Only two types of factors were considered, modeling generalized "active" and "inactive" complexes, associated with transcriptionally active euchromatin, and heterochromatin respectively. The simulated chromatin was "patterned" into active-binding, inactive-binding, and non-binding regions. Active regions were identified based on histone modification data (using "chromatin state" predictions from the hidden Markov model approach of (Ernst et al. 2011)), whereas inactive regions were identified simply by looking at GC content (as heterochromatin and gene-poor regions broadly correlate with low GC content). In other versions of the TF model, different combinations of histone modification, protein binding or DNA-accessibility data were used to label regions as active/repressed/ heterochromatic (Brackley et al. 2016a; Johnson et al. 2015).

A natural consequence of the bridging-induced attraction is that multivalent factors which bind specifically to different chromatin regions spontaneously segregate into "specialized" clusters (Brackley et al. 2016b). In this way, the active and inactive factors (and their cognate binding sites) cluster separately, and the model naturally generates the A (active) and B (inactive) compartments seen in Hi-C maps. The model is also consistent with microscopy which revealed that heterochromatin forms liquid-like phase-separated regions (Larson et al. 2017; Strom et al. 2017). As for the case of nuclear bodies, such regions do not coarsen indefinitely, so a more accurate term to describe this phenomenon is *microphase separation* – which is used in physics to describe self-assembly of clusters of self-limiting size. The same effect may be the mechanism underlying the formation of "specialized factories", which are protein clusters rich in either RNA polymerase II or III, but which not contain both (Papantonis and Cook, 2013; Xu and Cook, 2008).

As well as A/B compartmentalization, the TF model also gives a good prediction of the locations of TAD boundaries: for example, in Brackley et al. (2016b), 85% of the boundaries in chromosome 19 were correctly located to within 100

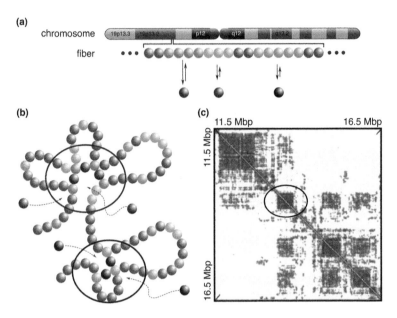

Figure 5.3 (a) Schematic of the TF model used in Brackley et al. (2016b) to study chromosome folding. Chromatin beads are colored according to the hidden Markov model described in Ernst et al. (2011), and to GC content (see text, and Brackley et al. (2016b) for more details). Active, euchromatic, factors (red spheres) bind strongly (specifically) to pink beads (essentially promoters and enhancers (Brackley et al. 2016b)) and weakly (non-specifically) to green beads (transcribed regions). Inactive, heterochromatic, factors (black spheres) bind weakly to grey beads (low GC content). Blue chromatin beads are non-binding. (b) Bridging-induced attraction drives the formation of separate red and black protein clusters, consistent with the microphase separation of chromosomes into active and inactive regions (see text). (c) Protein clusters lead to domains in the contact map. This zoom is from a region of human chromosome 19, which compares favorably to Hi-C data (see Brackley et al. 2016b). Adapted from Brackley et al. (2016b), with permission.

kbp. Also, some inter-domain interactions are correctly captured (as the off-diagonal blocks in the contact maps in Figure 5.3 match those in Hi-C maps, see Brackley et al. 2016b). This level of agreement is remarkable given the fitting-free nature of the model, where the only input is the 1D information on the protein binding landscape. A similar good agreement was found when considering the 3D folding of smaller regions at a higher resolution: simulations of the α and β globin loci in mouse gave a good prediction of the interaction profiles obtained from Capture C experiments (Brackley et al. 2016a), using a TF model where DNase hypersensitivity data was used to infer protein binding. Both of these studies (Brackley et al. 2016a,b) focused on active regions – predictions of interactions within inactive or repressed regions tend to compare slightly less favorably with Hi-C data, although the TF model still captures the overall interaction trends.

It is important to stress that the phenomenology described previously hinges critically on the assumption that the chromatin-binding proteins form complexes which are multivalent, so that they can form bridges. There are several examples of factors which have this property: an example of a heterochromatin-associated bridge is HP1α (Kilic et al. 2015), and repressive polycomb-related proteins include multimeric PRC1 (Wani et al. 2016); even a simple linker histone like H1 can potentially bind the genome in multiple places (Mack et al. 2015). Examples of multivalent euchromatin-associated bridges include complexes of polymerases and transcription factors, and the mediator complex, both of which are thought to be involved in enhancer–promoter looping.

We close this section mentioning another popular model for chromosome organization which is of a similar spirit to the TF model, namely the "block-copolymer" model (which was previously used to study the folding of *Drosophila* chromosomes (Jost et al. 2014)). In that model the chromatin beads interact attractively with each other directly, so bridging proteins are implied but not explicitly modeled. This approach is equivalent to the TF model if bridging proteins are abundant enough to saturate the binding sites; however, the two models differ in the regime where only some of the binding sites are occupied.

5.4 THE ACTIVE AND DIFFUSIVE LOOP EXTRUSION MODELS

As mentioned above, another popular model for 3D chromosome organization is loop extrusion. The idea that some factors bind to the chromatin at a single point and a loop is generated through some kind of tracking along the contour of the fiber, was first mooted in Nasmyth (2001) in the context of looping in mitotic chromosomes (Alipour and Marko, 2012), but has more recently been applied to chromatin organization during interphase (Fudenberg et al. 2016, Sanborn et al. 2015). A likely candidate for the extruding factor is the SMC complex cohesin (Uhlmann, 2016), which (as the model is described in Fudenberg et al. 2016 and Sanborn et al. 2015) uses a motor-like activity to reel in the chromatin and grow a loop using ATP; we therefore refer to this as the *active* LE model (to distinguish it from an alternative version based on bidirectional diffusion, which is discussed below).

Loop extrusion is an appealing model for two main reasons. First, if extrusion is halted when cohesin reaches a CTCF protein with its binding site oriented against the extrusion direction, then the model elegantly explains the strong bias towards convergent over divergent CTCF looping reported in Rao et al. (2014). This assumption is reasonable in the view of evidence that cohesin and CTCF interact in an orientation-dependent manner (Fudenberg et al. 2016, Xiao et al. 2011). Second, computer simulations have shown that the model can also give good predictions of the TAD patterns observed in Hi-C data (this requires a constant flux of extruders and a careful choice of parameters (Fudenberg et al. 2016)).

While extrusion can seemingly predict many of the interaction patterns observed in Hi-C data, the idea remains controversial. One crucial issue is that the model requires a fast motor with high processivity to extrude the loop. Although single-molecule experiments showed that the related condensin SMC complex is able to move unidirectionally and extrude loops on naked DNA *in vitro* (see below), under similar conditions cohesin has only been observed to move diffusively (Davidson et al. 2016, Kanke et al. 2016, Stigler et al. 2016). Motivated by this we recently proposed a *diffusive* loop extrusion model (Brackley et al. 2017a), asking whether extrusion is still a viable mechanism for generating chromatin loops if there is no motor activity. We imagine a simple scenario in which a pair of cohesin complexes are loaded at adjacent positions on a chromatin fiber in a handcuff configuration (see Figure 5.4a; an alternative arrangement in which a single cohesin ring embraces two chromatin fibers leads to similar results, and is briefly discussed in Brackley et al. (2018)). We then assume that each side of the handcuff can diffuse by sliding back and forth along the fiber, so that a loop grows and shrinks diffusively – i.e., cohesin works as a molecular slip-link

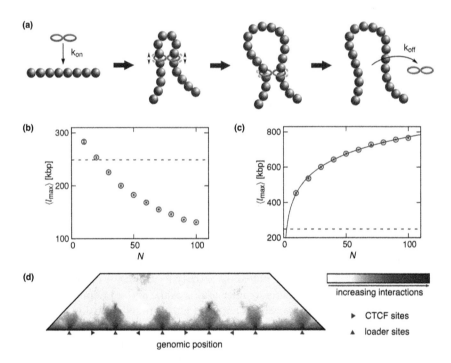

Figure 5.4 (a) Schematic description of the diffusive loop extrusion model. A cohesin dimer, with the topology of a hand-cuff (or a molecular slip-link), is loaded on a chromatin fiber at a rate k_{on}. It subsequently diffuses until it detaches, at a rate k_{off}. (b,c) Plots from Brackley et al. (2017a) showing how the mean size of the largest loop in a simplified 1D simulation depends on the number of cohesin handcuffs for a case with no loaders (b) and for the case with a loader in the middle (c). (d) A contact map (close to the diagonal) for a simulation with multiple chromatin segments, with CTCF-like proteins at each end functioning as barriers for diffusive loop extrusion, and with loaders in the middle of each segment. Adapted from Brackley et al. (2018, 2017a), with permission.

(Brackley et al. 2017a). The cohesin is later unloaded from the chromatin fiber at a constant rate. If cohesin interacts with CTCF directionally, for instance by forming a stable complex only when the CTCF site is oriented towards the diffusing cohesin, then again this is sufficient to explain the bias favoring convergent CTCF loop formation (Brackley et al. 2017a). Importantly, even though the motion is diffusive, the system is not in thermodynamic equilibrium: the cohesin rings are always loaded at adjacent points along the fiber (the loop length is initially zero), whereas the loop can be of any size when they detach. As a result, the system is not time-reversible (since a loop cannot form with finite size) and detailed balance is violated. The absence of thermodynamic equilibrium is both necessary to get the results discussed below, and consistent with experiments showing that cohesin needs ATP to both bind to and unbind from chromatin (Murayama and Uhlmann, 2015; Uhlmann, 2016).

While it might be difficult to design an experiment which can discriminate between the active and diffusive LE models *in vivo,* we can examine what would be required for each to generate the kind of loops which are observed in Hi-C. The residence time of cohesin on chromatin is about 20 min (Gerlich et al. 2006; Hansen et al. 2017; Ladurner et al. 2014), so to generate loops of 100 kbp the *active* LE model would require a motor which moves with speeds of 2–5 kbp/min (loops as large as 1 Mbp would require speeds 10-fold larger). Condensin was observed to move at ~3.6 kbp/min on DNA *in vitro* (Terakawa et al. 2017), and was able to extrude loops at on average 36 kbp/min (Ganji et al. 2018); whether cohesin can achieve similar motion, and indeed if either complex can perform similarly on chromatinized DNA, remains an open question. Another possibility is that a different motor protein pushes cohesin to facilitate extrusion; while faster motors are known to exist in bacteria, the magnitude of the speed required can be put into context by comparing it to that of RNA polymerase, the most processive motor currently known to be active during interphase, which is ~1 kbp/min. Polymerase and bacterial DNA-translocase enzymes move with a tracking motion, whereas the motion of condensin appears to proceed in steps – this might give an explanation for fast active extrusion if the cohesin can step from nucleosome to nucleosome along the chromatin fiber. Then again, this raises the question of how the motor can maintain the direction of extrusion to promote loop growth but not shrinking – an aspect of the process which is difficult to reconcile with a molecular stepping motion.

The *diffusive* LE model, on the other hand, is only a viable mechanism if the diffusive sliding is fast enough to generate loops of 100 kbp or more within the ~20 min cohesin residence time. Our work presented in (Brackley et al. 2017a) (some of whose results are reported in Figure 5.4) suggests that this may well be the case. A simple theoretical model puts a lower limit of 10 kbp^2/s on the required diffusion constant. If chromatin exists as a 30 nm fiber (having a compaction of ~100 bp/nm) then this equates to D ~0.001 μm^2/s as a minimum diffusion constant for viable loop generation. If it exists as a 10 nm fiber (compaction ~20 bp/nm) then the required diffusion constant becomes D ~0.025 μm^2/s. Recent *in vitro* experiments measured D = 0.2525 ± 0.0031μm^2/s for acetylated cohesin diffusing on chromatin fibers reconstituted in *Xenopus* egg extract. This value comfortably fulfils the requirements for diffusive LE to be viable, although it should be kept in mind that experiments in Kanke et al. (2016) were performed on stretched chromatin in a dilute solution, and in the absence of other chromatin-bound protein complexes. Other recent single-molecule experiments (Stigler et al. 2016) studied (non-acetylated) cohesin on DNA with nucleosome-like obstacles, and found that cohesin did not translocate over obstacles larger than 20 nm (possibly suggesting that acetylation greatly enhances cohesin diffusivity). The issue of cohesin having to negotiate obstacles on the chromatin is of course shared by the active LE model.

A third possibility is that cohesin might have a motor activity which acts to push it a short distance in one direction, before the directionality is lost. Active steps, or "kicks", back and forward would give diffusive motion at long time scales, but effectively with an increased diffusion rate. This then rescinds the requirement for a processive motor which keeps its direction.

Our work on the diffusive LE model also revealed some intriguing effects when multiple cohesin slip-links are present on the same chromatin segment. In Brackley et al. (2017a) we considered two scenarios for the binding kinetics of cohesin on chromatin: one where cohesin is loaded at random locations and one where there are preferred loading sites. The latter option is motivated by existing evidence that the cohesin-loading factor (NIPBL in humans, or Scc2 in yeast) binds at preferred genomic locations, and that cohesin may be loaded near the promoters of active genes (Kagey et al. 2010). The dynamics are very different in the two cases (Figures. 5.4b,c). For random loading, slip-links form many consecutive loops, which compete with each other for space: as a result, increasing the number of slip-links in a given segment leads to a *decrease* of the maximal loop size (Figure 5.4b, and Brackley et al. 2017a). In stark contrast, the presence of a loading site favors the formation of structures with nested loops, which creates a ratchet effect promoting loop growth. That is to say, increasing the number of slip-links yields an *increase* in the maximal loop size (Figure 5.4c). The effect can be understood as the presence of a loading site setting up an inhomogeneous density of slip-links on the fiber, leading to an osmotic pressure which favors loop growth over shrinking. Such a ratchet provides an avenue to boost the efficiency of diffusive loop extrusion even if the effective 1D diffusion coefficient of cohesin along the chromatin is small. Other authors reported that a similar effect operates in the case of a mixture of cohesin dimers (hand-cuffs) and monomers (where the latter are single rings which do not form loops), with the monomers creating the osmotic pressure (Yamamoto and Schiessel, 2017).

5.5 SOME CONSEQUENCES OF THE TF AND LE MODELS

As detailed above, the TF model naturally explains genome compartmentalization and the formation of protein clusters, and can correctly predict the locations of a large proportion of TAD boundaries; it cannot, however, explain the observations related to CTCF loops. The LE model elegantly explains the bias toward convergent CTCF, and gives good predictions of TAD patters; it cannot explain compartmentalization. The fact that each model explains different aspects of genome organization suggests that transcription factors and cohesin may *both* play important, and possibly complementary, roles in establishing chromosome structure. If this is the case, one would expect that experimentally disrupting either transcription factors or cohesin should have different effects. Overall this is the case, but such experiments are not at all straightforward, and sometimes give conflicting results. Particularly it remains unclear as to whether transcription drives chromatin conformation, or *vice versa*.

An immediate consequence of the TF model is that any change in expression (of regulated genes) should be accompanied by changes in chromosome interactions. Indeed, very high-resolution conformation studies of the globin loci using Capture C (Brackley et al. 2016a, Hughes et al. 2014) revealed completely different conformations in erythroid cells (where the globin genes are highly active) and embryonic stem cells (where they are inactive). More recently (Morgan et al. 2017), a dCas9 system was used where specific chromatin loops can be reversibly induced by the addition of the plant phytohormone S-(+)-abscisic acid (ABA). The authors of this work were able to increase β-globin expression by bringing the gene's promoter into physical contact with nearby active enhancers; the effect was context specific, as the expression was not altered by loop formation in cell lines where the β-globin locus was heterochromatinized. Interestingly, using the same system to force a loop between Oct4 and an upstream enhancer also induced looping to a downstream enhancer; additionally, when a loop was induced for more than 10 hours, it was found to persist even after ABA was removed. Together these results imply not only that TF looping can drive transcription, but transcription can also drive chromatin conformation. Experiments with the RNA polymerase inhibitor α-amanitin, as in (Hug et al. 2017), are more difficult to interpret: treatment with the drug did not have a major effect on the formation of TADs during *Drosophila* embryogenesis. However, the nature of the transcriptional inhibition is not clear, since the levels of polymerase II bound near promoters are only slightly affected.

The (active and diffusive) LE model predicts that eliminating chromatin-bound cohesin should lead to dramatic changes in 3D structure. To see this effect experimentally is very challenging, because cohesin is essential for proper sister chromatid cohesion in cycling cells, and eliminating all chromatin-bound cohesin in cell cycle–arrested cells is difficult. The most recent data come from experiments where deletion of the loader protein NIPBL (Schwarzer et al. 2017) or degradation of a cohesin subunit (Rao et al. 2017) was induced in non-dividing cells. This did indeed show that CTCF loops and loop-associated TADs disappear on cohesin removal. Earlier results reporting only minor changes may not have completely eradicated cohesin from the chromatin. Interestingly, cohesin removal was also shown to render A/B compartments more prominent, and to promote the formation of superenhancer hubs involving very long-range interactions: this suggests that there is a cross-talk between the two levels of organization (by TFs and by LE factors). Despite the clear changes in chromosome interactions revealed by the Hi-C data from these experiments, the changes in gene expression are rather modest; this surprising result leaves the function of TADs and CTCF loops as a mystery yet to be solved.

Looking ahead, an important task will be to clarify the interplay between cohesin and TF in directing chromosome organization during interphase, and theory and simulations are likely to be a fruitful device in this endeavor. There are several questions which can be addressed by a combined model. For instance, how is the efficiency of active and diffusive LE affected by clusters of chromatin-binding proteins assembled via the bridging-induced attraction? Are the two "organizers" working antagonistically or cooperatively? Can a combined model fully explain Hi-C data, and, if not, what else is missing from the picture? Can these models also tell us something more about the elusive transition between interphase and mitosis? Questions like these are just beginning to be asked, and it will be exciting to see how models and simulations improve our understanding of this fascinating branch of molecular biology in the future.

ACKNOWLEDGMENTS

This work was supported by ERC (CoG 648050, THREEDCELL- PHYSICS).

REFERENCES

Alipour, E. and Marko, J.F., 2012. Self-organization of domain structures by DNA-loop-extruding enzymes, *Nucleic Acids Research*, 40 (22), 11202–11212.

Barbieri, M., Chotalia, M., Fraser, J., Lavitas, L.M., Dostie, J., Pombo, A., and Nicodemi, M., 2012. Complexity of chromatin folding is captured by the strings and binders switch model, *Proceedings of the National Academy of Sciences USA*, 109 (40), 16173–16178.

Brackley, C., Johnson, J., Michieletto, D., Morozov, A., Nicodemi, M., Cook, P., and Marenduzzo, D., 2018. Extrusion without a motor: A new take on the loop extrusion model of genome organization, *Nucleus*, 9 (1), 95–103.

Brackley, C.A., Brown, J.M., Waithe, D., Babbs, C., Davies, J., Hughes, J.R., Buckle, V.J., and Marenduzzo, D., 2016a. Predicting the three-dimensional folding of cis-regulatory regions in mammalian genomes using bioinformatic data and polymer models, *Genome Biology*, 17 (1), 59.

Brackley, C.A., Johnson, J., Kelly, S., Cook, P.R., and Marenduzzo, D., 2016b. Simulated binding of transcription factors to active and inactive regions folds human chromosomes into loops, rosettes and topological domains, *Nucleic Acids Research*, 44 (8), 3503–3512.

Brackley, C.A., Johnson, J., Michieletto, D., Morozov, A.N., Nicodemi, M., Cook, P.R., and Marenduzzo, D., 2017a. Non-equilibrium chromosome looping via molecular slip-links, *Physical Review Letters*, 119, 138101.

Brackley, C.A., Liebchen, B., Michieletto, D., Mouvet, F.L., Cook, P.R., and Marenduzzo, D., 2017b. Ephemeral protein binding to DNA shapes stable nuclear bodies and chromatin domains, *Biophysical Journal*, 28, 1085.

Brackley, C.A., Taylor, S., Papantonisc, A., Cook, P.R., and Marenduzzo, D., 2013. Nonspecific bridging-induced attraction drives clustering of DNA-binding proteins and genome organization, *Proceedings of the National Academy of Sciences USA*, 110, E3605–E3611.

Chandra, T., Ewels, P.A., Schoenfelder, S., Furlan-Magaril, M., Wingett, S.W., Kirschner, K., Thuret, J.Y., Andrews, S., Fraser, P., and Reik, W., 2015. Global reorganization of the nuclear landscape in senescent cells, *Cell Reports*, 10 (4), 471–483.

Chiariello, A.M., Annunziatella, C., Bianco, S., Esposito, A., and Nicodemi, M., 2016. Polymer physics of chromosome large-scale 3d organisation, *Scientific Reports*, 6, 29775.

Criscione, S.W., Teo, Y.V., and Neretti, N., 2016. The chromatin landscape of cellular senescence, *Trends in Genetics*, 32 (11), 751–761.

Davidson, I.F., Goetz, D., Zaczek, M.P., Molodtsov, M.I., Huis In 't Veld, P.J., Weissmann, F., Litos, G., Cisneros, D.A., Ocampo-Hafalla, M., Ladurner, R., Uhlmann, F., Vaziri, A., and Peters, J.M., 2016. Rapid movement and transcriptional re-localization of human cohesin on DNA, *The EMBO Journal*, 35 (24), 2671–2685.

Dixon, J.R., Selvaraj, S., Yue, F., Kim, A., Li, Y., Shen, Y., Hu, M., Liu, J.S., and Ren, B., 2012. Topological domains in mammalian genomes identified by analysis of chromatin interactions, *Nature*, 485 (7398), 376–380.

Ernst, J., Kheradpour, P., Mikkelsen, T.S., Shoresh, N., Ward, L.D., Epstein, C.B., Zhang, X., Wang, L., Issner, R., Coyne, M., et al., 2011. Mapping and analysis of chromatin state dynamics in nine human cell types, *Nature*, 473 (7345), 43.

Fudenberg, G., Imakaev, M., Lu, C., Goloborodko, A., Abdennur, N., and Mirny, L.A., 2016. Formation of chromosomal domains by loop extrusion, *Cell Reports*, 15 (9), 2038–2049.

Ganji, M., Shaltiel, I.A., Bisht, S., Kim, E., Kalichava, A., Haering, C.H., and Dekker, C., 2018. Real-time imaging of DNA loop extrusion by condensin, *Science*, eaar7831.

Gerlich, D., Koch, B., Dupeux, F., Peters, J.M., and Ellenberg, J., 2006. Live-cell imaging reveals a stable cohesin-chromatin interaction after but not before DNA replication, *Current Biology*, 16 (15), 1571–1578.

Giorgetti, L., Galupa, R., Nora, E.P., Piolot, T., Lam, F., Dekker, J., Tiana, G., and Heard, E., 2014. Predictive polymer modeling reveals coupled fluctuations in chromosome conformation and transcription, *Cell*, 157 (4), 950–963.

Hansen, A.S., Pustova, I., Cattoglio, C., Tjian, R., and Darzacq, X., 2017. CTCF and cohesin regulate chromatin loop stability with distinct dynamics, *eLife*, 6, e25776.

Hsieh, T.H.S., Weiner, A., Lajoie, B., Dekker, J., Friedman, N., and Rando, O.J., 2015. Mapping nucleosome resolution chromosome folding in yeast by micro-c, *Cell*, 162 (1), 108–119.

Hug, C.B., Grimaldi, A.G., Kruse, K., and Vaquerizas, J.M., 2017. Chromatin architecture emerges during zygotic genome activation independent of transcription, *Cell*, 169, 216–228. e19.

Hughes, J.R., Roberts, N., McGowan, S., Hay, D., Giannoulatou, E., Lynch, M., Gobbi, M.D., Taylor, S., Gibbons, R., and Higgs, D.R., 2014. Analysis of hundreds of cis-regulatory landscapes at high resolution in a single, high-throughput experiment, *Nature Genetics*, 46, 205–212.

Johnson, J., Brackley, C., Cook, P., and Marenduzzo, D., 2015. A simple model for DNA bridging proteins and bacterial or human genomes: Bridging-induced attraction and genome compaction, *Journal of Physics: Condensed Matter*, 27 (6), 064119.

Jost, D., Carrivain, P., Cavalli, G., and Vaillant, C., 2014. Modeling epigenome folding: Formation and dynamics of topologically associated chromatin domains, *Nucleic Acids Research*, 42 (15), 9553–9561.

Kagey, M.H., Newman, J.J., Bilodeau, S., Zhan, Y., Orlando, D.A., van Berkum, N.L., Ebmeier, C.C., Goossens, J., Rahl, P.B., Levine, S.S., Taatjes, D.J., Dekker, J., and Young, R.A., 2010. Mediator and cohesin connect gene expression and chromatin architecture, *Nature*, 467, 430–435.

Kanke, M., Tahara, E., Huis In't Veld, P.J., and Nishiyama, T., 2016. Cohesin acetylation and wapl- pds5 oppositely regulate translocation of cohesin along DNA, *The EMBO Journal*, 35 (24), 2686–2698.

Kilic, S., Bachmann, A.L., Bryan, L.C., and Fierz, B., 2015. Multivalency governs hp1α association dynamics with the silent chromatin state, *Nature Communications*, 6, 7313.

Ladurner, R., Bhaskara, V., Huis in't Veld, P.J., Davidson, I.F., Kreidl, E., Petzold, G., and Peters, J.M., 2014. Cohesin's ATPase activity couples cohesin loading onto DNA with smc3 acetylation, *Current Biology*, 24 (19), 2228–2237.

Larson, A.G., Elnatan, D., Keenen, M.M., Trnka, M.J., Johnston, J.B., Burlingame, A.L., Agard, D.A., Redding, S., and Narlikar, G.J., 2017. Liquid droplet formation by hp1α suggests a role for phase separation in heterochromatin, *Nature*, 547, 236–240.

Le, T.B., Imakaev, M.V., Mirny, L.A., and Laub, M.T., 2013. High-resolution mapping of the spatial organization of a bacterial chromosome, *Science*, 342 (6159), 731–734.

Lieberman-Aiden, E., van Berkum, N.L., Williams, L., Imakaev, M., Ragoczy, T., Telling, A., Amit, I., Lajoie, B.R., Sabo, P.J., Dorschner, M.O., Sandstrom, R., Bernstein, B., Bender, M.A., Groudine, M., Gnirke, A., Stamatoyannopoulos, J., Mirny, L.A., Lander, E.S., and Dekker, J., 2009. Comprehensive mapping of long-range interactions reveals folding principles of the human genome, *Science*, 326 (5950), 289–293.

Lupianez, D.G., Spielmann, M., and Mundlos, S., 2016. Breaking tads: How alterations of chromatin domains result in disease, *Trends in Genetics*, 32 (4), 225 – 237.

Mack, A., Schlingman, D., Salinas, R., Regan, L., and Mochrie, S., 2015. Condensation transition and forced unravelling of DNA-histone H1 toroids: A multi-state free energy landscape, *Journal of Physics: Condensed Matter*, 27 (6), 064106.

Mao, Y.S., Zhang, B., and Spector, D.L., 2011. Biogenesis and function of nuclear bodies, *Trends in Genetics*, 27 (8), 295–306.

Mifsud, B., Tavares-Cadete, F., Young, A.N., Sugar, R., Schoenfelder, S., Ferreira, L., Wingett, S.W., Andrews, S., Grey, W., Ewels, P.A., et al., 2015. Mapping long-range promoter contacts in human cells with high-resolution capture Hi-C, *Nature Genetics*, 47 (6), 598.

Morgan, S.L., Mariano, N.C., Bermudez, A., Arruda, N.L., Wu, F., Luo, Y., Shankar, G., Jia, L., Chen, H., Hu, J.F., Hoffman, A.R., Huang, C.C., Pitteri, S.J., and Wang, K.C., 2017. Manipulation of nuclear architecture through crispr-mediated chromosomal looping, *Nature Communications*, 8, 15993.

Murayama, Y. and Uhlmann, F., 2015. DNA entry into and exit out of the cohesin ring by an interlocking gate mechanism, *Cell*, 163, 1628–1640.

Nagano, T., Lubling, Y., Stevens, T.J., Schoenfelder, S., Yaffe, E., Dean, W., Laue, E.D., Tanay, A., and Fraser, P., 2013. Single-cell Hi-C reveals cell-to-cell variability in chromosome structure, *Nature*, 502 (7469), 59.

Nagano, T., Lubling, Y., Varnai, C., Dudley, C., Leung, W., Baran, Y., Mendelson-Cohen, N., Wingett, S., Fraser, P., and Tanay, A., 2017. Cell-cycle dynamics of chromosomal organisation at single-cell resolution, *Nature*, 547, 61–67.

Nasmyth, K., 2001. Disseminating the genome: joining, resolving, and separating sister chromatids during mitosis and meiosis, *Annual Review of Genetics*, 35 (1), 673–745.

Naumova, N., Imakaev, M., Fudenberg, G., Zhan, Y., Lajoie, B.R., Mirny, L.A., and Dekker, J., 2013. Organization of the mitotic chromosome, *Science*, 342 (6161), 948–953.

Papantonis, A. and Cook, P.R., 2013. Transcription factories: genome organization and gene regulation, *Chemical Reviews*, 113 (11), 8683–8705.

Rao, S.S., Huang, S.C., St Hilaire, B.G., Engreitz, J.M., Perez, E.M., Kieffer-Kwon, K.R., Sanborn, A.L., Johnstone, S.E., Bascom, G.D., Bochkov, I.D., et al., 2017. Cohesin loss eliminates all loop domains, *Cell*, 171 (2), 305–320.

Rao, S.S., Huntley, M.H., Durand, N.C., Stamenova, E.K., Bochkov, I.D., Robinson, J.T., Sanborn, A.L., Machol, I., Omer, A.D., Lander, E.S., and Lieberman-Aiden, E., 2014. A 3d map of the human genome at kilobase resolution reveals principles of chromatin looping, *Cell*, 159 (7), 1665–1680.

Sanborn, A.L., Rao, S.S.P., Huang, S.C., Durand, N.C., Huntley, M.H., Jewett, A.I., Bochkov, I.D., Chinnappan, D., Cutkosky, A., Lia, J., Geeting, K.P., Gnirke, A., Melnikov, A., McKenna, D., Stamenova, E.K., Lander, E.S., and Aiden, E.L., 2015. Chromatin extrusion explains key features of loop and domain formation in wild-type and engineered genomes, *Proceedings of the National Academy of Sciences USA*, 112, E6456–E6465.

Sati, S. and Cavalli, G., 2017. Chromosome conformation capture technologies and their impact in understanding genome function, *Chromosoma*, 126 (1), 33–44.

Schwarzer, W., Abdennur, N., Goloborodko, A., Pekowska, A., Fudenberg, G., Loe-Mie, Y., Fon seca, N.A., Huber, W., Haering, C.H., Mirny, L., et al., 2017. Two independent modes of chromatin organization revealed by cohesin removal, *Nature*, 551 (7678), 51.

Serra, F., Di Stefano, M., Spill, Y.G., Cuartero, Y., Goodstadt, M., Bau, D., and Marti-Renom, M.A., 2015. Restraint-based three-dimensional modeling of genomes and genomic domains, *FEBS Letters*, 589 (20PartA), 2987–2995.

Sexton, T., Yaffe, E., Kenigsberg, E., Bantignies, F., Leblanc, B., Hoichman, M., Parrinello, H., Tanay, A., and Cavalli, G., 2012. Three-dimensional folding and functional organization principles of the *Drosophila* genome, *Cell*, 148, 458–472.

Stigler, J., Camdere, G., Koshland, D.E., and Greene, E.C., 2016. Single-molecule imaging reveals a collapsed conformational state for DNA-bound cohesin, *Cell Reports*, 15 (5), 988–998.

Strom, A.R., Emelyanov, A.V., Mir, M., Fyodorov, D.V., Darzacq, X., and Karpen, G.H., 2017. Phase separation drives heterochromatin domain formation, *Nature*, 547, 241–245.

Terakawa, T., Bisht, S., Eeftens, J.M., Dekker, C., Haering, C.H., and Greene, E.C., 2017. The condensin complex is a mechanochemical motor that translocates along DNA, *Science*, 358 (6363), 672-676.

Tiana, G., Amitai, A., Pollex, T., Piolot, T., Holcman, D., Heard, E., and Giorgetti, L., 2016. Structural fluctuations of the chromatin fiber within topologically associating domains, *Biophysical Journal*, 110 (6), 1234–1245.

Uhlmann, F., 2016. SMC complexes: From DNA to chromosomes, *Nature Reviews Molecular Cell Biology*, 17.

van de Werken, H.J.G., Landan, G., Holwerda, S.J.B., Hoichman, M., Klous, P., Chachik, R., Splinter, E., Valdes-Quezada, C., Oz, Y., Bouwman, B.A.M., Verstegen, M.J.A.M., de Wit, E., Tanay, A., and de Laat, W., 2012. Robust 4c-seq data analysis to screen for regulatory DNA interactions, *Nature Methods*, 9, 969–972.

Wani, A.H., Boettiger, A.N., Schorderet, P., Ergun, A., Miinger, C., Sadreyev, R.I., Zhuang, X., Kingston, R.E., and Francis, N.J., 2016. Chromatin topology is coupled to polycomb group protein subnuclear organization, *Nature communications*, 7, 10291.

Xiao, T., Wallace, J., and Felsenfeld, G., 2011. Specific sites in the c terminus of CTCF interact with the SA2 subunit of the cohesin complex and are required for cohesin-dependent insulation activity, *Molecular and Cellular Biology*, 31 (11), 2174–2183.

Xu, M. and Cook, P.R., 2008. Similar active genes cluster in specialized transcription factories, *Journal of Cell Biology*, 181 (4), 615–623.

Yamamoto, T. and Schiessel, H., 2017. Osmotic mechanism of the loop extrusion process, *Physical Review E*, 96, 030402(R).

Yu, M. and Ren, B., 2017. The three-dimensional organization of mammalian genomes, *Annual Review of Cell and Developmental Biology*, 33 (1), 265–289.

Zirkel, A., Nikolic, M., Sofiadis, K., Mallm, J.P., Brant, L., Becker, C., Altmueller, J., Franzen, J., Koker, M., Gusmao, E.G., et al., 2017. Topological demarcation by HMGB2 is disrupted early upon senescence entry across cell types and induces CTCF clustering, *bioRxiv*, 127522, 1–34.

6

Introducing Supercoiling into Models of Chromosome Structure

FABRIZIO BENEDETTI, DUSAN RACKO, JULIEN DORIER, AND ANDRZEJ STASIAK

6.1 INTRODUCTION

Numerous studies have indicated that portions of DNA in interphase chromosomes are torsionally stressed and this results in supercoiling of implicated chromatin fibers (Naughton et al., 2013, Kouzine et al., 2013, Baranello et al., 2018). Since RNA polymerases are prevented from encircling transcribed DNA in the dense nuclear milieu but still need to follow the DNA helix, it is the transcribed DNA that is forced to undergo axial rotation with respect to the relatively immobile RNA polymerase (Cook, 1999, Cook, 2009). Therefore, as originally proposed by Liu and Wang (1987), negative supercoiling is generated behind transcribing RNA polymerases, whereas ahead of transcribing polymerases it is positive supercoiling that is generated. As topoisomerases associated with RNA polymerase preferentially relax positive supercoils (Baranello et al., 2016), transcription effectively injects negative supercoiling into chromosomes (Naughton et al., 2013). Since negative supercoiling facilitates DNA strand separation, the role of negative supercoiling was primarily seen as a facilitator of DNA replication and transcription as these processes require progressive global or local strand separation, respectively (Bates and Maxwell, 2005). Studies showing that supercoiling is able to increase the frequency of intra-molecular contacts in DNA molecules (Vologodskii and Cozzarelli, 1996, Liu et al., 2001) suggested that this contact stimulation may also be an important role of supercoiling in eukaryotic chromosomes (Benedetti et al., 2014b). The observation that eukaryotic chromosomes are composed of linearly arranged chromatin blocks with increased frequency of internal contacts (Dixon et al., 2012, Nora et al., 2012) suggested their similarity to topological domains in bacterial chromosomes, which are believed to consist of supercoiled DNA loops (Postow et al., 2004). Therefore, chromatin blocks with increased frequency of internal contacts were given the name of topologically associated domains TADs (Nora et al., 2012). However, the question of whether TADs indeed consist of supercoiled chromatin fibers is not answered yet. Experimental approaches that detect supercoiling, such as psoralen photobinding are complex and require even more complex control experiments. This complexity contributed to a partial inconsistency between the results of different groups (Naughton et al., 2013, Kouzine et al., 2013). The experimental methods determining contact maps in chromosomes are currently the most advanced methods that provide structural information about interphase chromosomes (Grob and Cavalli, 2018). In addition, these methods are highly reproducible among different laboratories. However, contact maps do not tell directly whether TADs are supercoiled or not. One needs to use various simulation methods (Tiana and Giorgetti, 2017) to be able to interpret contact data and evaluate, for example, whether experimental contact maps are recapitulated better or worse by modeled chromosome fragments in which chromatin fibers are supercoiled or not. To this end, one needs to be able to model supercoiled chromatin fibers. This is somewhat difficult though as standard software packages for molecular dynamics simulations, that can be used without any modifications to model non-supercoiled chromatin fibers, were not foreseen to model the effects of supercoiling. We present here how standard modeling software packages such as HooMD (Anderson et al., 2008, Glaser et al., 2015) or ESPResSo (Limbach et al., 2006) can be modified to simulate supercoiled chromatin fibers. We also present our results suggesting that chromatin fibers forming TADs are supercoiled.

6.2 HOW TO INTRODUCE TORSIONAL RIGIDITY INTO FREELY SWIVELING STANDARD BEADED CHAIN MODELS

A standard beaded chain model is frequently sufficient for a coarse-grained approach to model equilibrium behavior of thermally fluctuating, torsionally unstressed polymeric chains subject to confinement (Reith et al., 2012, Dorier and Stasiak, 2013, Racko and Cifra, 2013). Usually, the diameter of beads, denoted as σ, is assumed to correspond to the mean diameter of modeled polymeric chains. In the case of DNA in a physiological solution, this diameter amounts to about 3 nm and in the case of decondensed chromatin fibers, it amounts to about 10 nm. Once it is decided what physical distance corresponds to σ, this sets the length scale of the model, which in turn permits various mechanical properties of that model to be set accordingly. For example, the bending persistence length (Lp) of DNA, which is known to be about 50 nm, is c. 17 larger than the effective diameter of DNA at physiological conditions. Therefore, in a beaded chain model of DNA, its bending rigidity should be set so that it will result in its persistence length being 17 times larger than the diameter of its beads.

The persistence length of chromatin fibers is less well defined than that of protein-free DNA and it can vary depending on conditions. Chromatin persistence length was reported to range from about 30 nm, as measured *in vitro* by single-molecule methods (Cui and Bustamante, 2000), to about 200 nm, as estimated using live imaging techniques (Bystricky et al., 2004). Following a theoretical approach to persistence length of chromatin (Mirny, 2011), we operate with an intermediate value of chromatin persistence length corresponding to about 50 nm.

As already mentioned, the effective diameter of modeled polymers is frequently used to set the scale of the model and thus the size of beads in a beaded chain model used for simulations. Decondensed chromatin takes the structure of 10-nm-thick fibers, known as 10 nm fibers (Brackley et al., 2015). These fibers are therefore conveniently modeled as beaded chains where individual beads have 10 nm diameter and each correspond to 400–600 bp (Fudenberg et al., 2016). This level of coarse-graining is suitable if one wants to model the behavior of relatively short chromatin fibers (up to 0.5 Mb). However, the modeling of larger systems, like specific chromosome regions or even entire chromosomes confined within a nucleus of a mammalian cell, for example, would require prohibitively long computation times, if one would maintain this level of coarse-graining. Efficient modeling of larger system requires a coarser coarse-graining. At least *in vitro*, 10 nm chromatin fibers can be induced to condense into 30 nm fibers (Thoma et al., 1979) and the linear density of these fibers is known. Therefore, coarser coarse-graining approaches used to model chromatin are frequently based on 30 nm beads where each bead corresponds to about 4000 bp (Benedetti et al., 2014a) and part of the simulations presented here use this coarse-graining.

Standard beaded chain models permit free swiveling, which makes them unable to maintain torsional tension and thus not suitable to model the effects of supercoiling. More complex models are needed to have chains that are elastically deformed in response to torsional tension resulting in plectonemic coiling (Brackley et al., 2014). Figure 6.1 shows the construction of our model that can maintain and react to torsional tension. The crucial elements needed to maintain the torsional tension in our model are quintuples of accessory beads that are placed between each pair of the main beads of the chain. These accessory beads together with the bonds keeping them in place have no excluded volume and thus do not affect the bending of the main chain even in configurations where some of

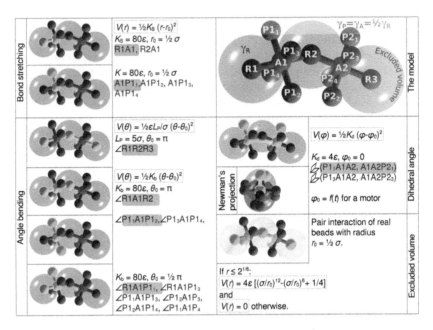

Figure 6.1 Inner workings of the beaded chain model of torsionally constrained elastic filaments used to simulate chromatin fibers. Beads and bonds forming our model are shown in the top-right panel together with their indexing that is used to describe various applied potentials. These potentials account for the resistance of modeled bonds to deviate from their intrinsic length, their rest bending angle or their rest dihedral angle that approximates a twist angle, respectively, as well as for the resistance of main chain beads to overlap with each other. Beads and bonds colored in yellow relate to those listed as first examples (highlighted in yellow) to which a given potential applies. The equations of the respective potentials contain values of parameters used during our simulations. In order to explain better the dihedral potential used to introduce resistance to twisting, a Newman's projection is shown where the angle between two planes defining the dihedral potential is seen directly and shows a similarity to the twist angle.

these phantom beads clash with each other or with other beads in the chain. The accessory beads serve to define dihedral angles that approximate twist angles and also specify dihedral potentials that maintain torsional constraint and oppose twisting deformation in our model. As dihedral angles are angles between intersecting planes, and we want to use them to approximate twist angles, we are interested here in the planes that intersect with each other along the line connecting the central beads of sequential quintuples, i.e., beads A1 and A2 and where one of these planes contains the bond connecting A1 with P1$_1$, and where the second plane contains then the bond A2P2$_1$. When one projects the two planes along the line of their intersection, one can perceive the dihedral angle directly and appreciate its similarity to the twist angle. The projection on which the dihedral angle can be perceived directly is known as Newman-projection and we show it in the panel of Figure 6.1 that contains the dihedral angle description. The dihedral angle potential V_d in the form $V_d(\varphi) = K_d (\varphi - \varphi_0)^2$ ensures that sequential quintuples can't freely rotate, along the chain axis with respect to each other. K_d is the force constant that determines the torsional stiffness of the model and $\varphi - \varphi_0$ is the dihedral angle approximating the twist angle (see Figure 6.1). For each two sequential quintuples of accessory beads, two dihedral angles are calculated: In one case it involves two intersecting planes that each contain one of the periaxial beads with the subscript 1 and in the second case each contains one periaxial bead with the subscript 3 (see Figure 6.1). Taking the dihedral angles measured with respect to two perpendicularly oriented bonds (the bond P1$_1$A1 is quasi-perpendicular to the bond P1$_3$A1) into account minimizes possible errors in the estimation of local torsional stress acting between sequential quintuples of accessory beads.

For the calculations of dihedral angles, we only use three beads of each quintuple but all beads in the quintuples are used to emulate rotational hydrodynamic drag experienced by chromatin fibers. Correctly accounting for rotational hydrodynamic drag is especially important when one models out-of-equilibrium processes such as the generation of supercoils (Racko et al., 2015). Further discussion of hydrodynamic effects will follow in the sections devoted to simulations of dynamic supercoiling.

By placing the quintuples of accessory beads in the middle between hinge points of the chain, and not immediately after these hinge points (Brackley et al., 2014), we avoided the difficulty with determining dihedral angles in case of strong bends in the chain. For example, a placement of quintuples at the start of each segment (where segments connect the sequential main beads in the chain) in the case of 90° bending between sequential segments of the chain, would make it that the plane determined by one of the two sequential quintuples becomes coplanar with respect to a straight line connecting the centers of these quintuples. In such a case, the dihedral angle is not defined. Another problem with placing quintuples at the start of each segment arises when the bend between sequential segments becomes larger than 90°. If that happens there is a sudden 180° jump of dihedral angle and this creates additional difficulties in the evaluation of dihedral angles. For quintuples placed in the middle between hinge points, the dihedral angle measured along the line connecting the centers of sequential quintuples is defined for all bending angles up to 180°, whereas the largest bends in a self-avoiding beaded chain can only reach 120°.

6.3 THE CONTROVERSY ABOUT TORSIONAL RIGIDITY OF CHROMATIN FIBERS

Magnetic tweezers experiments have shown that the torsional stiffness of chromatin fibers depends on whether the fiber is overwound or underwound (Celedon et al., 2009, Bancaud et al., 2006). Already, a low level of overwinding is sufficient to start the structural phase transition in chromatin fibers (Celedon et al., 2009, Bancaud et al., 2006). Before the transition, the incoming and outgoing DNA linkers of individual nucleosomes form left-handed crossings, whereas after the transition these crossings are right-handed (Bancaud et al., 2006). Once over-wound chromatin fibers start this structural transition their resistance to further torsional deformations becomes very low, resulting in the torsional persistence length of only 5 nm (Bancaud et al., 2006).

A completely different situation applies, though, to negatively supercoiled chromatin fibers. Celedon et al. have shown that when one starts to underwind chromatin fibers they oppose it with a rapid increase of counteracting torque (Celedon et al., 2009). As shown by Celedon et al. in their Figures 6.3 and 6.4, for the same number of introduced left-handed rotations and the same stretching force of 0.3 pN, the measured torque changes are larger in chromatin fibers than in the DNA molecules of the same size (Celedon et al., 2009).

Although Bancaud et al. popularized the notion that the torsional persistence length of chromatin is only about 5 nm, their own data show that this applies only to chromatin fibers that are overwound and as such are undergoing a nucleo-some switch transition (see Figure 6.2 in Bancaud et al.). The chromatin fragment tested by Bancaud et al. was constructed to contain 38 nucleosomes, therefore, knowing that each nucleosome introduces ΔLk of -1 (Bates and Maxwell, 2005) the ΔLk at which this fragment is expected to be torsionally relaxed is about -38, as compared to naked DNA of the same size. Indeed, Figure 6.2 in Bancaud et al. shows that as soon as the manipulated chromatin fragment starts to experience underwinding, it starts to form plectonemes and thus is not in the regime of high torsional plasticity (Bancaud et al., 2006).

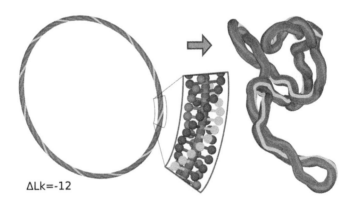

$\Delta Lk=-12$

Figure 6.2 Starting configuration of the modeled, torsionally stressed chroma-tin circle and an equilibrated configuration that diminished its elastic energy by forming plectonemes. Inset with a detailed view of the inner workings shows sequential levels of periaxial beads defining dihedral potential. Colored stripes in less-detailed presentations of the simulated chromatin rings permit us to follow the twisting of periaxial beads. Torsional stress acting through the dihedral potential decreases when twisting of periaxial beads decreases. However, when the dihedral circuit is closed the reduction of twisting can be only achieved when modeled molecules acquire writhe and form plectonemes. Notice that in the plectonemically wound configuration the twisting of peri-axial beads is much smaller than in the starting configuration.

In our modeling studies, we start with the assumption that transcription-induced positive supercoiling is quickly relaxed by DNA topoisomerases (Baranello et al., 2016), whereas negative supercoiling is long-lived (Naughton et al., 2013). Therefore, we are interested only in modeling the effect of negative supercoiling on chromatin fibers and consequently set their torsional persistence length similar to that of naked DNA.

In addition to bending and torsional rigidity, there are many other parameters that are also important in specifying the model that we use to simulate chromatin fibers. Such Parameters such as bond-length elasticity or excluded volume are also important. However, these other parameters are more standard and are not specific for models that simulate supercoiling of chromatin fibers. Therefore, we do not discuss them separately but just list them in Figure 6.1 together with detailed drawings presenting the construction of the beaded chain models used by us to model supercoiling of chromatin fibers.

6.4 SETTING THE DESIRED ΔLK

The effect of torsional stress on chromatin fibers depends not only on their torsional rigidity, as expressed by their torsional persistence length, but also on their supercoiling density. Since the level of supercoiling density in various portions of chromosomes is not well established, one needs to be able to simulate chromatin fibers with various supercoiling densities to evaluate which density recapitulates best the experimental data. Figure 6.2 shows how we prepare starting

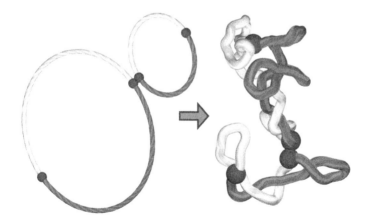

Figure 6.3 Starting and equilibrated configurations of the system permitting us to model supercoiling in sequential TADs that do not form closed loops. The lightly colored halves of circles serve only an accessory role permitting us to close the dihedral circuits and thus to maintain supercoiling in the entire circles, including the intensely colored halves. In each circle, we can set the desired ΔLk individually. After equilibration, we continue simulations of thermally fluctuating chains to register intra- and inter-TAD contacts needed for contact maps and determination of the α-exponent (see Figure 6.4). Only intensely colored chains are taken into account in the statistics of contacts.

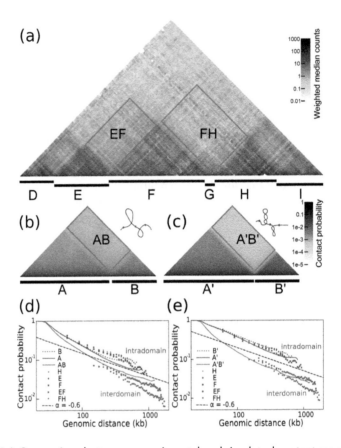

Figure 6.4 Comparison between experimental and simulated contact maps as well as contact probability profiles. (a) experimentally determined contact map within a portion of the X chromosome in male embryonic stem cells of mice. The map is of the genomic region shown by Nora et al. in their Figure 6.1c and was generated here using the deposited data (Nora et al., 2012). (b) and (c) contact maps obtained in simulations of systems composed of two neighboring chromatin loops (see Figure 6.3). In (b) and (c) the set density of supercoiling is ΔLk = −1 per 400 kb and ΔLk = −8 per 400 kb, respectively. (d) and (e) show comparisons between the rates of contact decrease with increasing genomic distance that were determined experimentally by Nora et al. for the genomic region shown in (a) and obtained in simulations presented in (b) and (c) respectively. The experimental results are shown as individual data points, whereas simulated results from continuous profiles. The straight dashed line shows the slope, which would correspond to the rate of contact decrease characterized by the α exponent of −0.6. One capital letter descriptions indicate to which experimental (D-E) or simulated (A, B) TADs the data refer to. Two capital letter descriptions indicate contacts between regions located in two different, correspondingly indicated TADs. Red and blue colors relate to intra- and inter-TAD contacts, respectively. Notice that simulations give the contact probability directly whereas in experimental data we only have relative contact probability and on log/log plots the experimental data can be all shifted vertically by the same factor during the fitting procedure.

configurations of modeled chromatin fibers with a given ΔLk, as measured with respect to torsionally relaxed chromatin fibers of the same size. We start with a perfectly circular configuration (Figure 6.2a). To introduce ΔLk of –12, for example, we calculate first what twist angle between sequential quintuples of accessory beads is needed to effectuate 12 left-handed rotations as one progresses through all sequential quintuples along the chain. Subsequently, without yet activating the dihedral potential, we rotate each sequential quintuple around the axis of the chain to match the desired twist angle (see inset in Figure 6.2). Once the integer number of rotations is introduced, we switch on the dihedral potential as well as all other potentials important for our model and start the equilibration of modeled chains. Since the dihedral potential circuit is closed, the decrease of all dihedral angles is only possible when the axis of modeled chromatin rings acquire writhe. On the other hand, writhing is opposed by the bending potential. The elastic resistance to bending and to torsional deformations both grow with the square of the deformation angle. Therefore, the starting configurations, in which we input large torsional deformations, will initially decrease their overall elastic energy by decreasing their torsional deformations while increasing their bending deformations but then will reach an equilibrium near the point where the further decrease of torsional energy is opposed by an equivalent increase of bending and other interaction energies. Figure 6.2b shows a simulation snapshot of an equilibrated, thermally fluctuating 80 kb large, circular chromatin fiber with ΔLk = –12. The colored stripes indicate how the introduced ΔLk is redistributed. The lower the rate of twisting of the colored stripes the smaller the torsional elastic energy of modeled chromatin fibers is. Before equilibration, all ΔLk is visible as a twist, whereas after equilibration the rate of twisting is decreased when modeled molecules acquire writhe and form plectonemes. However, the bending energy is increased compared to the perfectly circular starting configuration.

6.5 MODELING TADs AS PORTIONS OF SUPERCOILED CHROMATIN RINGS

Knowing that bacterial chromosomes are organized into supercoiled loops (Postow et al., 2004) it is natural to consider the possibility that individual TADs are supercoiled chromatin loops, having their borders bound to the chromosomal or nuclear matrix. How can supercoiling in portions of chromatin fibers delimited by two attachment sites be modeled? One could fix in space the first and last quintuple of each loop and this would allow us to set various ΔLk for each chromatin portion between the two attachment sites. However, by doing this one would need to arbitrarily set some physical separation distance between the attachment sites, whereas in reality these distances may fluctuate. Figure 6.3 shows our initial approach that allowed us to introduce supercoiling into individual TADs (Benedetti et al., 2014a). We used accessory chains that circularized individual loops and allowed us to close dihedral circuits for each individual loop. To minimize the effects of accessory chains on the interaction frequency of main chain beads, the accessory chains were not entered into the statistics of contacts presented in simulated contact maps (see Figure 6.4). In addition, we connected neighboring loops with linker chains that corresponded to chromatin portions between border elements of neighboring TADs. In a physical sense, the chains entering into statistics corresponded to a continuous chain that can extend along the whole chromosome and where individual TADs can form independent supercoiled loops, as would be the case if the borders of these loops were nuclear matrix attachment sites (Keaton et al., 2011). The genomic position of DNA sequences that constitute border elements in individual TADs is defined in every chromosome (Rao et al., 2014, Sanborn et al., 2015). However, the physical distances between matrix regions that bind borders of TADs may differ in different cells and may also fluctuate in the same cell. In addition, that physical distance may depend on the level of supercoiling as more supercoiled chromatin loops being more compact are likely to have their border elements closer to each other than chromatin loops with a similar genomic size but with a lower magnitude of supercoiling. To account for the crowding of chromatin within chromosomal territories, we adjusted the size of the simulation box with the periodic boundary conditions so that modeled chains with two neighboring TADs occupied 20% of the volume of the simulation box (Benedetti et al., 2014a). Periodic boundary conditions make it that the simulated molecules can take the same shape as in an unrestricted space but experience the effects of crowding. During simulations, all contacts occurring within individual copies were summed up whereas inter-copy contacts were not entered into the statistics.

6.6 COMPARISON BETWEEN EXPERIMENTAL AND SIMULATED CONTACT MAPS

Figure 6.4a shows an experimental map of contacts occurring within a specific region of the X chromosome in male embryonic stem cells of mice. This map corresponds to the contact map shown in Figure 6.1c by Nora et al. and was generated here using the deposited data (Nora et al., 2012).

It is interesting to evaluate to what extent a simple model composed of just two neighboring supercoiled chromatin loops, having sizes corresponding to 800 and 400 kb (as presented in Figure 6.3), is able to recapitulate the experimental data. Although several recent papers indicated that negative supercoiling is generated by transcription (Naughton et al., 2013, Kouzine et al., 2013, Baranello et al., 2016) the level of supercoiling *in vivo* is not determined yet. Therefore, we proceeded with simulations in which we set various levels of supercoiling. Figures 6.4 b and c show how the contact probabilities within simulated systems change as the magnitude of supercoiling density increases from the $\Delta Lk = -1$ per 400 kb (**b**) to $\Delta Lk = -8$ per 400 kb (**c**). It is visible that as the magnitude of supercoiling increases the ratio between intra- and inter-TADs contacts also increases. This is shown more clearly in Figures 6.4d and e, which compare simulated contact probabilities (continuous lines) with the experimental data (scatter plots). In the simulated system with a low magnitude of supercoiling ($\Delta Lk = -1$ per 400 kb) the ratio between intra-TAD and inter-TAD contact for loci separated by the same genomic distance is only about 1.5 (d), whereas in reality this difference is about 3. It was required to increase the magnitude of supercoiling density of simulated systems to $\Delta Lk = -8$ per 400 kb to approach the experimentally determined ratio between intra-TAD and inter-TAD contacts (e).

Interestingly, the same density of supercoiling ($\Delta Lk = -8$ per 400 kb) reproduced reasonably well the experimentally determined rate with which the contact probability between various pairs of loci located in the same TAD decreases with their genomic distance. That rate of the contact decrease with genomic distance is known as the α exponent. It was shown by Nora et al. that for loci located in the same TAD the α exponent is about -0.6 (Nora et al., 2012).

The fact that the same supercoiling density of modeled chromatin loops reproduced two different experimentally determined chromatin characteristics, i.e., the intra/inter TAD contact ratio as well as the decay rate of intra-TAD contacts with the genomic distance suggests that the local, supercoiling-induced compaction of chromatin portions forming TADs in our simulations is likely to be similar to this occurring *in vivo*. It should be mentioned though, that although we impose in our simulations a certain ΔLk, i.e., $\Delta Lk = -8$ per 400 kb, its effect on the overall chromatin conformation depends on the ratio between bending and torsional rigidity of the modeled chromatin fibers. As mentioned earlier, there is a substantial uncertainty in these values. If the torsional persistence length of chromatin fibers is significantly lower than what we assumed, one would require a higher magnitude of ΔLk to reach the same overall compaction as obtained in our models.

Figure 6.5 shows how our simulated TADs look. As already mentioned, simulations of pairs of TADs were performed under conditions mimicking the effect of high chromatin crowding within chromosomal territories. For purpose of better visualization, one of the periodic copies composed of two neighboring TADs that are both supercoiled is shown partially "extracted" from the crowded simulated system. It is clear that each of two TADs, shown in blue and red color, respectively, are self-compacted by supercoiling. That image is consistent with the notion that supercoiling of individual TADs promotes their self-compaction.

6.7 DYNAMIC SUPERCOILING OF CHROMATIN FIBERS

The simulations presented previously were all of "static" systems where modeled chromatin rings, in which we set various ΔLk, were simply reaching their thermal equilibrium. Biologically more relevant and computationally more challenging are the dynamic situations where at some regions supercoiling is actively introduced into chromatin by ongoing transcription and in other regions supercoiling is dissipated by DNA topoisomerases.

In Figure 6.1, we presented a construction of our beaded chain model where sequential periaxial beads serve to define the dihedral angle, which in turn is needed to set dihedral potential that introduces torsional rigidity into modeled chromatin fibers. The same periaxial beads serve to construct torsional motors. Individual periaxial beads can be compared to wrenches attached perpendicularly to the central axis of modeled chromatin fibers. By acting on even one of these wrenches with a certain force one induces torque acting on the axis of modeled chromatin fibers. When the dihedral potential is active the torque induced

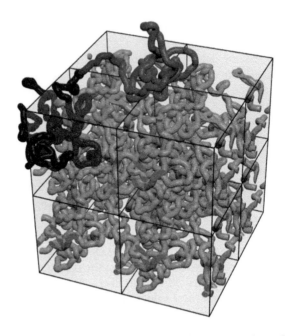

Figure 6.5 Structure of supercoiled TADs. Simulation snapshot of an equilibrated system with highly crowded, supercoiled TADs. The supercoiling density was set to ΔLk = −8 per 400 kb. For better visibility, one periodic copy, representing two neighboring TADs with sizes corresponding to 800 and 400 kb is presented upon removing copies that obstruct the view. 800 and 400 kb TADs are colored in blue and red, respectively. The accessory chains needed to close the dihedral circuits are not shown.

at one point is transmitted along the modeled fiber so that all sequential sets of periaxial beads also start to rotate and thus the entire chromatin portion can be induced to undergo axial rotation. RNA polymerases are known to be powerful torsional motors (Herbert et al., 2008) and since they are prevented from encircling transcribed DNA during transcription in the dense nuclear milieu, it is the transcribed DNA that is forced to undergo axial rotation (Cook, 1999, Cook, 2009). As a result of enforced axial rotation of DNA in the dense nuclear milieu, one has the situation where negatively supercoiled regions are generated behind transcribing RNA polymerase, whereas positively supercoiled regions accumulate ahead of RNA polymerase (Liu and Wang, 1987). Recent studies revealed that topoisomerases are associated with RNA polymerases and they preferentially relax positive supercoiling arising ahead of RNA polymerases (Baranello et al., 2016). Figure 6.6 shows our simulation of a circular chromatin loop with two RNA polymerases converging toward each other. Since we are only interested here in the topological consequences of transcription, we do not model newly synthesized RNA chains but only the effect of the torque introduced by each RNA polymerase. RNA polymerases are schematically presented as spinning top-like symbols pointing in the direction of transcription, whereas the directions of torque imposed by each RNA polymerase is indicated by circular

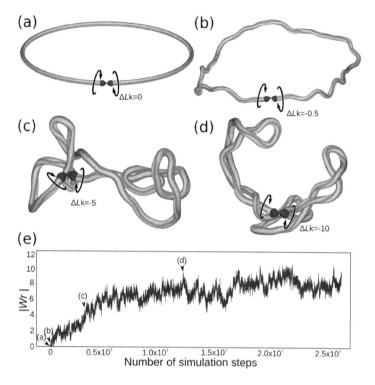

Figure 6.6 TADs modeled as circular chromatin loops progressively acquire negative supercoiling when positive supercoiling generated ahead of transcribing RNA polymerases is dissipated. Spinning top-like symbols indicate the positions and directions of the modeled RNA polymerases and more precisely the positions of periaxial beads (see Figure 6.1) that serve to introduce torque into the modeled chromatin rings. The dihedral potential value was set to 0 just ahead of modeled RNA polymerases. This setting accounts for the action of DNA topoisomerases that dissipates positive torque and thus positive supercoiling that would be otherwise produced ahead of RNA polymerase. Since the dihedral potential has usual values in the rest of the modeled chains, there is an accumulation of negative supercoiling that is generated behind modeled RNA polymerases. The circular arrows indicate directions of rotations, which were applied only to periaxial beads at positions corresponding to modeled RNA polymerases. The colored stripes permit us to trace positions of all periaxial beads (see Figure 6.2) and show how the torque is transferred along modeled chromatin fibers. (a) Starting configuration before equilibration and before application of torque. (b-d) Simulation snapshots obtained after Brownian dynamics simulation accounting for the effects of thermal motion in a solution was started at the same time as torsional motors. The number of rotation introduced by torsional motors is indicated for the corresponding snapshots. (e) The magnitude of writhe, which is a measure of supercoiling, initially grows but later is stabilized after torsional motors have introduced about 10 rotations. Torsional motors modeled here are constant torque motors that stall when the opposing torsional stress approaches 2 pN·nm. Arrowheads indicate when configurations shown in panels b–d were generated during the simulation and what was their writhe as well as that of the starting configuration (a).

arrows. The imposed torque values were set to 2 pN·nm, which corresponded to c. 50% of RNA polymerase stalling torque (Herbert et al., 2008). The dihedral potential defining the torsional rigidity of the modeled chromatin portions was acting along the entire chain with the exception of bonds placed between converging RNA polymerases. The freedom of swiveling at this region was to mimic the action of DNA topoisomerase that relaxes positive supercoiling arising ahead of transcribing RNA polymerases (Baranello et al., 2016). As shown in Figure 6.6, once the simulation has started, the torque introduced by modeled RNA polymerases progressively supercoiled chromatin ring. The profile of writhe, which is a measure of supercoiling, shows that the magnitude of supercoiling initially grows and then stabilizes. This stabilization occurs in our simulations when the counteracting torque resulting from supercoiling becomes as large as the torque of the two torsional motors.

6.8 TADS IN CHROMOSOMES OF FISSION YEAST CORRESPOND TO DOMAINS WITH DIVERGENT TRANSCRIPTION

In 2014, Mizuguchi et al. reported that chromosomes of fission yeast (*S. pombe*) are organized into TAD-like self-interacting domains and that genomic positions of these TADs correspond to positions of domains with divergent transcription (Mizuguchi et al., 2014). This latter observation prompted us to test by simulation whether sequential domains with divergent transcription can self-organize into TADs just due to negative supercoiling generated between diverging RNA polymerases. To this aim, we the simulated behavior of long chromatin fibers containing ten domains with divergent transcription. In each domain with divergent transcription, we placed two torsional motors that recapitulate the effect of divergent transcription and lead to accumulation of negative supercoiling between diverging RNA polymerases. To account for the fact that type I topoisomerases associated with RNA polymerases preferentially relax positive supercoiling generated ahead of transcribing RNA polymerases (Baranello et al., 2016), we set the dihedral potential to zero in portions of the chain located between RNA polymerases converging toward the borders of domains with divergent transcription. Portions of the chain with dihedral potential set to zero can freely swivel and this mimics the action of eukaryotic type I DNA topoisomerases (Seol and Neuman, 2016). In addition, motivated by studies showing that type II DNA topoisomerases are located at borders of TADs (Uuskula-Reimand et al., 2016), we introduced at the borders between domains with divergent transcription short "phantom" regions that showed no excluded volume and therefore could permit other regions of modeled chains to pass through these regions. Such passages correspond to passages that are mediated by type II DNA topoisomerases (Seol and Neuman, 2016).

Figure 6.7a shows a snapshot of our simulated system containing ten domains with divergent directions of transcription. Each domain is colored differently. Torsional motors are presented in the same way as in Figure 6.6 and bonds that

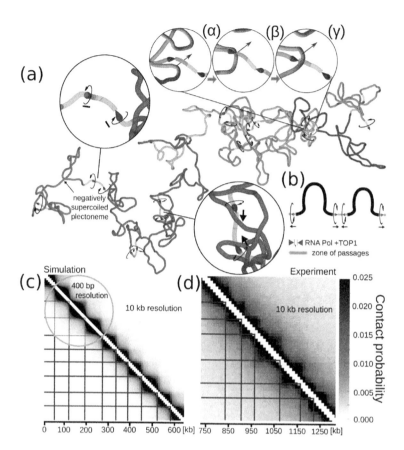

Figure 6.7 Simulations of the chromosomal portion with 10 divergently transcribed domains produce contact maps resembling experimental contact maps of S. pombe chromosomes. a. Simulation snapshot of chromosome fragments containing 10 divergently transcribed domains. Each domain is shown in different color. Insets offer a detailed view on borders between domains with the position and direction of action of torsional motors, preceding them free swivels and phantom-like zones of free passages. Insets α, β and γ show an intra-chain passage occurring during simulation. b. Scheme of the overall construction indicating positions of modeled RNA polymerases, preceding them are free swivels and zones of intra-chain passages accounting for the action of type II DNA topoisomerases. c. contact map obtained in simulations of constructs presented in a and b. d. The experimental contact map of a portion of chromosome 2 of S. pombe. Hi-C contact map of that chromosome portion was shown in Figure 6.1f by Mizuguchi et al. and we generated it here using the deposited data (Mizuguchi et al., 2014). Blue vertical lines in panel c and d indicate positions of borders between domains with divergent transcription in simulated and real biological systems, respectively. Figure 6.7 is based on Figure 6.2 in (Benedetti et al., 2017).

allow free rotations are placed just ahead of spinning top symbols indicating the orientation of RNA polymerases, which induce rotational motion of transcribed regions. The regions that allow intrachain passages are shown as semi-transparent. Figure 6.7b schematically presents the overall construction of modeled chromosome fragments including placements of special regions acting as torsional motors, free swivels, and chain portions permitting intrachain passages. Insets α, β, and γ show three selected snapshots visualizing the process of intrachain passage occurring during the simulation at one of the semi-transparent regions. Our simulations show that individual domains with divergent transcription show moderate self-compaction due to the negative supercoiling accumulated between diverging RNA polymerases. Figure 6.7c presents contact maps obtained in ongoing simulations of chromosome fragments with ten domains with diverging directions of transcription. That simulated contact map shows that each domain with diverging directions of transcription corresponds to a TAD-like region showing an increased frequency of internal contacts. In our simulations, the size of individual domains was randomly chosen from the experimentally observed size distribution of TAD-like domains in *S. pombe* and does not correspond to a particular genomic region. However, when our simulated contact map (Figure 6.7c) is compared to experimentally determined contact map (Figure 6.7d) of a particular genomic region studied by Mizuguchi et al. (Mizuguchi et al., 2014), one can appreciate the qualitative similarity of these contact maps. This similarity, together with the fact that in *S. pombe* chromosomes the location of TADs coincides with the location of chromosomal domains with divergent directions of transcription, supports our proposal that transcription-induced supercoiling organizes chromosomes into TADs.

6.9 TRANSCRIPTION-INDUCED SUPERCOILING CAN DRIVE CHROMATIN LOOP EXTRUSIONS

Several recent studies provided evidence that TADs in chromosomes of higher eukaryotes are formed by chromatin loop extrusion (Rao et al., 2014, Sanborn et al., 2015, Fudenberg et al., 2016). It was proposed that cohesin in the form of individual rings or co-joined two rings binds to chromatin fiber in such a way that it induces the formation of a small chromatin loop with chromatin fibers passing through cohesin rings (Sanborn et al., 2015, Fudenberg et al., 2016). These small chromatin loops were proposed to grow rapidly so that after about 20 min, which is the average life-time of cohesin rings bound to chromatin fibers (Hansen et al., 2017), these loops can reach the size of TADs, which can be as large as a megabase. Cohesins spanning each individual loop were proposed to slide with respect to embraced chromatin fibers until they reach and bind correctly oriented CTCF proteins bound to specific DNA sequences known as CTCF binding sites. CTCF binding sites determine the location of TADs borders (Rao et al., 2014, Sanborn et al., 2015, Fudenberg et al., 2016). It is now established that both cohesin and CTCF are essential for the formation of TADs and cells without cohesin do not form TADs (Rao et al., 2017), whereas in cells without CTCF the formed chromatin loops are frequently larger as their growth is not stopped at the CTCF binding sites (Nora et al., 2017). Although new studies have provided a growing number of details about the process of chromatin loop extrusion (Rao et al., 2017, Nora et al., 2017), it is not yet known how the process is driven. Cohesin is an ATPase, therefore, in principle, it could use the energy of ATP hydrolysis to actively slide along chromatin fibers. However, various biochemical tests only indicated that the sole role of ATPase activity in cohesin is during its binding and unbinding from chromatin fibers (Stigler et al., 2016). In the absence of evidence supporting the possibility of active sliding of cohesin along chromatin, other possibilities were considered. For example, other motor proteins acting as DNA translocases may be needed to push cohesin rings along chromatin fibers. Experiments have shown, for example, that FtsK, which is a bacterial DNA translocase, is perfectly able to push cohesin rings along DNA (Stigler et al., 2016). However, a proposal that other DNA translocases push cohesin rings needs to explain how these translocases would "know" in which direction they should push cohesin rings, so that chromatin loops would grow and not shrink with time.

In vitro observations of individual cohesin rings on DNA and reconstituted chromatin fibers revealed that individual cohesin rings embrace chromatin fibers very tightly (Stigler et al., 2016). These observations rather excluded the possibility that two chromatin fibers can be embraced by the same cohesin ring and supported the proposal that co-joined cohesin rings in form of handcuffs are needed during chromatin loop extrusion (Stigler et al., 2016). These *in vitro* observations revealed also that due to the tightness of cohesin rings there is a large drag opposing diffusion of cohesin rings along embraced by them chromatin fibers.

Recent studies of TADs organization in human cells revealed that topoisomerase TopIIB is associated with CTCF proteins bound at TADs borders (Uuskula-Reimand et al., 2016). This observation suggests that there is a flux of transcription-induced supercoiling from the source of its generation, transcribing RNA polymerases to sites where it can be dissipated, i.e., sites of action of TopIIB at TADs borders. This flux can be realized by the axial rotation of chromatin fibers located between transcribing polymerases and the borders of TADs. We, therefore, tested by simulation what difference it makes when chromatin fibers undergoing transcription can freely rotate as compared to a situation where their rotation is limited by the presence of cohesin handcuffs.

Figure 6.8 shows our toy models of individual TADs, in which we placed motors introducing negative supercoiling as this reflects the combined topological effect of transcribing RNA polymerase associated with DNA topoisomerase Top1. There is a net production of negative supercoiling during transcription as RNA polymerase produces negative supercoiling behind and positive supercoiling ahead of its actual position but positive supercoiling is preferentially relaxed by DNA topoisomerase Top1 positioned ahead of RNA polymerase (Baranello et al., 2016). Motors producing negative supercoiling are simply modeled by us as a step in the dihedral circuit where the rest angle needed for the calculation of the dihedral angle potential is progressively changing with each step of simulations. In our toy models of TADs, we also accounted for the action of TopIIB, which is known to be positioned at TADs borders (Uuskula-Reimand et al., 2016). Since the action of TopIIB can be very local, permitting passages between incoming and outgoing linkers from the same nucleosome (Salceda et al., 2006), we accounted for this action by placing free swivels at the TAD borders. To account for the non-local action of TopIIB permitting passages of distal chromatin portion through each other, we placed several beads devoid of self-avoidance near simulated TAD borders.

The simulation snapshots shown in the left column in Figure 6.8 show what happens when supercoiling generated at modeled sites of transcription can freely diffuse till modeled sites of TopIIB action. As could be predicted, in such a situation there is no accumulation of supercoiling. The right column of Figure 6.8 shows what happens when the dissipation of supercoiling is limited by tight

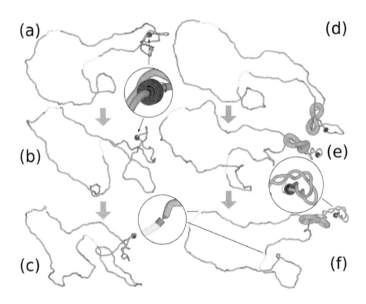

Figure 6.8 Accumulation of transcription-induced supercoiling in chromatin loops spanned by cohesin handcuffs. a–c. Snapshots from the simulation of our modeled TAD under conditions where the diffusion of continuously introduced transcription-induced supercoiling is not inhibited. Supercoiling generated by transcribing RNA polymerase can freely diffuse to sites of supercoiling relaxation by DNA topoisomerases and there is no accumulation of supercoiling. d–f. Simulation snapshots showing progressive accumulation of transcription-induced supercoiling in chromatin loops spanned by cohesin handcuffs that limit free rotations of modeled chromatin fibers. Two insets show positions of torsional motors that introduce negative supercoiling and thus mimic the combined action of RNA polymerase with associated Top1 that preferentially relaxes positive supercoiling generated ahead of RNA polymerase but does not act on negative supercoiling generated behind transcribing polymerase (Baranello et al., 2016). Another inset shows positions of sites that permit free swiveling (these sites are presented as sharp tips in contact with flat surfaces) and sites that permit intersegmental passages (these sites are presented as semitransparent regions of modeled chromatin fibers). Figure 6.8 is a modified version of Figure 6.2 in Racko et al. (2017).

cohesin handcuffs. To model the effect of cohesin handcuffs, we added to our simulations a chain forming a figure-of-eight arrangement and placed it on a starting configuration of modeled TAD so that a short loop containing a motor was spanned by cohesin handcuffs (see Figure 6.8d). To account for a large hydrodynamic drag opposing rotational and translational movement of chromatin fibers with respect to enclosing the cohesin rings (Stigler et al., 2016), we

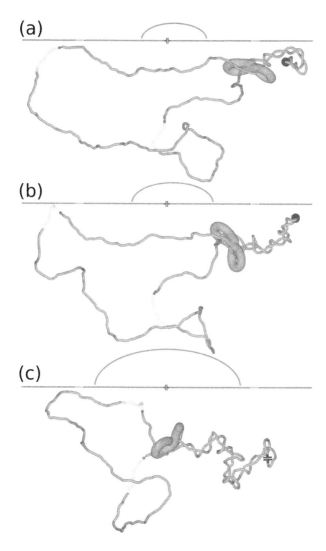

Figure 6.9 Transcription-induced supercoiling drives chromatin loop extrusion by pushing cohesin handcuffs. a–c. Simulation snapshots showing further progression of the situation presented in Figure 6.8 f. Growing plectoneme pushes cohesin handcuffs towards sites where supercoiling is dissipated. Schematics above snapshots illustrate how the length of chromatin loops spanned by cohesin handcuffs (green arc) grows with time.

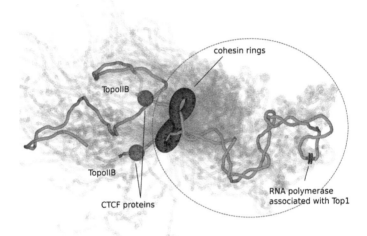

Figure 6.10 The mechanism of chromatin loop extrusion shown in one image resulting from the superposition of many simulation snapshots. In addition to elements discussed in Figures 6.8 and 6.9, the simulations presented in Figure 6.10 contained beads that were too large to pass through openings in cohesin handcuffs. These beads were intended to represent CTCF proteins that block the movement of cohesin handcuffs and limit the size of extruded chromatin loops. Positions of TopIIB, which are normally associated with CTCF (Uuskula-Reimand et al., 2016) are indicated. See the main text for the explanation of how the directionality of CTCF binding sites can be included in the model.

associated this large hydrodynamic drag with the quintuples of additional beads (see Figure 6.1) of modeled chromatin fibers that were passing through the cohesin handcuffs and thus at a given step of simulation were recognized as the closest to centers of both cohesin rings forming the handcuffs (Racko et al., 2017). This updating procedure permitted the cohesin handcuffs to move, while still limiting free supercoil diffusion via axial rotation of modeled chromatin fibers. Figures 6.8e and f show that transcription-induced supercoiling starts to accumulate in the chromatin loop that contains the source of supercoiling and is spanned by cohesin handcuffs.

Figure 6.9a–c shows a further evolution of the system presented in Figure 6.8d–f. As supercoiling continues to be generated, the grooving plectoneme pushes the cohesin handcuffs towards TADs borders. This process is formally analogous to the chromatin loop extrusion as shown on schematic diagrams above the snapshots. Our simulations, therefore, show that transcription induced supercoiling can drive chromatin loop extrusion.

An important part of the loop extrusion mechanism is that the loop extrusion process should stop when the cohesin rings reach the correctly oriented CTCF proteins at the TAD borders. To model such a stopping mechanism it is sufficient to introduce bulky beads at the sites corresponding to TAD borders. These bulky beads representing CTCF proteins possibly bound with other proteins such as TopIIB should be larger than the openings of modeled cohesin rings. When pushed cohesin rings reach such bulky beads they will be stopped there as shown in our concluding Figure 6.10. To account for the observation that stable loops are only observed between convergent CTCF sites (Rao et al., 2014) one would need still additional modifications of the model. For example, one can include in the model stabilization of cohesin handcuffs beyond their average 20 min life-time (Hansen et al., 2017) occurring only when the orientation of the CTCF protein with respect to contacting cohesin permits the C-terminal end of CTCF protein to bind cohesin (Xiao et al., 2011).

ACKNOWLEDGMENTS

This work is supported in part by Swiss National Science Foundation (Grant 31003A_166684) and the Leverhulme Trust (grant RP2013-K-017).

REFERENCES

Anderson, J. A., Lorenz, C. D. & Travesset, A. 2008. General purpose molecular dynamics simulations fully implemented on graphics processing units. *Journal of Computational Physics*, 227, 5342–5359.

Bancaud, A., Conde E Silva, N., Barbi, M., Wagner, G., Allemand, J. F., Mozziconacci, J., Lavelle, C., Croquette, V., Victor, J. M., Prunell, A. & Viovy, J. L. 2006. Structural plasticity of single chromatin fibers revealed by torsional manipulation. *Nat. Struct. Mol. Biol.*, 13, 444–450.

Baranello, L., Levens, D. & Kouzine, F. 2018. DNA supercoiling(omics). In: Lavelle, C. & Victor, J. M. (eds.) *Nuclear Architecture and Dynamics.* Chennai, India: Elsevier Inc.

Baranello, L., Wojtowicz, D., Cui, K., Devaiah, B. N., Chung, H. J., Chan-Salis, K. Y., Guha, R., Wilson, K., Zhang, X., Zhang, H., Piotrowski, J., Thomas, C. J., Singer, D. S., Pugh, B. F., Pommier, Y., Przytycka, T. M., Kouzine, F., Lewis, B. A., Zhao, K. & Levens, D. 2016. RNA polymerase II regulates topoisomerase 1 activity to favor efficient transcription. *Cell*, 165, 357–371.

Bates, A. D. & Maxwell, A. 2005. *DNA Topology*, Oxford, Oxford University Press.

Benedetti, F., Dorier, J., Burnier, Y. & Stasiak, A. 2014a. Models that include supercoiling of topological domains reproduce several known features of interphase chromosomes. *Nucleic Acids Res.*, 42, 2848–2855.

Benedetti, F., Dorier, J. & Stasiak, A. 2014b. Effects of supercoiling on enhancer–promoter contacts. *Nucleic Acids Res.*, 42, 10425–10432.

Benedetti, F., Racko, D., Dorier, J., Burnier, Y. & Stasiak, A. 2017. Transcription-induced supercoiling explains formation of self-interacting chromatin domains in S. pombe. *Nucleic Acids Res.*, 45, 9850–9859.

Brackley, C. A., Allan, J., Keszenman-Pereyra, D. & Marenduzzo, D. 2015. Topological constraints strongly affect chromatin reconstitution in silico. *Nucleic Acids Res.*, 43, 63–73.

Brackley, C. A., Morozov, A. N. & Marenduzzo, D. 2014. Models for twist-able elastic polymers in Brownian dynamics, and their implementation for LAMMPS. *J. Chem. Phys.*, 140, 135103.

Bystricky, K., Heun, P., Gehlen, L., Langowski, J. & Gasser, S. M. 2004. Long-range compaction and flexibility of interphase chromatin in budding yeast analyzed by high-resolution imaging techniques. *Proc. Natl. Acad. Sci. U S A*, 101, 16495–16500.

Celedon, A., Nodelman, I. M., Wildt, B., Dewan, R., Searson, P., Wirtz, D., Bowman, G. D. & Sun, S. X. 2009. Magnetic tweezers measurement of single molecule torque. *Nano Lett.*, 9, 1720–1725.

Cook, P. R. 1999. The organization of replication and transcription. *Science*, 284, 1790–1795.

Cook, P. R. 2009. A model for all genomes: The role of transcription factories. *J. Mol. Biol.*, 395, 1–10.

Cui, Y. & Bustamante, C. 2000. Pulling a single chromatin fiber reveals the forces that maintain its higher-order structure. *Proc. Natl. Acad. Sci. U S A*, 97, 127–132.

Dixon, J. R., Selvaraj, S., Yue, F., Kim, A., Li, Y., Shen, Y., Hu, M., Liu, J. S. & Ren, B. 2012. Topological domains in mammalian genomes identified by analysis of chromatin interactions. *Nature*, 485, 376–380.

Dorier, J. & Stasiak, A. 2013. Modelling of crowded polymers elucidate effects of double-strand breaks in topological domains of bacterial chromosomes. *Nucleic Acids Res.*, 41, 6808–6815.

Fudenberg, G., Imakaev, M., Lu, C., Goloborodko, A., Abdennur, N. & Mirny, L. A. 2016. Formation of chromosomal domains by loop extrusion. *Cell Rep.*, 15, 2038–2049.

Glaser, J., Nguyen, T. D., Anderson, J. A., Liu, P., Spiga, F., Millan, J. A., Morse, D. C. & Glotzer, S. C. 2015. Strong scaling of general-purpose molecular dynamics simulations on GPUs. *Computer Physics Communications*, 192, 97–107.

Grob, S. & Cavalli, G. 2018. Technical review: A Hitchhiker's guide to chromosome conformation capture. *Methods Mol. Biol.*, 1675, 233–246.

Hansen, A. S., Pustova, I., Cattoglio, C., Tjian, R. & Darzacq, X. 2017. CTCF and cohesin regulate chromatin loop stability with distinct dynamics. *Elife*, 6, e25776.

Herbert, K. M., Greenleaf, W. J. & Block, S. M. 2008. Single-molecule studies of RNA polymerase: motoring along. *Annu. Rev. Biochem.*, 77, 149–176.

Keaton, M. A., Taylor, C. M., Layer, R. M. & Dutta, A. 2011. Nuclear scaffold attachment sites within ENCODE regions associate with actively transcribed genes. *PLoS One*, 6, e17912.

Kouzine, F., Gupta, A., Baranello, L., Wojtowicz, D., Ben-Aissa, K., Liu, J., Przytycka, T. M. & Levens, D. 2013. Transcription-dependent dynamic supercoiling is a short-range genomic force. *Nat. Struct. Mol. Biol.*, 20, 396–403.

Limbach, H. J., Arnold, A., Mann, B. A. & Holm, C. 2006. ESPResSo: An extensible simulation package for research on soft matter systems. *Comp. Phys. Comm.*, 174, 704–727.

Liu, L. F. & Wang, J. C. 1987. Supercoiling of the DNA template during transcription. *Proc. Natl. Acad. Sci. U S A*, 84, 7024–7027.

Liu, Y., Bondarenko, V., Ninfa, A. & Studitsky, V. M. 2001. DNA supercoiling allows enhancer action over a large distance. *Proc. Natl. Acad. Sci. U S A*, 98, 14883–14888.

Mirny, L. A. 2011. The fractal globule as a model of chromatin architecture in the cell. *Chromosome Res.*, 19, 37–51.

Mizuguchi, T., Fudenberg, G., Mehta, S., Belton, J. M., Taneja, N., Folco, H. D., Fitzgerald, P., Dekker, J., Mirny, L., Barrowman, J. & Grewal, S. I. 2014. Cohesin-dependent globules and heterochromatin shape 3D genome architecture in S. pombe. *Nature*, 516, 432–435.

Naughton, C., Avlonitis, N., Corless, S., Prendergast, J. G., Mati, I. K., Eijk, P. P., Cockroft, S. L., Bradley, M., Ylstra, B. & Gilbert, N. 2013. Transcription forms and remodels supercoiling domains unfolding large-scale chromatin structures. *Nat. Struct. Mol. Biol.*, 20, 387–395.

Nora, E. P., Goloborodko, A., Valton, A. L., Gibcus, J. H., Uebersohn, A., Abdennur, N., Dekker, J., Mirny, L. A. & Bruneau, B. G. 2017. Targeted Degradation of CTCF Decouples Local Insulation of Chromosome Domains from Genomic Compartmentalization. *Cell*, 169, 930–944, e22.

Nora, E. P., Lajoie, B. R., Schulz, E. G., Giorgetti, L., Okamoto, I., Servant, N., Piolot, T., Van Berkum, N. L., Meisig, J., Sedat, J., Gribnau, J., Barillot, E., Bluthgen, N., Dekker, J. & Heard, E. 2012. Spatial partitioning of the regulatory landscape of the X-inactivation centre. *Nature*, 485, 381–385.

Postow, L., Hardy, C. D., Arsuaga, J. & Cozzarelli, N. R. 2004. Topological domain structure of the Escherichia coli chromosome. *Genes Dev.*, 18, 1766–1779.

Racko, D., Benedetti, F., Dorier, J., Burnier, Y. & Stasiak, A. 2015. Generation of supercoils in nicked and gapped DNA drives DNA unknotting and postreplicative decatenation. *Nucleic Acids Res.*, 43, 7229–7236.

Racko, D., Benedetti, F., Dorier, J. & Stasiak, A. 2017. Transcription-induced supercoiling as the driving force of chromatin loop extrusion during the formation of TADs in interphase chromosomes. *Nucleic Acids Res.*

Racko, D. & Cifra, P. 2013. Segregation of semiflexible macromolecules in nanochannel. *J. Chem. Phys.*, 138, 184904.

Rao, S. S., Huntley, M. H., Durand, N. C., Stamenova, E. K., Bochkov, I. D., Robinson, J. T., Sanborn, A. L., Machol, I., Omer, A. D., Lander, E. S. & Aiden, E. L. 2014. A 3D map of the human genome at kilobase resolution reveals principles of chromatin looping. *Cell*, 159, 1665–1680.

Rao, S. S. P., Huang, S. C., Glenn st Hilaire, B., Engreitz, J. M., Perez, E. M., Kieffer-Kwon, K. R., Sanborn, A. L., Johnstone, S. E., Bascom, G. D., Bochkov, I. D., Huang, X., Shamim, M. S., Shin, J., Turner, D., Ye, Z., Omer, A. D., Robinson, J. T., Schlick, T., Bernstein, B. E., Casellas, R., Lander, E. S. & Aiden, E. L. 2017. Cohesin Loss Eliminates All Loop Domains. *Cell*, 171, 305–320, e24.

Reith, D., Cifra, P., Stasiak, A. & Virnau, P. 2012. Effective stiffening of DNA due to nematic ordering causes DNA molecules packed in phage capsids to preferentially form torus knots. *Nucleic Acids Res.*, 40, 5129–5137.

Salceda, J., Fernandez, X. & Roca, J. 2006. Topoisomerase II, not topoisomerase I, is the proficient relaxase of nucleosomal DNA. *Embo. J.*, 25, 2575–2583.

Sanborn, A. L., Rao, S. S., Huang, S. C., Durand, N. C., Huntley, M. H., Jewett, A. I., Bochkov, I. D., Chinnappan, D., Cutkosky, A., Li, J., Geeting, K. P., Gnirke, A., Melnikov, A., Mckenna, D., Stamenova, E. K., Lander, E. S. & Aiden, E. L. 2015. Chromatin extrusion explains key features of loop and domain formation in wild-type and engineered genomes. *Proc. Natl. Acad. Sci. U S A*, 112, E6456–E6465.

Seol, Y. & Neuman, K. C. 2016. The dynamic interplay between DNA topoisomerases and DNA topology. *Biophys. Rev.*, 8, 101–111.

Stigler, J., Camdere, G. O., Koshland, D. E. & Greene, E. C. 2016. Single-molecule imaging reveals a collapsed conformational state for DNA-bound cohesin. *Cell Rep.*, 15, 988–998.

Thoma, F., Koller, T. & Klug, A. 1979. Involvement of histone H1 in the organization of the nucleosome and of the salt-dependent superstructures of chromatin. *J. Cell. Biol.*, 83, 403–427.

Tiana, G. & Giorgetti, L. 2017. Integrating experiment, theory and simulation to determine the structure and dynamics of mammalian chromosomes. *Curr. Opin. Struct. Biol.*, 49, 11–17.

Uuskula-Reimand, L., Hou, H., Samavarchi-Tehrani, P., Rudan, M. V., Liang, M., Medina-Rivera, A., Mohammed, H., Schmidt, D., Schwalie, P., Young, E. J., Reimand, J., Hadjur, S., Gingras, A. C. & Wilson, M. D. 2016. Topoisomerase II beta interacts with cohesin and CTCF at topological domain borders. *Genome Biol.*, 17, 182.

Vologodskii, A. & Cozzarelli, N. R. 1996. Effect of supercoiling on the juxtaposition and relative orientation of DNA sites. *Biophys. J.*, 70, 2548–2556.

Xiao, T., Wallace, J. & Felsenfeld, G. 2011. Specific sites in the C terminus of CTCF interact with the SA2 subunit of the cohesin complex and are required for cohesin-dependent insulation activity. *Mol. Cell Biol.*, 31, 2174–2183.

7

Structure and Microrheology of Genome Organization: From Experiments to Physical Modeling

ANDREA PAPALE AND ANGELO ROSA

The mechanisms beyond chromosome folding within the nuclei of eukaryotic cells have fundamental implications in important processes like gene expression and regulation. Yet, they remain widely unknown. Unveiling the secrets of nuclear processes requires a cross-disciplinary approach combining experimental techniques to theoretical, mathematical and physical modeling. In this review, we discuss our current understanding of the generic aspects of genome organization during interphase in terms of the conceptual connection between the large-scale structure of chromosomes and the physics beyond the crumpled structure of entangled ring polymers in solution. Then, we employ this framework to discuss recent experimental and theoretical results for microrheology of Brownian nanoprobes dispersed in the nuclear medium.

7.1 OUTLINE OF THE REVIEW

An accurate description of the mechanisms underlying the regulation of the genome in eukaryotes inevitably involves the study of the genetic code contained in the DNA string. Obviously fundamental, this information represents nonetheless only a small part of the intricate puzzle that determines the correct functioning of the entire cell. Each strand of DNA contained in a single chromosome is, in fact, part of the cell nucleus, and the way each chromosome is individually bent within the nucleus and in relation to the other chromosomes and the other nuclear structures is crucial to the future of the whole cell. In other words, the proper functioning of the genome of each organism is based not only on the alphabet contained in the sequence (genome in one dimension, or 1D genome), but also on how this sequence is folded and moves within the cell nucleus (genome "in space and time", or 4D genome [1, 2]).

The intricate relationship between genome structure and function within the nucleus can be now systematically explored owing to the development of high-resolution experimental techniques providing more and more accurate data for chromosome positioning and interactions [3, 4], chromosome mobilities [5, 6] and the viscoelastic properties of the nucleus and the cytoplasm [7]. At the same time, the amount of experimental data is growing so fast as to require the additional input provided by sophisticate quantitative tools such as rigorous statistical methods [8], machine learning [9] and physical models [10–13] of the three-dimensional structure and dynamics of chromosomes and the nuclear and cellular environments.

In this review article, we focus on recent experimental progress concerning nuclear chromosome structure and dynamics and the motion of nuclear bodies and their interpretation in terms of theoretical concepts borrowed from generic polymer and soft matter physics. In particular, we highlight two fundamental aspects: the physical origin of chromosome organization explained in terms of the slow relaxation of large polymers subjected to topological constraints and the impact of nuclear structure on the Brownian diffusion of nanoprobes microinjected within the nucleus (microrheology).

The material of the review is organized as follows: Sections 7.2.1 and 7.2.2 provide the necessary introduction to the phenomenology of chromosome organization and single-particle tracking applied to the exploration of the nucleus. In Sections 7.3.1 and 7.3.2 we present the general concepts and applications of polymer theory to model nuclear architecture and microrheology. Conclusions with highlights on future research topics are sketched in Section 4. Technical details on the general principles of microrheology and the physics of ring polymers which may be skipped at first reading are organized in specific sections (boxes) throughout of the article.

7.2 NUCLEAR ORGANIZATION AND GENOME STRUCTURE

7.2.1 From DNA to chromosomes

The cells of eukaryotes are partitioned into distinct compartments (Figure 7.1A), each of which is delimited by a "wall" made of a single or double lipid

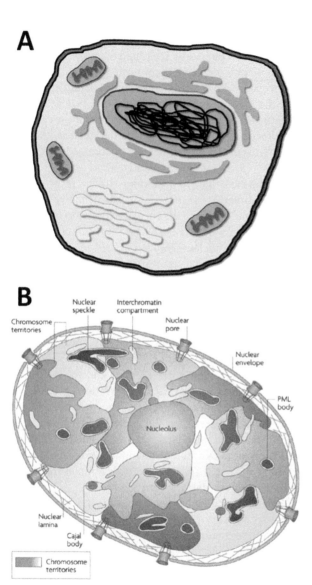

Figure 7.1 (A) Schematic illustration of the typical eukaryotic cell, showing the peculiar division into physically separated compartments. The nucleus is one of these compartments, it is shown in light red at the cell's center. The black rope inside represents DNA. (B) More detailed representation of the nucleus, showing its own compartmentalization. Inside, there exists regions void of chromatin ("interchromatin" compartments) and chromosomes condense into "territories" (see Sec. 7.2). Reproduced with permission from Ref. [14].

layer membrane [15]. In general, each compartment has evolved to fulfill a well-defined function.

The nucleus constitutes one of these compartments (Figure 7.1B): It consists of a roughly spherical region physically separated from the rest of the cell by the nuclear envelope, whose external layer is connected to the cytoplasm, while the internal layer connects to the nuclear lamina. Importantly, the structure of the envelope helps the nucleus to sustain its shape. A fundamental role of the nucleus is to isolate the DNA double-helix from the rest of the cell and to protect it from physical and/or chemical damage. Inside the nucleus, the genetic information carried by DNA is decoded and then post-processed to fulfill the cellular processes.

It is now well established that, in a manner similar to proteins who must acquire a unique three-dimensional shape (a so-called "native" state) in order to accomplish their functional role [15], the correct expression of the genetic information encoded in the linear sequence of DNA is the result of appropriate folding of the double-helix inside the nucleus [16, 17].

A vivid example of the intriguing connection between genome structure and function is provided by the nuclear architecture of the rod photoreceptor cells in nocturnal *against* diurnal mammals [18]. The rods of the diurnal retinas show the typical architecture of nearly all eukaryotic cells, with most heterochromatin (a tightly packed form of chromatin) close to the nuclear periphery and euchromatin (a gene-rich, lightly packed form of chromatin) concentrated toward the nuclear interior. Instead, the rods of nocturnal retinas display the reverse pattern of the heterochromatin nearby the nuclear center and the euchromatin closer to the nuclear envelope. The two opposite configurations are the results of the best adaptation of the corresponding species to the environment. At the same time, chromosome "misfolding" is typically associated with severe pathologies: for instance, fibroblasts of individuals affected by premature aging due to the Hutchinson–Gilford progeria syndrome show massive chromatin decondensation not observed in healthy cells [19–21]. Further examples include some forms of cancer [22] and other genetic dysfunctions [23].

In a typical human nucleus, about two meters (corresponding to $\approx 6 \times 10^9$ base-pairs (bp)) of DNA are packed into distinct chromosomes, each chromosome made of a unique filament of chromatin fiber. Chromatin results from the association of the double-helix with specific protein complexes (Figure 7.2). Approximately 147 bp of DNA wrap around the nucleosome complex (an octamer of core histone proteins (H2A, H2B, H3, H4) [25]), forming a 10-nm-wide and 6-nm-thick nucleosome-core particle (ncp) [26]. Consecutive ncp's are linearly connected into the so-called "10 nm" fiber by ≈ 50 bp of "linker" DNA [24], making the typical distance between the centers of neighboring core particles of the order of "10 nm + 50 bp/(3 bp/nm) = 25 nm". The contour length density of the 10 nm fiber is hence "200 bp/(25 nm) = 8 bp/nm", which is ≈ 3 times more compact than bare DNA. In spite of the considerable experimental work of the last decades, there is little consensus concerning how chromatin folds above the 10 nm fiber.

In general, *in vitro* studies of reconstituted nucleosomal arrays have pointed out [24] the role of nucleosome–nucleosome interactions in mediating the

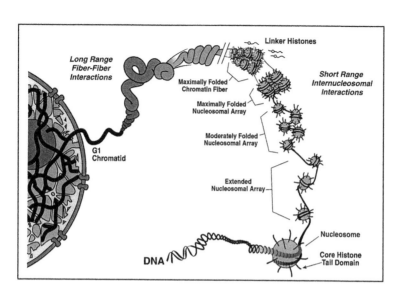

Figure 7.2 Schematic illustration of DNA and chromatin fiber structure in nuclei of eukaryotic cells. The chromatin fiber originates from the wrapping of DNA around the nucleosome complex which produces the necklace-like structure known as the 10 nm fiber, and the folding of 10 nm fibers into 30 nm fibers. The nature and very existence of the latter remain highly debated. Reproduced with permission from Ref. [24].

formation of helical-like structures with a diameter in the range 30–40 nm and a contour length density of ≈ 100 bp/nm, *i.e.*, ≈ 30 times more compact than bare DNA. This so-called "30 nm" fiber has been proposed as an essential element of the three-dimensional structures of interphase and mitotic chromosomes *in vivo*. Yet, its true existence remains highly controversial.

In fact, recent experimental studies by Maeshima and coworkers based on small-angle X-ray scattering (SAXS) on HeLa cells [27] in combination with computational modeling, essentially detected no structural features beyond the 10 nm fiber. Based on these results, the authors proposed [28] an alternative model where chromosomes in interphase nuclei look like an interdigitated polymer melt of nucleosome fibers lacking the 30 nm chromatin structure (Figure 7.3A). Very recently, these results have been substantially confirmed by chromEMT [29], a novel high-resolution experimental technique combining electron microscopy tomography (EMT) with a labeling method (ChromEM) that selectivity enhances the contrast of DNA. ChromEMT supports the picture where chromatin fibers form disordered structures packed together at different concentrations in the nucleus (Figure 7.3B). Interestingly, although chromatin compaction is locally changing in time, measurements of density fluctuations at high-resolution reveal that nuclear chromatin behaves like a compact and dynamically "stable" *fractal medium* [30].

Although the distribution of chromatin fibers seems to display, to some extent, some degree of randomness, other notable features emerge which suggest

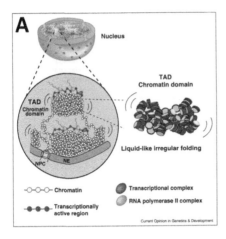
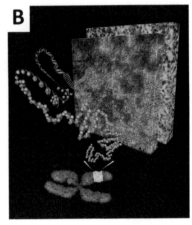

Figure 7.3 (A) Polymer melt-like model by Maeshima and coworkers [27, 28] of the eukaryotic nucleus filled by interdigitated 10 nm chromatin fibers. Topologically associating domains (TADs) partition the genome into regions where chromatin–chromatin contacts (see also Figure 7.4(C)) are more intense between the elements of the same region than between elements belonging to different regions. Reproduced with permission from [28]. (B) ChromEMT [29] reveals that chromatin forms a disordered 3d structure with regions of variable concentrations from high (red) to low (blue). Reproduced with permission from [29].

that some order at the nuclear level does exist [33]. First, chromosome mapping by "fluorescence in situ hybridization" (FISH) reveals the presence of distinct and moderately overlapping [34] regions termed "chromosome territories", see Figure 7.4A,B. Moreover, the spatial distance of each territory from the nuclear envelope is non-random, with gene-poor/rich chromosomes being systematically closer to the nuclear envelope/center [35]. Second, chromatin–chromatin contacts detected by Hi-C [4] have shown spatial segregation of chromosome sequences the size of a few megabasepairs (Mbp) termed "A/B sub-compartments" (Figure 7.4C). The data suggest that sub-compartments tend to interact more if they are alike than if they are not, and that A/B compartments correlate with (tissue-dependent) active/inactive chromatin. Third, chromosomes were partitioned into *topologically associating domains* (TADs, Figure 7.4C) of linear size $\leq$ 1 Mbp: within a TAD, DNA sequences interact more frequently with each other than with sequences outside the TAD [32]. Remarkably, TADs appear well conserved across tissues *within* the same species [36] and even *between* different species [37].

In summary, nuclear chromatin fibers form an intricate polymer-like network at small chromatin scales, with "vague" echoes of ordered structures starting from intermediate to large spatial scales (TADs $\longrightarrow$ A/B compartments $\longrightarrow$ territories).

The next question is how such intricacy affects and is affected by another important ingredient of nuclear organization, the presence of macromolecular

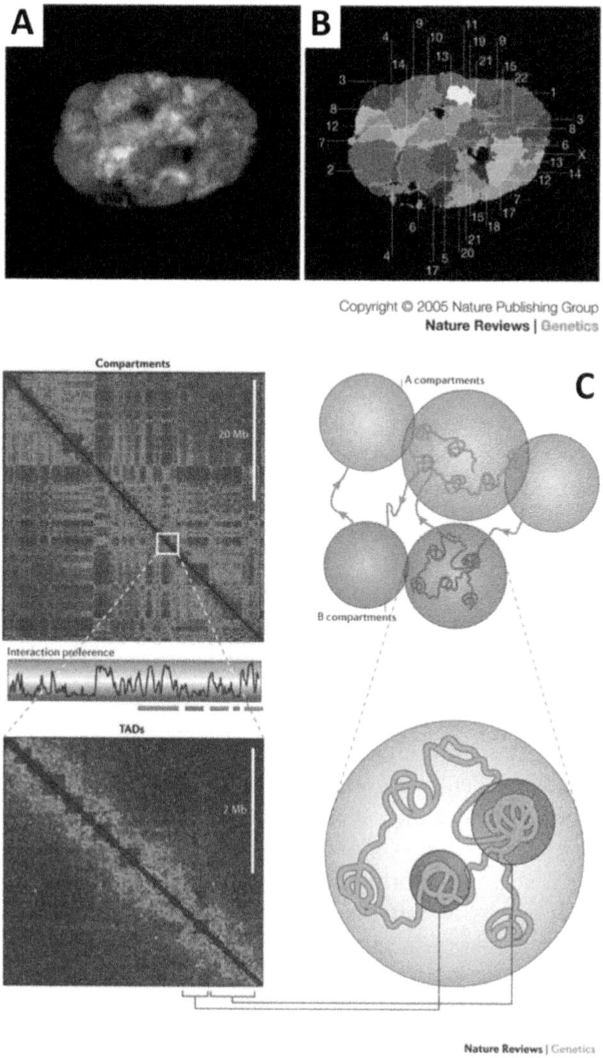

Nature Reviews | Genetics

Nature Reviews | Genetics

Figure 7.4 Chromosome organization is hierarchical from territories down to macro-domains and TADs. (A,B) Territorial organization of the human nucleus is visualized by FISH by using a combination labeling scheme in which each chromosome is labeled with a different set of fluorochromes. In this way, each chromosome territory can be identified by the corresponding combination of different colors and, then, appropriately annotated by its corresponding number. Reproduced with permission from [31]. (C) Hi-C shows an extensive network of chromatin–chromatin contacts within the genome. These interactions can be represented in the form of matrices showing a characteristic patterning into tissue-specific macro-domains (≥ 1 megabasepairs (Mbp)) of active/inactive (A/B) chromatin [4], and tissue-independent micro-domains (≤ 1 Mbp) termed TADs [32]. DNA interacts more frequently intra-TAD than inter-TAD. Reproduced with permission from [10].

complexes and enzymes which move through the nucleus directed towards specific DNA target-binding sequences [38, 39]. In the next section, we discuss the connection between chromatin folding and the diffusion of nuclear complexes from the point of view of *microrheology*, one of the most versatile and powerful experimental tools available today.

7.2.2 Microrheology of the nucleus

Microrheology is based on the tracking of the Brownian motion of fluorescent nanoprobes injected inside the cytoplasm or the nucleus. From nanoprobe motion, one extracts the time (τ) mean-square displacement (MSD) of the probe, $\langle \Delta x^2(\tau) \rangle$, which is used as a proxy for the viscoelastic properties of the embedding medium (see Box 7.1 and Refs. [7, 40–43] for details).

In fact, the MSD constitutes an important source of information concerning the nature of the environment [44]. For instance, in a thermally fluctuating, purely viscous medium, nanoprobe motion is described by standard diffusion with $\langle x^2(\tau) \rangle \sim D\tau$ where D is the diffusion coefficient. Instead, in complex and disordered media [44–46], nanoprobes can behave quite differently: in general, $\langle x^2(\tau) \rangle \sim D_a \tau^a$ with $\alpha \neq 1$ and D_a is the "generalized" diffusion coefficient.

Particularly relevant to the cellular context is the case of *subdiffusion* with $0 < \alpha < 1$. In fact, a growing number of experimental studies employing single-particle tracking of fluorescently labeled chromatin loci [5, 6] has demonstrated that loci dynamics is typically subdiffusive [47] and, at least in some cases, ATP-dependent [48, 49]. From the physical point of view, subdiffusion can be ascribed either to the macromolecular crowding of the nucleus [50, 51] which obstructs free chromatin motion or to the polymer-like nature of the chromatin fiber [52], or, most likely, to a combination of both.

Subdiffusion is also an important feature emerging in microrheological studies of tracked nanoprobes within the cytoplasm or the nucleus. However, the literature on this topic is surprisingly much more limited than on single-particle tracking of chromosome loci.

To our knowledge, the first microrheological studies in live cells are ascribable to Tseng et al. [40, 53] who measured the viscoelastic properties of the cytoplasm and the intranuclear region of mouse cells (Swiss 3T3 fibroblasts). Yellow-green fluorescent spherical nanoprobes of diameter = 100 nm were microinjected within the cytoplasm and their trajectories tracked inside the nucleus and the perinuclear region of the cytoplasm. Important differences between the two situations were reported. Nanospheres fluctuating in the crowded nuclear region displayed non-overlapping trajectories with "caged-and-escape" motion. On the contrary, nanospheres moving inside the cytoplasm showed extensive overlap. The corresponding MSDs reflect these differences. The MSD of nanoprobes diffusing inside the nucleus grows with τ on short time scales (0–0.1 s), then shows a plateau (0.1–1 s), and finally grows again at large lag-times, in agreement with the "caged-and-escape" motion between confining domains of average linear size ≈ 290 nm. Conversely, the plateau displayed by cytoplasmic nanospheres

BOX 7.1: Principles of particle-tracking microrheology.

Microrheology exploits the erratic (Brownian) motion of fluorescent nanoprobes (see Figure 7.5) carefully injected inside the cytoplasm or the nucleus as a proxy for the viscoelastic properties of the embedding medium [7, 40–43]. Compared to standard (bulk) rheology, microrheology grants systematic screening over wide ranges of length and time scales for the feasibility of designing trackable nanoprobes of linear sizes ranging from only a few nanometers [100] to hundreds of nanometers [40] and microns [54]. Microrheology is nowadays especially suitable for studies of biological materials [43] since, being minimally invasive, it allows to perform experiments *in vivo* and with very small samples [41].

Experimental data for microrheology can be obtained by various means, such as dynamic light scattering (DLS) [101]. More commonly, the motion of the probe in the form of its spatial coordinates (Figure 7.5) can be recorded through direct imaging and transformed into the

time-mean-square displacement, $\left\langle \Delta x^2(\tau) \right\rangle \equiv \frac{1}{T-\tau} \int_0^{T-\tau} \left(\bar{x}(t+\tau) - \bar{x}(t) \right)^2 dt,$

where T is the measurement time and τ the lag-time [44]. Then, the viscoelasticity of the embedding medium and nanoprobe motion are connected by the following mathematical relation [102]:

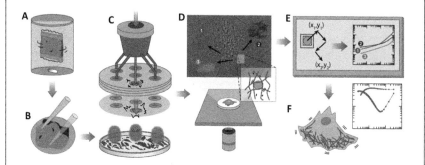

Figure 7.5 Sequential steps in microrheology: (A) After the initial preparation of the sub-micron fluorescent probes, (B) the beads are spread on a grid and (C) ballistically injected inside the cytoplasm where they rapidly disperse. (D) The cells are then placed under a fluorescence microscope and the random motion of the probes is monitored with high spatial and temporal resolutions. Examples of three trajectories are shown in red (1), blue (2), green (3). (E) The recorded time-dependent coordinates, $\bar{x}(t)$, of the probes are transformed into time-lag mean-square displacements (MSDs). (F) Finally, the MSDs of the probes are used to derive the local values of the frequency-dependent storage ($\hat{G}'(\omega)$) and loss ($\hat{G}''(\omega)$) moduli of the cytoplasm and/or the nucleoplasm. Reproduced with permission from [7].

(Continued)

$$\hat{G}(\omega) = -i \frac{2\kappa_B T}{\pi d\omega \left\langle \Delta \hat{x}^2(\omega) \right\rangle}. \qquad (7.1)$$

Here: $i = \sqrt{-1}$ is the imaginary unit; κ_B is the Boltzmann constant; T is the absolute temperature; d is the nanoprobe diameter; $\left\langle \Delta \hat{x}^2(\omega) \right\rangle$ is the Laplace-Fourier (LF) transform of $\left\langle \Delta x^2(\tau) \right\rangle$ (ω is the frequency). $\hat{G}(\omega) \equiv \hat{G}'(\omega) + i\hat{G}''(\omega)$ is the complex *shear modulus* of the medium: its real ($\hat{G}'(\omega)$) and imaginary ($\hat{G}''(\omega)$) parts correspond to the storage (elastic) and loss (viscous) moduli [74], respectively.

To illustrate the method, we consider the general situation where nano-probe diffusion is power-law-like [45]: $\left\langle \Delta x^2(\tau) \right\rangle = 6D_\alpha \tau^\alpha$, where D_α is the (generalized) diffusion coefficient ($0 \leq \alpha \leq 1$) and τ is the lag-time. With the corresponding LF-transform given by $\left\langle \Delta \hat{x}^2(\omega) \right\rangle = 6D_\alpha \Gamma(\alpha+1)(i\omega)^{-(\alpha+1)}$, $\hat{G}'(\omega)$ and $\hat{G}''(\omega)$ are expressed by the simple formulas:

$$\hat{G}'(\omega) = \frac{\kappa_B T}{3\pi d} \frac{\cos(\pi a / 2)}{D_\alpha \Gamma(a+1)} \omega^a, \hat{G}'(\omega) = \frac{\kappa_B T}{3\pi d} \frac{\sin(\pi a / 2)}{D_\alpha \Gamma(a+1)} \omega^a. \qquad (7.2)$$

The two "special" limits of $\alpha = 0$ and $\alpha = 1$ correspond, respectively, to the well-known cases of $\hat{G}(\omega) = \hat{G}' = const = \frac{\kappa_B T}{3\pi d D_0}$ and $\hat{G}(\omega) = i\hat{G}''(\omega) = i\frac{\kappa_B T}{3\pi d D_1} \omega \equiv i\eta\omega$. In the former case, the medium responds as an elastic (Hookean) solid, while in the latter its behavior is as of a classical fluid with "bulk" viscosity $= \eta$. In the intermediate case of $0 < \alpha < 1$ both, $\hat{G}'(\omega)$ and $\hat{G}''(\omega)$, are non-zero and the medium displays intermediate (solid/liquid) properties.

takes a higher value and reflects the restricted motion inside the cell. Finally, MSDs were used (Box 7.1) to calculate the complex shear modulus whose real ($\hat{G}'(\omega)$) and imaginary ($\hat{G}''(\omega)$) parts correspond to the storage and loss moduli of the medium embedding the nanoprobes. Qualitatively, the curves for the cytoplasm and the nucleoplasm have similar shapes. Quantitatively, by comparing the plateau values for $G'(\omega)$ under shear the nucleoplasm is ≈twice stiffer than the cytoplasm. Moreover, the low viscosity of the cytoplasm compared to the nucleus should facilitate the transport of proteins and molecules from and to the nucleus. At the same time, nuclear viscosity, higher if compared to cytoplasm, might play an active role in chromosome reorganization during interphase.

While the work by Tseng et al. focuses on passive diffusion within the cytoplasm or the nucleus, the motion of a large number of macromolecular nuclear bodies and subnuclear organelles like transcription compartments (TCs), promyelocytic leukemia (PML) nuclear bodies or Cajal bodies (CBs) which are involved in transcriptional regulation or RNA processing results from the combination of both, passive and active (*i.e.*, energy-consuming), processes [55–57]. Moreover, recent work in bacteria [48] suggests that consumption of ATP increases the mobility of cellular bodies and chromatin more steeply with temperature in untreated cells than in ATP-depleted cells.

In order to understand the role of active processes on nuclear dynamics and the motion of nuclear bodies, Hameed et al. [54] compared the passive motion of nanoprobes to the driven motion of transcription compartments (TCs). TCs are chromatin domains with an open chromatin structure which partially colocalize to active "transcription factories". During this process and at physiological temperatures (37 °C), they undergo directed movements which are influenced by ATP-dependent chromatin remodeling processes [58], and which are suppressed at lower, non-physiological temperatures.

To characterize the motion of TCs, Hameed et al. tracked tens of nanoprobes of linear size = 1 μm microinjected within the nuclei of HeLa cells at 25°C by using a protocol similar to the one by Tseng et al. (Figure 7.6A). The results are in quantitative agreement with those reported in previous work, in particular, the nanoprobe motion is caged within domains of linear size ≈250 nm (Figure 7.6B), a value remarkably close to the one (≈290 nm) measured by Tseng et al. in murine fibroblasts. Furthermore, single trajectories can be clustered into two groups according to the long-term behaviors of corresponding MSDs: in the first group, MSDs are plateauing after a long time while in the second they steadily increase (Figure 7.6C–E). The analysis is finally completed by computing the storage and loss moduli, $\hat{G}'(\omega)$ and $\hat{G}''(\omega)$ (Figure 7.6F): the nucleus behaves like a "power-law" solid ($\hat{G}' > \hat{G}''$) at low frequencies (again, in qualitative agreement with the experiments by Tseng et al.) crossing over to viscous-like behavior at large frequencies. The procedure was then repeated at 37 °C with analogous results.

Next, passive nanoprobe motion was compared to the motion of TCs at the same two temperatures. As anticipated above, at the non-physiological temperature of 25 °C, TC motion loses directionality and becomes similar to the passive motion of nanoprobes with analogous confinement and dispersion of MSD curves (Figures 7.6G,H). Conversely, trajectories taken at 37 °C display "mixed" behavior of confined motion and jump between close-by cages (Figures 7.6I,J, analogous to the results for passive nanoprobes in murine fibroblasts discussed before and significantly larger mobility (Figure 7.6K). Accordingly (Figure 7.6L), curves for storage and loss moduli at 25 °C are qualitatively similar to the ones for passive nanoprobes, while at the higher temperature they show a drastic change with the nuclear environment becoming sensibly much softer to TC motion. The temperature dependent behavior is dramatically affected by ATP depletion and perturbations to chromatin remodeling processes [54], suggesting that TC motion is partially stimulated by an active component.

Interestingly, the dynamic behavior of TCs contrasts analogous results [56] for the motion of Cajal bodies (CBs) in healthy (normal) and ATP-depleted nuclei. CBs are dynamic structures implicated in RNA-related metabolic processes. They can diffuse inside the nucleus, merge or split to form larger or smaller CBs and even associate/dissociate with/from specific genomic loci [56]. These processes were investigated in normal cells and in ATP-depleted cells in order to quantify the role of ATP in CB dynamics (see Figure 7.7). Typically, CBs show anomalous diffusion while moving within the interchromatin nuclear compartment. Quite unexpectedly, upon ATP depletion CBs tend to diffuse faster and they are no

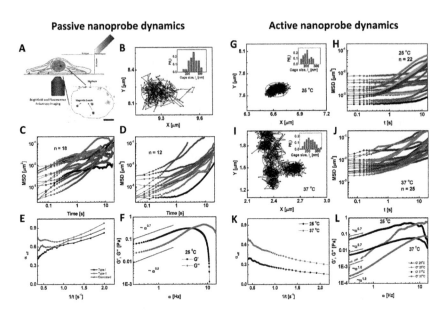

Figure 7.6 Microrheology of mammalian nuclei (live HeLa cells (human)): passive (A–F) versus active (G–L) dynamics. (A) Schematic illustration of the experimental setup used for single-particle tracking. Inset: focus on microinjected probes tracked by fluorescence microscopy. (B) Typical trajectory of a nanoprobe at 25 °C showing diffusion in a confined cage. Inset: Histogram of cage sizes l_c. (C,D) Time mean-square displacements (MSDs) for different nanoprobes, displaying behavior I (C, plateauing at large times) and II (D, monotonically increasing). (E) Mean effective exponents α_{eff} (MSD$(\tau) \sim \tau^{\alpha_{eff}}$) as a function of inverse time for trajectories I and II and their combination. (F) Storage and loss moduli, $\hat{G}'(\omega)$ and $\hat{G}''(\omega)$ as functions of frequency ω. At low ω's the nucleus is elastic $(\hat{G}' > \hat{G}'')$ while becoming increasingly viscous at higher ω's. (G) Typical trajectory of a transcription compartment (TC) at 25 °C showing diffusion in a confined cage as for microinjected beads. Inset: Histogram of cage sizes l_c. (H) Time MSDs for different TCs at 25 °C. (I) Typical trajectory of a TC at 37 °C showing diffusion in confined cages intermitted with jumps even across long distances. (J) Time MSDs for different TCs at 37 °C. (K) Mean effective exponents α_{eff} as functions of inverse time for the two temperatures. (L) Storage and loss moduli, $\hat{G}'(\omega)$ and $\hat{G}''(\omega)$, as functions of frequency ω for the two temperatures. Reproduced from [54] under Creative Commons License.

longer associated with dense chromatin regions. In conclusion, the association between CB and chromatin is an active process needing ATP.

To summarize, these results illustrate the prominent role of microrheology in the characterization of nuclear organization and how this influences the motion of nuclear bodies that participate in the correct functioning of cellular processes. In the next section, we discuss the connection between the physics of solutions of crumpled polymers and chromosome structure and dynamics, and illustrate its implications in the theoretical description of nuclear microrheology.

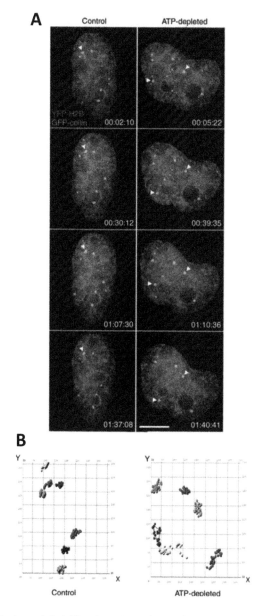

Figure 7.7 Diffusion of Cajal bodies (CBs) through the nuclear interchromatin space is an APT-dependent process. (A) Examples of consecutive temporal frames of nuclei of HeLa live cells: a healthy (control) nucleus (left) versus an ATP-depleted nucleus (right). CBs are stained green, while chromosomal DNA is stained red. Consecutive positions of CBs are indicated by the white arrowheads. CBs in ATP-depleted nuclei show higher mobility and they are no longer associated with dense chromatin regions. Scale bar = 10 µm. (B) Reconstructed trajectories of individual CBs. Different colors correspond to different CBs. Reproduced with permission from [56].

7.3 POLYMER MODEL OF NUCLEAR CHROMOSOME ORGANIZATION

7.3.1 "Topological" origin of chromosome territories

In spite of their intrinsic complexity (discussed in Sections 7.2), the general behavior of interphase chromosomes is remarkably well described by generic polymer physics [11–13, 66–71].

To explain these ideas, we start from a set of experiments dated back to the 60–70s featuring very accurate estimates of nuclear volumes (NV, in micrometers^3(μm^3)) from different organisms compared to the sizes of the corresponding genomes (GS, in basepairs (bp)). The data are summarized in Table 7.1 (animals) and Table 7.2 (plants). Remarkably, the data fit well (see Figure 7.8) to the *linear* relationship "$GS \sim NV$". Moreover, this law appears to be the same for animals and plants including the prefactor which, within statistical fluctuations (see Figure 7.8, inset), suggests a rather robust DNA density of $\rho \approx (0.054 \pm 0.028)$bp / nm^3. This value corresponds to a volume occupancy from $\approx$7% (for DNA being modeled as a cylinder of 2.2 nm of diameter with linear density of $\approx$3 bp/nm [15]) up to $\approx$25% (for chromatin being modeled as a cylinder of 30 nm of diameter with linear density of $\approx$100 bp/nm, see Section 7.2).

Under these conditions and supported by experimental observations on the polymer-like nature of the chromatin fiber (Section 7.2), the theory of semi-dilute polymer solutions [74] represents a good starting point for a quantitative description of chromosome organization inside the nucleus.

At the beginning of interphase, each chromosome evolves from its initial, compact mitotic conformation and starts swelling inside the nucleus (Figure 7.9A). Rosa and Everaers [72] argued that the time to reach the *complete* mixing of all chromosomes starting from the fully *unmixed* state can be estimated by assuming ordinary reptation dynamics [74, 75] for linear polymers in concentrated solutions: $\tau_{mix} \approx \tau_e (\frac{L_c}{L_e})^3$ where $\tau_e \approx 32$ seconds and $L_e \approx 0.12$ megabasepairs (Mbp) are, respectively, the entanglement time and entanglement length of the chromatin fibers solution (Table 7.3). With typical mammalian chromosomes of total contour length L_c of the order of 10^{2}Mbp (Table 7.1), τ_{mix} is exceeding by orders of magnitude the typical cell lifetime. As a consequence, the spatial structures of chromosomes remain effectively stuck into territorial-like conformations retaining the topological "memory" of the initial mitotic state.

These considerations were adapted into a generic bead-spring polymer model [72], taking into account the density, stiffness and local topology conservation of the chromatin fiber (Table 7.3). Extensive Molecular Dynamics computer simulations then showed that the swelling of model mitotic-like chromosomes (Figure 7.9B) leads to compact territories with physical properties akin to crumpled conformations of ring polymers in entangled solutions (Box 7.2). The analogy between chromosome territories and ring polymers motivated the formulation of the efficient multiscale algorithm described in [73] which is capable of generating hundreds of putative chromosome conformations (see Figure 7.9C for a single example) in negligible computer time. The polymer model was shown

Table 7.1 List of nuclear volumes (NV, in micrometers^3 (μm^3)), nuclear radii ($\equiv (\frac{3}{4\pi} NV)^{1/3}$, in μm, genome sizes (GS, in basepairs (bp)) and genome densities ($\equiv \frac{GS}{NV}$) for different animal species or different cell types of the same species. Corresponding sources are indicated at the top of each sub-panel

Organism	Animals			
	Nuclear volume [μm^3]	Nuclear radius [μm]	Genome size [×10⁹ bp]	Genome density [bp/nm³]
Anurans, liver parenchymal cells [59]				
A. obstetricans	253	3.92	20.54	0.081
X. laevis	125	3.10	7.34	0.059
B. marinus	221	3.75	9.49	0.043
B. viridis	122	3.08	10.47	0.086
B. fowleri	157	3.35	12.32	0.079
B. bufo	231	3.81	13.89	0.060
B. calamita	123	3.09	9.00	0.073
B. americanus	136	3.19	10.56	0.078
H. squirella	137	3.20	10.17	0.074
H. septentrionales	107	2.95	4.30	0.040
R. pipiens	168	3.42	14.67	0.087
R. catesbiana	225	3.77	14.87	0.066
R. temporaria	129	3.14	8.61	0.067
R. esculenta	196	3.60	13.79	0.070
Salamanders, liver parenchymal cells [59, 60]				
N. maculosus	1784	7.52	192.47	0.108
P. anguinus	1223	6.63	102.79	0.084
A. tigrinum	1104	6.41	83.42	0.076
A. mexicanum	943	6.08	75.31	0.080
A. means	3852	9.73	188.56	0.049
N. viridescens	943	6.08	91.15	0.097
T. granulosa	716	5.55	66.50±5.87	0.093±0.008
T. cristatus	697	5.50	51.35	0.074
T. vulgaris	768	5.68	69.44	0.090
T. alpestris	730	5.59	73.55	0.101
D. fuscus	523	5.00	35.21	0.067
E. bislineata	1236	6.66	73.37	0.059
P. ruber	579	5.17	48.90±1.96	0.085±0.003

(Continued)

Table 7.1 (Continued) List of nuclear volumes (NV, in micrometers³ (µm³)), nuclear radii (≡ ($\frac{3}{4\pi}$NV)$^{1/3}$, in µm), genome sizes (GS, in basepairs (bp)) and genome densities (≡ $\frac{GS}{NV}$) for different animal species or different cell types of the same species. Corresponding sources are indicated at the top of each sub-panel

Organism	Animals			
	Nuclear volume [µm³]	Nuclear radius [µm]	Genome size [×10⁹ bp]	Genome density [bp/nm³]
Other organisms [61]				
N. viridescens (lens)	4174	9.99	93.50	0.022
T. cristatus (heart)	1748	7.47	58.29	0.033
R. pipiens (embryo)	627	5.31	14.02	0.022
X. laevis (kidney)	294	4.13	7.51	0.026
X. laevis (heart)	307	4.19	7.51	0.025
S. holbrooki (heart)	197	3.61	3.70	0.019
P. crinitus (lung)	153	3.32	6.16	0.040
M. musculus	435	4.70	13.50	0.031
H. sapiens (lymphocytes)	232	3.81	6.10	0.026
H. sapiens (lung)	170	3.44	6.10	0.036
H. sapiens (HeLa, cervix)	374	4.47	10.45	0.028
C. sabaeus (kidney)	421	4.65	13.65	0.032
C. griseus (ovary)	188	3.55	5.87	0.031
G. gallus domesticus (embryo)	210	3.69	2.62	0.013
T. pyriformis	678	5.45	15.36	0.023
D. melanogaster (imaginal disc)	78	2.65	0.29	0.004
S. cerevisiae	3.3	0.92	0.02	0.005

to reproduce the experimentally observed behavior of (sequence-averaged [71]) properties of interphase chromosomes: these include chromosomes spatial positions measured by FISH, chromatin–chromatin interaction data and time mean-square displacements of chromosome loci [71, 72, 78, 79].

For illustration purposes, the single chromosome structure is described through the structure factor [74] $S(q) \equiv \left\langle e^{i\vec{q}\cdot(\vec{r}_i-\vec{r}_j)} \right\rangle$ as a function of the norm of the wave vector $q \equiv |\vec{q}|$ (Figure 7.11A). $\vec{r}_i$ are the spatial positions of chromosome loci and the average is taken over all chromosome conformations. For wave vectors $q \lesssim \frac{2\pi}{d_T}$ where $d_T \approx 245$ nm is the tube diameter of the chromatin fiber [72, 78] $S(q) \sim q^{-3}$, which corresponds to the expected result for a compact, scale-free polymer.

While $S(q)$ provides information on single-chain properties, it is instructive to look at the spatial relationship between different territories. To this end, we consider the average DNA density at spatial distance r from the chromosome

Table 7.2 Notation is as in Table 7.1

Organism	Nuclear volume [μm^3]	Nuclear radius [μm]	Genome size [$\times 10^9$ bp]	Genome density [bp/nm³]
Plants				
Higher plants [62]				
K. daigremontiana	105.0±3.8	2.93±0.04	10.76±0.98	0.103±0.013
R. sativus	111.0±2.9	2.98±0.03	4.89±0.98	0.044±0.010
R. sanguineus	120.0±3.9	3.06±0.03	4.89±0.98	0.041±0.010
T. majus	152.0±4.4	3.31±0.03	10.76±0.98	0.071±0.009
V. angustifolia	186.0±5.8	3.54±0.04	9.78±0.98	0.053±0.007
R. stenophyllus	212.0±6.2	3.70±0.04	11.74±0.98	0.055±0.006
R. obtusifolius	217.0±6.9	3.73±0.04	7.82±0.98	0.036±0.006
R. longifolius	249.0±9.3	3.90±0.05	12.71±1.96	0.051±0.010
C. nipponicum	270.0±10.2	4.01±0.05	43.03±4.89	0.159±0.024
G. sp. HV mansoer	281.0±8.5	4.06±0.04	5.87±0.98	0.021±0.004
H. annuus	293.0±10.7	4.12±0.05	15.65±1.96	0.053±0.009
C. jackmannii	347.0±15.0	4.36±0.06	21.52±0.98	0.062±0.006
Chrysanthemum sp. I	360.0±10.9	4.41±0.05	36.19±6.85	0.101±0.022
N. demascena	392.0±12.8	4.54±0.05	27.38±1.96	0.070±0.007
C. yezoense	478.0±16.4	4.85±0.06	28.36±1.96	0.059±0.006
T. blossfeldiana	481.0±16.8	4.86±0.06	39.12±2.93	0.081±0.009
V. faba	521.0±14.1	4.99±0.05	43.03±7.82	0.083±0.017
N. tazetta	579.0±20.6	5.17±0.06	30.32±1.96	0.052±0.005
A. cepa HV excel	621.0±24.6	5.29±0.07	52.81±5.87	0.085±0.013
T. sp. golden harvest	844.0±23.4	5.86±0.05	70.42±7.82	0.083±0.012

(Continued)

Table 7.2 Notation is as in Table 7.1

| Organism | Plants | | | |
	Nuclear volume [μm^3]	Nuclear radius [μm]	Genome size [$\times 10^9$ bp]	Genome density [bp/nm^3]
S. sibirica HV alba	908.0±42.8	6.01±0.09	71.39±20.54	0.079±0.026
Tradescantia sp. I	916.0±27.6	6.02±0.06	57.70±4.89	0.063±0.007
T. paludosa	947.0±36.6	6.09±0.08	52.81±5.87	0.056±0.008
L. squamigera	1017.0±24.0	6.24±0.05	125.18±11.74	0.123±0.015
Chrysanthemum sp. II	1183.0±44.9	6.56±0.08	77.26±9.78	0.065±0.011
T. virginiana	1324.0±38.8	6.81±0.07	113.45±7.82	0.086±0.008
L. longiflorum I	1347.0±44.8	6.85±0.08	103.67±13.69	0.077±0.013
C. lacustre	1663.0±53.8	7.35±0.08	138.88±7.82	0.084±0.007
T. paludosa	1767.0±96.7	7.50±0.14	115.40±16.63	0.065±0.013
L. longiflorum II	2809.0±158.1	8.75±0.16	173.11±19.56	0.062±0.011
Herbaceous angiosperms [63, 64]				
A. thaliana	32±3	1.97±0.06	0.59±0.27	0.018±0.010
L. maritima	49±5	2.27±0.08	1.08±0.04	0.022±0.003
C. arietinum	96±5	2.84±0.05	1.86	0.019±0.001
N. lutea	139±7	3.21±0.05	1.89	0.014±0.001
S. oleracea	156±7	3.34±0.05	2.01	0.013±0.001
A. pulsatilla	435±27	4.70±0.10	34.03	0.078±0.005
T. navicularis	552±27	5.09±0.08	53.97±3.65	0.098±0.011
C. majalis	710±17	5.53±0.04	33.22	0.047±0.001
F. lanceolata	1466±115	7.05±0.19	89.24	0.061±0.005
F. camschatcensis	1824±103	7.58±0.14	109.78	0.060±0.003
L. longiflorum	3273±167	9.21±0.16	68.85	0.021±0.001

(Continued)

Table 7.2 Notation is as in Table 7.1

Organism	Nuclear volume [μm³]	Nuclear radius [μm]	Genome size [×10⁹ bp]	Genome density [bp/nm³]
		Plants		
S. formosissima	4638±262	10.35±0.20	128.01	0.028±0.002
Gymnosperms [65, 64]				
P. strobus I	1468±94	7.05±0.15	50.17	0.034±0.002
P. strobus II	1259±51	6.70±0.09	50.17	0.040±0.002
P. glauca I	1137±52	6.48±0.10	31.59	0.028±0.001
A. balsamea	1114±62	6.43±0.12	32.08	0.029±0.002
L. laricina	1110±55	6.42±0.11	18.58	0.017±0.001
P. ponderosa	1095±51	6.39±0.10	47.34	0.043±0.002
P. resinosa	1084±59	6.37±0.12	46.55	0.043±0.002
Ts. canadensis I	1077±67	6.36±0.13	36.38	0.034±0.002
P. abies	1023±61	6.25±0.12	39.14	0.038±0.0002
P. glauca II	1014±47	6.23±0.10	31.59	0.031±0.002
P. pungens	977±38	6.16±0.08	35.50	0.036±0.001
P. glauca III	953±37	6.11±0.08	31.59	0.033±0.001
Ts. canadensis II	852±41	5.88±0.09	36.38	0.043±0.002
L. leptolepis	844±42	5.86±0.10	25.82	0.031±0.002
T. media I	645±28	5.36±0.08	22.01±0.39	0.034±0.002
P. douglasii	742±29	5.62±0.07	37.26	0.050±0.002
Ta. canadensis	677±28	5.45±0.08	22.69	0.034±0.001
T. media II	493±23	4.90±0.08	22.01±0.39	0.045±0.003
S. giganteum	431±21	4.69±0.08	19.41	0.045±0.002
T. occidentalis	358±19	4.41±0.08	24.17	0.068±0.004

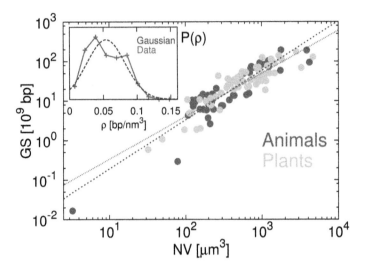

Figure 7.8 Scatter plot of genome-size (GS) versus nuclear volume (NV). Detailed data for animals and plants families are summarized in Tables 7.1 and 7.2, respectively. Lines correspond to best fits of the two sets of data to $GS = \alpha NV^{\gamma}$, which give $\gamma_{animals} = 1.25 \pm 0.09$ and $\gamma_{plants} = 1.08 \pm 0.07$. Inset: Probability distribution function for genome density $P(\rho)$ (solid curve) is well described by the Gaussian function with same average and standard deviation (dashed curve).

center of mass and its two components (Figure 7.11B): the self-density contribution from the given chromosome ($\rho_{DNA}^{self}(r)$) and the external contribution from the surrounding chromosomes ($\rho_{DNA}^{ext}(r)$). The plots demonstrate that chromosomes are rather "soft". As for common polymer systems [80], the core of each chromosome contains a significant amount of DNA protruding from close-by chains. In summary, territoriality is not a preclusion for chromosome strands to intermingle with each other, in agreement with cryo-FISH experiments [34].

7.3.2 Microrheology of the nucleus

It is indeed quite remarkable that, separately, Tseng et al. [40] and Hameed et al. [54] reported a consistent value of 250–290 nm value for nanoprobe-caging domains within nuclei of different types of cells and organisms (Section 7.2.2): this suggests a common origin for the domains. As noticed by Valet and Rosa [81], this value is also surprisingly close to the nominal tube diameter, $d_T \approx 245$nm, predicted by the "topological" polymer model describing chromosome territories (Sec. 7.3.1).

Topological constraints by polymer fibers are likely to induce confinement of dispersed nanoprobes of diameter d if $d \gtrsim d_T$ [82–84]. Motivated by this phenomenon, Valet and Rosa [81] employed large-scale numerical simulations to study the effect of polymer entanglement on the diffusion of nanoprobes of diameter d, and therefore obtain quantitative information for the viscoelastic properties of the nucleoplasm modelled as a semi-dilute solution of chromatin fibers. Different

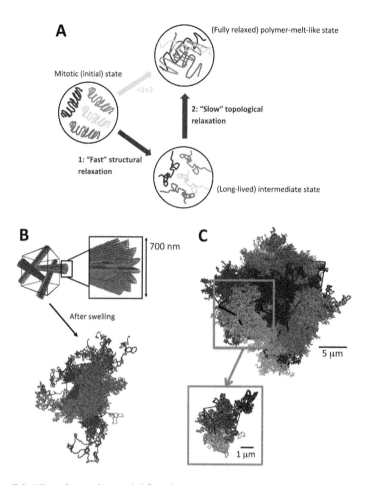

Figure 7.9 "Topological" model for chromosome territories. (A) At the beginning of interphase, condensed mitotic chromosomes start swelling. The path to full relaxation and complete mixing cannot take place on natural time scales due to "slow" relaxation of the topological degrees of freedom [72]. Chromosome structures thus remain effectively quenched into separate territories which retain "memory" of the initial conformations. (B) Numerical implementation of the model by Molecular Dynamics (MD) computer simulations [72]. Model chromosomes are initially prepared into non-overlapping mitotic-like structures. MD simulations show the rapid relaxation of polymer length scales up to the tube diameter d_T, while larger length scales fold into a crumpled structure resembling the behavior of ring polymers (Box 7.2). Each color corresponds to a single model chromosome. Reproduced from [72] under Creative Commons License. (C) The analogy between solutions of ring polymers and chromosome territories can be systematically exploited owing to an efficient coarse-grain protocol [73] which can produce hundreds of independent model conformations of mammalian-sized chromosomes. The model can be mapped to real-time and length scales (see bars) with no free parameters [72, 73]. The snapshot here provides a typical view for a model human nucleus. Reproduced with permission from [73].

Table 7.3 ρ_{DNA}, DNA density calculated for an "average" human nucleus of 5 μm radius; ρ_{30nm}, 30 nm chromatin fiber density assuming fiber compaction of 100 bp/nm (Sec. 7.2.1); ℓ_K, Kuhn length of the 30 nm fiber [76]; ξ, average distance from a monomer on one chain to the nearest monomer on another chain ("correlation length" [74]); L_e, entanglement length obtained from the condition of "optimal packing" of 20 chains per entanglement volume by Kavassalis and Noolandi [77]; d_T, average spatial distance between entanglements ("tube diameter" [74]); τ_e, time scale marking the onset of entanglement effects

Physical parameters of the "bead-spring" polymer model by Rosa and Everaers [72, 73].	
ρ_{DNA}	$0.012 \text{ bp} / \text{nm}^3$
ρ_{30nm}	$1.2 \cdot 10^{-4} / \text{nm}^2$
ℓ_K (30 nm fiber)	$300 \text{ nm} = 30 \text{ kbp}$
ξ	90 nm
$\dfrac{L_e}{\ell_K} = \left(\dfrac{20}{\rho_{30nm}\, \ell_K^2} \right)^2$	$4\ell_K = 1200 \text{ nm} = 0.12 \text{ Mbp}$
$d_T = \sqrt{\dfrac{\ell_K L_e}{6}}$	245 nm
τ_e	32 s

BOX 7.2: Structure and microrheology of entangled ring polymers in solution.

Structure – Ring polymers in entangled solutions have to respect global topological invariance requiring that all chains remain permanently *unlinked* at the expense of entropic loss [103]. Consequently, topological constraints between close-by rings induce chain conformations to fold into compact (*i.e.*, "territorial" [104]) structures which are reminiscent of the "crumpled" (or "fractal" [4]) globule [105–108]. Recent numerical work [73, 109, 110] has confirmed this conjecture and, thus, demonstrated that the typical end-to-end mean-square spatial distance between chain monomers with contour separation L is given by:

$$\langle R^2(L) \rangle \approx \begin{cases} L^2, & L \leq \ell_K \\ \ell_K L, & \ell_K \leq L \leq L_e \\ \ell_K L_e \left(\dfrac{L}{L_e} \right)^{2/3}, & L \geq L_e \end{cases} \tag{7.3}$$

ℓ_K is the Kuhn length of the polymer describing chain stiffness [74]. L_e, which depends on ℓ_K and on the solution density [77, 111], is the so-called *entanglement* length marking the onset of entanglement effects. The corresponding end-to-end spatial distance between entanglement strands $d_T \approx (\ell_K L_e)^{1/2}$ (the "entanglement distance") is also called the "tube diameter" by analogy to systems of linear chains [75]. Among their noticeable features and in spite of compactness, rings do not expel close-by rings: on average, in fact, their surface remains "rough" and shares many contacts with neighbors [108, 110, 112, 113]. Indeed, rings interpenetrate as "threading" conformations [114, 115] (Figure 7.10A) akin to interacting "branched structures" [73] with long-range (loose) loops [116, 117].

Microrheology – Depending on nanoprobe diameter three regimes (Figure 7.10B) can be distinguished [82–84]:

I) Small nanoprobes, $d < \xi$, where ξ (the "correlation length") is the average distance from a monomer on one chain to the nearest monomer on another chain [74]. Nanoprobes interact only with the solvent, and their motion remains diffusive:

$$\langle \Delta x^2 (\tau) \rangle \sim D_s \, \tau \sim \frac{\kappa_B T}{\eta_s \, d} \, \tau. \tag{7.4}$$

D_s is the diffusion coefficient and η_s is the viscosity of the solvent.

II) Intermediate nanoprobes, $\xi \lesssim d \lesssim d_T$. Nanoprobe motion is now affected by the polymers, showing three regimes:

$$\langle \Delta x^2 (\tau) \rangle \sim \begin{cases} D_s \tau, & \tau < \tau_\xi \sim \dfrac{\eta_s \xi^3}{\kappa_B T} & \text{(II.a)} \\[2ex] D_s \tau_\xi \left(\dfrac{\tau}{\tau_\xi} \right)^{1/2}, & \tau_\xi < \tau < \tau_d \sim \tau_\xi \left(\dfrac{d}{\xi} \right)^4 & \text{(II.b)} \\[2ex] D_s \left(\dfrac{\xi}{d} \right)^2 \tau, & \tau > \tau_d & \text{(II.c)} \end{cases} \tag{7.5}$$

In (II.a), nanoprobe motion is driven only by random collisions with the solvent, as in (I). This regime stops at τ_ξ, the relaxation time of a polymer strand of spatial size ξ. Then (II.b), the nanoprobe experiences a time-dependent viscosity $\eta(\tau) \sim \eta_s \, n_{str}(\tau) \equiv \eta_s (\tau / \tau_\xi)^{1/2}$, where $n_{str}(\tau)$ is the number of strands which have relaxed at time τ. This regime stops at time τ_d, the relaxation time of a larger polymer strand of spatial size $= d = \xi \sqrt{n_{str}(\tau_d)}$. Above τ_d (II.c), nanoprobe motion becomes diffusive again with effective viscosity $\sim \eta_s \, \eta_{str}(\tau_d)$, which is $\sim (d / \xi)^2$ times *larger* than the value in pure solvent.

III) Large nanoprobes, $d > d_T$. Regime (II.a) still holds, while regime (II.b) stops at $\tau_{d=d_T} = \tau_\xi (d_T / \xi)^4$. Above τ_{d_T}, entanglements affect nanoprobe motion. By scaling arguments [74], the time-dependent friction

(Continued)

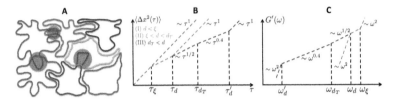

Figure 7.10 (A) Schematic illustration of ring polymers in dense solution. Shaded areas highlight threadings between close-by chains. (B) Time mean-square displacement $\langle \Delta x^2(\tau) \rangle$ of nanoprobes in solutions of ring polymers. (C) Corresponding predictions (obtained by using Eqs. 7.1 and 7.2) for the storage modulus $G'(\omega)$ as a function of frequency ω. Each crossover frequency is the inverse of the corresponding crossover time in panel (B) multiplied by "2π". An analogous plot for the loss modulus $\hat{G}''(\omega)$ can be constructed. Plots are in log-log scales.

$\eta = \eta(\tau) \approx \tau G(\tau)$ where: $G(\tau) \sim \dfrac{\kappa_B T}{v^{2/3} L(\tau)}$ is the stress relaxation modulus, v is the monomer volume and $L(\tau) \sim L_e(\tau / \tau_e)^{1/\gamma}$ ($\gamma = 2.33\text{–}2.57$ [116, 118, 119]) is the contour length of the polymer strand with relaxation time τ. Therefore,

$$\langle \Delta x^2(\tau) \rangle \sim \frac{\kappa_B T}{\eta(L(\tau))d} \tau \sim \frac{v^{2/3} L_e}{d} \left(\frac{\tau}{\tau_e}\right)^{1/\gamma}, \qquad (7.6)$$

and nanoprobe diffusion is anomalous with exponent $1/\gamma = 0.39\text{–}0.43$. This regime breaks down at $\tau_d' \sim \tau_e (d^2 / (\ell_\kappa L_e))^{3\gamma/2} = \tau_e (d^2 / (\ell_\kappa L_e))^{3.50\text{–}3.86}$, the relaxation time of a ring strand of spatial extension $\approx d$ in the compact regime (Equation 7.3 for $L > L_e$). For $\tau > \tau_d'$, nanoprobe diffusion is normal with $\langle \Delta x^2(\tau) \rangle \sim \dfrac{\kappa_B T}{\eta(L(\tau_d'))d} \tau$.

By applying Equations 7.1 and 7.2 (Box 7.1), the shapes of the storage and loss moduli can be then recovered (Figure 7.10C).

nanoprobes were considered, with d ranging from 30 nm (the fiber diameter) to 300 nm (slightly above d_T).

Figure 7.12 reports the main results, in terms of: (A) nanoprobe time mean-square displacement, $\langle \Delta x^2(\tau) \rangle$; (B) time-dependent diffusion coefficient, $D(\tau) \equiv \dfrac{\langle \Delta x^2(\tau) \rangle}{6\tau}$; (C) time-dependent viscosity $\eta(\tau) \equiv \dfrac{\kappa_B T}{3\pi d D(\tau)}$; (D) asymptotic diffusion coefficient ($D_\infty \equiv D(\tau \to \infty)$) and viscosity ($\eta_\infty \equiv \eta(\tau \to \infty)$) versus nanoprobe diameter. The data demonstrate that for d smaller than the polymer correlation length (Box 7.2) $\xi \approx 90$ nm and neglecting the short-time ballistic regime ($\langle \Delta x^2(\tau) \rangle \sim \tau^2$), nanoprobe motion is not or only weakly coupled to chromosome dynamics, implying $\langle \Delta x^2(\tau) \rangle \equiv 6 D_\infty \tau$ with "standard" behaviors $D_\infty \sim d^{-1}$ and $\eta_\infty \sim d^0$. Viceversa, for $d > \xi$ coupling to chromosome dynamics

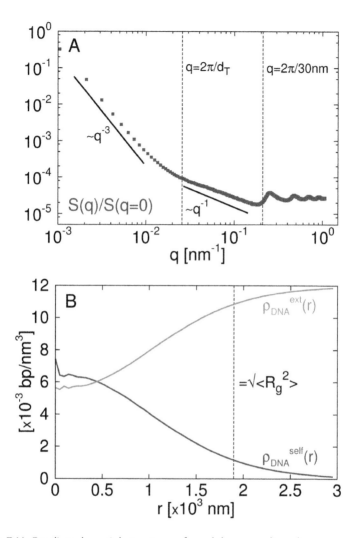

Figure 7.11 Predicted spatial structure of model mammalian chromosomes [72, 73]. (A) Structure factor $S(\bar{q})$ versus the norm of the wave vector $q \equiv |\bar{q}|$. The two regimes q^{-1} (rod-like) and q^{-3} (compact-like) are for spatial scales, respectively, below and above the tube diameter, $d_T \approx 245\text{nm}$, of the chromatin fiber. The wavy behavior at large q is an artifact due to the discrete bead-spring nature of the model. (B) Average DNA density at spatial distance r from the chromosome center of mass: $\rho_{DNA}^{self}(r)$, the self-density contribution from the given chromosome; $\rho_{DNA}^{ext}(r)$, the external contribution from the surrounding chromosomes. The sum of the two equals the average DNA density $= 0.012$ bp/nm^3 (Table 7.3). For reference, the dashed line corresponds to the predicted average size of a single chromosome territory.

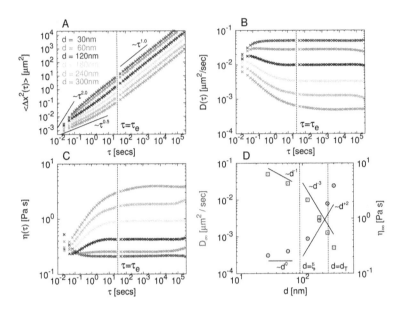

Figure 7.12 Viscoelasticity of model interphase chromosomes analyzed by microrheology. (A) Time mean-square displacement, $\Delta x^2(\tau)$, of nanoprobes with varying diameter d. Vertical dashed lines mark the position of chromatin entanglement time $\tau_e \approx 32$ seconds [72]. (B) Time-dependent diffusion coefficient, $D(\tau) \equiv \left\langle \Delta x^2(\tau) \right\rangle / 6\tau$. (C) Time-dependent viscosities, $\eta(\tau) \equiv \dfrac{\kappa_B T}{3\pi d\, D(\tau)}$. (D) Asymptotic diffusion coefficient ($\square$), $D_\infty \equiv D(\tau \to \infty)$, and particle viscosity ($\circ$), $\eta_\infty \equiv \eta(\tau \to \infty)$, as functions of nanoprobe diameter, d. Solid lines are for theoretical predictions in the non-entangled regime (Box 7.2). Polymer-mediated effects start at nanoprobe diameter $d \approx \xi \approx 90$ nm (Box 7.2). The largest nanoprobe diameter is of the order of the tube diameter, $d_T \approx 245$ nm, of the chromatin solution. Reproduced with permission from [81].

induces nanoprobe subdiffusion ($\left\langle \Delta x^2(\tau) \right\rangle \sim \tau^{1/2}$) at small τ and consequent "anomalous" behaviors $D_\infty \sim d^{-3}$ and $\eta_\infty \sim d^2$, in agreement with theoretical predictions (see discussion in Box 7.2).

Figure 7.13 completes the previous analysis by showing (A, B) the distribution functions for D_∞ and (C–E) the distribution functions for particle displacements $\Delta x(\tau) \equiv |x(\tau' + \tau) - x(\tau')|$ at different lag-times τ (see caption for details). In general, D_∞ and $\Delta x(\tau)$ appear Gaussian-distributed (black lines). With one notable exception: for $\tau \ll \tau_e$ and large nanoprobes, $P(\Delta x(\tau))$ appears significantly different from the Gaussian function. This follows from the presence of surrounding polymers exerting constraints and inducing spatial correlations [81] on nanoprobe displacement.

Finally, by using the fundamental relation of microrheology connecting the complex shear modulus to the nanoprobe mean-square displacement (Equation 7.1 in Box 7.1), theoretical predictions for the storage and loss moduli at frequency

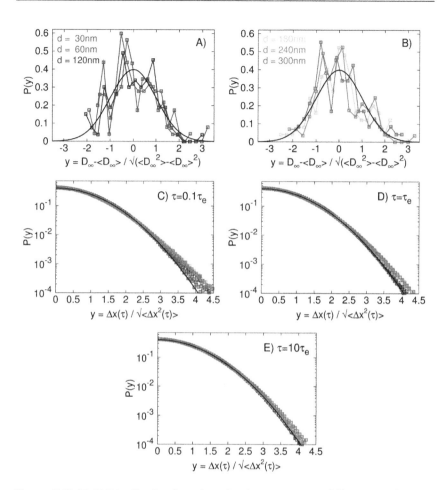

Figure 7.13 (A,B) Distribution functions for the asymptotic diffusion coefficients, D_∞. The shape of the distributions compares well to the Gaussian function (black line). (C,D,E) Distribution functions for one-dimensional nanoprobe displacements, $\Delta x(\tau) \equiv |x(t+\tau) - x(t)|$, at different lag-times τ (see corresponding captions). The x-axis has been rescaled according to the corresponding standard deviations $\sqrt{\langle \Delta x^2(\tau) \rangle}$ and the curves compared to the normal form of the Gaussian function describing ordinary diffusion [85]. At $\tau \ll \tau_e$, $P(\Delta x(\tau))$ shifts from Gaussian to non-Gaussian behavior at increasing nanoprobe diameters. Universal Gaussian behavior is recovered at all d's at $\tau \gg \tau_e$. Color code is as in Figure 7.12.

ω can be extracted and then compared to available experimental results, see Table 7.4. In spite of its simplicity, the polymer model is in reasonable agreement with experiments. The main difference is that experiments predict nuclei with $\hat{G}' > \hat{G}''$ (i.e., more solid- than liquid-like, see Box 7.1), while the polymer model predicts the opposite. Since nanoprobes with diameters larger than d_T should experience a more solid-like behavior (Box 7.2), this difference can be ascribed to the size of the

Table 7.4 Microrheology of the nucleus: theoretical modeling (top) versus experiments (bottom). Asymptotic diffusion coefficients (D_∞), viscosities (η_∞) and selected values for storage ($\hat{G}'(\omega)$) and loss ($\hat{G}''(\omega)$) moduli. The symbol "–" means no data are available

Microrheology of the nucleus: Theoretical predictions from Ref. [81]								
			0.1 Hz		1 Hz		10 Hz	
$d[nm]$	$D_\infty\left[\times 10^{-3}\,\mu m^2 \cdot s^{-1}\right]$	$\eta_\infty\,[Pa \cdot s]$	$\hat{G}'\,[Pa]$	$\hat{G}''\,[Pa]$	$\hat{G}'\,[Pa]$	$\hat{G}''\,[Pa]$	$\hat{G}'\,[Pa]$	$\hat{G}''\,[Pa]$
30	50.0	0.21	–	0.0214	–	0.2045	–	2.0651
60	28.0	0.25	0.0004	0.0238	0.0079	0.2532	–	2.2061
120	10.0	0.42	–	0.0423	0.0712	0.3948	0.7009	2.3778
180	3.3	0.92	0.0086	0.0882	0.2260	0.6865	1.6126	2.8336
240	1.3	1.81	0.0349	0.1469	0.4959	1.0163	2.6690	3.5608
300	0.5	3.86	0.0747	0.2972	0.8674	1.4476	3.7901	4.5105

Microrheology of the nucleus: experiments				
$d[nm]$	Organism	Frequency [Hz]	$\hat{G}'\,[Pa]$	$\hat{G}''\,[Pa]$
100	Swiss 3T3 fibroblasts (mouse) [40]	1 – 10	≈ 10	≈ 3 – 10
1000	HeLa cells (human) [54]	1	≈ 0.1	≈ 0.05

simulated nanoprobes which is just about d_T. It would be interesting to test then if larger nanoprobes would go more towards the observed experimental behavior.

At the same time, other factors which have not been taken into account because of the initial intention to keep the polymer model as simple as possible could contribute to explain deviations from experiments as well. In the next section, we will comment briefly on these issues and highlight possible directions for future work.

7.4 CONCLUSIONS AND FUTURE DIRECTIONS

In this review, we have summarized the results of our efforts to understand chromosome folding and nuclear structure in terms of generic polymer physics. In particular, we have discussed the physical origin of:

1.   Chromosome territories. In our framework, the territorial organization of the nucleus (Section 7.2.1) is explained in terms of the slow Brownian relaxation of non-overlapping long polymer chains subjected to topological constraints (Section 7.3.1). As a consequence, average chromosome conformations in eukaryotes resemble ring polymers (Figure 7.8) with a crumpled yet intermingling structure (Figure 7.9).

2.   Microrheology of the nucleus. The viscoelastic properties of the nucleus (Sec. 7.2.2) have been compared to the dynamic behavior of nanoprobes immersed in the ring polymers solution (Section 7.3.2). The model is in good quantitative agreement with theoretical expectations (Figure 7.10) and in qualitative agreement with the available experiments for nuclear microrheology (Table 7.4).

Obviously, due to the complexity of the genome and the simplicity of the model, it is no surprise that there is still much work ahead which remains to be done in order to arrive at a satisfactory picture of genome organization in terms of polymer physics. In this spirit, in this section, we discuss a few promising directions which should be undertaken to make the model more coherent with experimental data.

First, an evident inconsistency between the outcome of microrheology experiments and the results of our polymer model in the latter shows no sign of a plateau in the time MSD or the storage modulus $\hat{G}'(\omega)$ (compare Figure 3A,B of Ref. [40] to Figure 7.10A), in spite of the very similar sizes of nanoprobes used. It should be possible to "level" this discrepancy though, by introducing a fixed amount of long-lived or permanent *crosslinks* between chromatin fibers. Crosslinks are known to quench polymer dynamics without altering significantly the average polymer $3d$ structure [86] and may affect nanoprobe diffusive behavior when its size becomes larger than the polymer tube diameter [87]. On the biological side, there exists a conspicuous number of experimental observations [4, 32, 88] proving the existence of functionally relevant protein bridges between sequence-distant chromatin loci which, indeed, may act as effective crosslinks. In this respect then, numerical investigations of nanoprobe dynamics could help to estimate the specific amount of cross-links present in the genome and elucidating their role in chromatin organization.

Second, a few recent studies [48, 54] have demonstrated that chromosome activity and chromosome dynamics consist of the subtle interplay between *passive* thermal diffusion and *active*, ATP-dependent motion triggered by chromatin remodeling and transcription complexes. Stimulated by active processes, chromatin dynamics influences also the motion of dispersed nanoprobes [54]. Taken together, these results suggest that the standard picture adopted so far where chromatin is modeled as a passive polymer is an oversimplification. In recent years, non-equilibrium physics of active systems [89] and active polymers [90] has received considerable attention for being at the interface between statistical, soft matter and biological physics. To our knowledge, the first attempt to include activity in a numerical polymer model for eukaryotic chromosomes was made by Ganai et al. [91] who argued that non-random chromosome segregation is the result of differences in *non-equilibrium* activity across chromosomes originating in the inhomogeneous distribution of ATP-dependent chromatin remodeling and transcription machinery on each chromosome. In their model, each monomer is characterized by a given transcription level whose fluctuations are suitably taken into account by a local "effective" temperature: "hot" (respectively, "cold") monomers are associated with active and gene-rich (respectively, inactive and gene-poor) monomers. Through a similar approach and rigorous theoretical considerations, Smrek and Kremer [92] showed that entangled polymer solutions where single chains have different temperatures undergo a *non-equilibrium* phase separation similar to the classical *equilibrium* phase separation observed in polymer mixtures [74]. As in the aforementioned case of transcription or in the recently proposed looping extrusion mechanism [93], active processes play a fundamental role in chromosome organization. It would be interesting, then, to explore to which extent the viscoelastic properties of chromatin fibers are changed by the presence of non-equilibrium mechanisms.

To conclude, we hope to have convinced the reader that polymer physics represents a fundamental tool to describe and predict chromosome structure during the different stages of the cell cycle. In general, the conspicuous amount of experimental data currently being published is causing the field to boom and many different polymer models (see, for instance, Ref. [11]) are available at present. The specificity of the point of view adopted here (and in our work [72, 73, 78, 81] on which this review is based) consists of the assumption that topological constraints are an *essential* feature to be retained in all minimal polymer models. In particular, a quantitative understanding of this "null model" with the inclusion of proper mapping [72] to real-time and length scales is a prerequisite for attempts [94–99] to reconstruct or predict the three-dimensional chromosome structure and the dynamics of entire cell nuclei, and provide then a reliable description of the large-scale structure and dynamics of nuclear compartmentalization.

REFERENCES

1. Dekker, J., Belmont, A. S., Guttman, M., Leshyk, V. O., Lis, J. T., Lomvardas, S., Mirny, L. A., O'Shea, C. C., Park, P. J., Ren, B., Politz, J. C. R., Shendure, J., Zhong, S., and the 4D Nucleome Network. The 4d nucleome project. *Nature* 549(7671), 219–226 (2017).

2. http://www.4dnucleome.eu/.

3. Cremer, T. and Cremer, C. Chromosome territories, nuclear architecture and gene regulation in mammalian cells. *Nat. Rev. Genet.* 2, 292–301 (2001).

4. Lieberman-Aiden, E. et al. Comprehensive mapping of long-range interactions reveals folding principles of the human genome. *Science* 326(5950), 289–293 (2009).

5. Gasser, S. M. Nuclear architecture – Visualizing chromatin dynamics in interphase nuclei. *Science* 296, 1412–1416 (2002).

6. Bystricky, K. Chromosome dynamics and folding in eukaryotes: Insights from live cell microscopy. *FEBS Lett.* 589(20, Part A), 3014–3022 (2015).

7. Wirtz, D. Particle-tracking microrheology of living cells: Principles and applications. *Annu. Rev. Biophys.* 38, 301–326 (2009).

8. Forcato, M., Nicoletti, C., Pal, K., Livi, C. M., Ferrari, F., and Bicciato, S. Comparison of computational methods for Hi-C data analysis. *Nat. Methods* 14, 679, 06 (2017).

9. Libbrecht, M. W. and Noble, W. S. Machine learning applications in genetics and genomics. *Nat. Rev. Genet.* 16(6), 321–332 (2015).

10. Dekker, J., Marti-Renom, M. A., and Mirny, L. A. Exploring the three-dimensional organization of genomes: Interpreting chromatin interaction data. *Nat. Rev. Genet.* 14, 390–403 (2013).

11. Rosa, A. and Zimmer, C. Computational models of large-scale genome architecture. *Int. Rev. Cell Mol. Biol.* 307, 275–350 (2014).

12. Chiariello, A. M., Annunziatella, C., Bianco, S., Esposito, A., and Nicodemi, M. Polymer physics of chromosome large-scale 3d organisation. *Sci. Rep.* 6, 29775 (2016).

13. Jost, D., Vaillant, C., and Meister, P. Coupling 1d modifications and 3d nuclear organization: data, models and function. *Curr. Opin. Cell Biol.* 44, 20–27 (2017).

14. Lanctôt, C., Cheutin, T., Cremer, M., Cavalli, G., and Cremer, T. Dynamic genome architecture in the nuclear space: Regulation of gene expression in three dimensions. *Nat. Rev. Genet.* 8, 104, 02 (2007).

15. Alberts, B. et al. *Molecular Biology of the Cell*. Garland Science, New York, 5th edition, (2007).

16. Cremer, T. and Cremer, M. Chromosome territories. *Cold Spring Harbor Perspectives in Biology* 2(3), 1 (2010).

17. Gibcus, J. H. and Dekker, J. The hierarchy of the 3d genome. *Molecular Cell* 49(5), 773–782 (2013).

18. Solovei, I., Kreysing, M., Lanctôt, C., Kösem, S., Peichl, L., Cremer, T., Guck, J., and Joffe, B. Nuclear architecture of rod photoreceptor cells adapts to vision in mammalian evolution. *Cell* 137(2), 356 (2009).

19. Columbaro, M., Capanni, C., Mattioli, E., Novelli, G., Parnaik, V., Squarzoni, S., Maraldi, N., and Lattanzi, G. Rescue of heterochromatin organization in Hutchinson-Gilford progeria by drug treatment. *Cell. Mol. Life Sci.* 62(22), 2669–2678 (2005).

20. McCord, R. P., Nazario-Toole, A., Zhang, H., Chines, P. S., Zhan, Y., Erdos, M. R., Collins, F. S., Dekker, J., and Cao, K. Correlated alterations in genome organization, histone methylation, and DNA-lamin A/C interactions in Hutchinson-Gilford progeria syndrome. *Genome Res.* 23(2), 260–269 (2013).

21. Chandra, T., Ewels, P. A., Schoenfelder, S., Furlan-Magaril, M., Wingett, S. W., Kirschner, K., Thuret, J.-Y., Andrews, S., Fraser, P., and Reik, W. Global reorganization of the nuclear landscape in senescent cells. *Cell reports* 10(4), 471–483 (2015).

22. Fritz, A. J., Stojkovic, B., Ding, H., Xu, J., Bhattacharya, S., Gaile, D., and Berezney, R. Wide-scale alterations in interchromosomal organization in breast cancer cells: Defining a network of interacting chromosomes. *Hum. Mol. Genet.* 23(19), 5133–5146 (2014).

23. Norton, H. K. and Phillips-Cremins, J. E. Crossed wires: 3d genome misfolding in human disease. *J. Cell Biol.* 216(11), 3441 (2017).

24. Hansen, J. C. Conformational dynamics of the chromatin fiber in solution: Determinants, mechanisms, and functions. *Annu. Rev. Biophys. Biomol. Struct.* 31(1), 361–392 (2002).

25. Luger, K., Mader, A. W., Richmond, R. K., Sargent, D. F., and Richmond, T. J. Crystal structure of the nucleosome core particle at 2.8[thinsp]a resolution. *Nature* 389(6648), 251–260 (1997).

26. Luger, K. and Hansen, J. C. Nucleosome and chromatin fiber dynamics. *Curr. Opin. Struct. Biol.* 15(2), 188–196 (2005).

27. Maeshima, K., Imai, R., Hikima, T., and Joti, Y. Chromatin structure revealed by x-ray scattering analysis and computational modeling. *Methods* 70(2), 154–161 (2014).

28. Maeshima, K., Ide, S., Hibino, K., and Sasai, M. Liquid-like behavior of chromatin. *Curr. Opin. Genet. Dev.* 37(Supplement C), 36–45 (2016).

29. Ou, H. D., Phan, S., Deerinck, T. J., Thor, A., Ellisman, M. H., and O'Shea, C. C. ChromEMT: Visualizing 3D chromatin structure and compaction in interphase and mitotic cells. *Science* 357(6349), ii (2017).

30. Récamier, V., Izeddin, I., Bosanac, L., Dahan, M., Proux, F., and Darzacq, X. Single cell correlation fractal dimension of chromatin. *Nucleus* 5(1), 75–84 (2014).

31. Speicher, M. R. and Carter, N. P. The new cytogenetics: Blurring the boundaries with molecular biology. *Nat. Rev. Genet.* 6(10), 782–792 (2005).

32. Dixon, J. R., Selvaraj, S., Yue, F., Kim, A., Li, Y., Shen, Y., Hu, M., Liu, J. S., and Ren, B. Topological domains in mammalian genomes identified by analysis of chromatin interactions. *Nature* 485(7398), 376–380 (2012).

33. Marshall, W. F. Order and disorder in the nucleus. *Curr. Biol.* 12(5), R185–R192 (2002).

34. Branco, M. R. and Pombo, A. Intermingling of chromosome territories in interphase suggests role in translocations and transcription-dependent associations. *Plos Biol.* 4(5), e138 (2006).

35. Bolzer, A., Kreth, G., Solovei, I., Koehler, D., Saracoglu, K., Fauth, C., Muller, S., Eils, R., Cremer, C., Speicher, M., and Cremer, T. Three-dimensional maps of all chromosomes in human male fibroblast nuclei and prometaphase rosettes. *Plos Biol.* 3(5), 826–842 (2005).

36. Dixon, J. R., Jung, I., Selvaraj, S., Shen, Y., Antosiewicz-Bourget, J. E., Lee, A. Y., Ye, Z., Kim, A., Rajagopal, N., Xie, W., Diao, Y., Liang, J., Zhao, H., Lobanenkov, V. V., Ecker, J. R., Thomson, J. A., and Ren, B. Chromatin architecture reorganization during stem cell differentiation. *Nature* 518(7539), 331–336 (2015).

37. Lonfat, N., Montavon, T., Darbellay, F., Gitto, S., and Duboule, D. Convergent evolution of complex regulatory landscapes and pleiotropy at Hox loci. *Science* 346(6212), 1004 (2014).

38. Bancaud, A., Huet, S., Daigle, N., Mozziconacci, J., Beaudouin, J., and Ellenberg, J. Molecular crowding affects diffusion and binding of nuclear proteins in heterochromatin and reveals the fractal organization of chromatin. *Embo. J.* 28(24), 3785 (2009).

39. Hihara, S., Pack, C.-G., Kaizu, K., Tani, T., Hanafusa, T., Nozaki, T., Takemoto, S., Yoshimi, T., Yokota, H., Imamoto, N., Sako, Y., Kinjo, M., Takahashi, K., Nagai, T., and Maeshima, K. Local nucleosome dynamics facilitate chromatin accessibility in living mammalian cells. *Cell Reports* 2(6), 1645–1656 (2012).

40. Tseng, Y., Lee, J. S. H., Kole, T. P., Jiang, I., and Wirtz, D. Micro-organization and visco-elasticity of the interphase nucleus revealed by particle nanotracking. *J. Cell Sci.* 117(10), 2159–2167 (2004).

41. Cicuta, P. and Donald, A. Microrheology: A review of the method and applications. *Soft Matter* 3, 1449–1455 (2007).

42. Wu, P.-H., Hale, C. M., Chen, W.-C., Lee, J. S. H., Tseng, Y., and Wirtz, D. High-throughput ballistic injection nanorheology to measure cell mechanics. *Nat. Protocols* 7(1), 155–170 (2012).

43. Waigh, T. A. Advances in the microrheology of complex fluids. *Rep. Prog. Phys.* 79(7), 074601 (2016).

44. Metzler, R., Jeon, J.-H., Cherstvy, A. G., and Barkai, E. Anomalous diffusion models and their properties: non-stationarity, non-ergodicity, and ageing at the centenary of single particle tracking. *Phys. Chem. Chem. Phys.* 16, 24128–24164 (2014).

45. Saxton, M. J. Anomalous diffusion due to obstacles: a Monte Carlo study. *Biophys. J.* 66, 394–401 (1994).

46. Bouchaud, J.-P. and Georges, A. Anomalous diffusion in disordered media: Statistical mechanisms, models and physical applications. *Phys. Rep.* 195(4), 127–293 (1990).

47. Bronshtein, I., Kanter, I., Kepten, E., Lindner, M., Berezin, S., Shav-Tal, Y., and Garini, Y. Exploring chromatin organization mechanisms through its dynamic properties. *Nucleus* 7(1), 27–33 (2016).

48. Weber, S. C., Spakowitz, A. J., and Theriot, J. A. Nonthermal ATP-dependent fluctuations contribute to the in vivo motion of chromosomal loci. *Proc. Natl. Acad. Sci. USA* 109(19), 7338–7343 (2012).

49. Zidovska, A., Weitz, D. A., and Mitchison, T. J. Micron-scale coherence in interphase chromatin dynamics. *Proc. Natl. Acad. Sci. USA* 110(39), 15555–15560 (2013).

50. Marenduzzo, D., Finan, K., and Cook, P. R. The depletion attraction: An underappreciated force driving cellular organization. *J. Cell Biol.* 175(5), 681–686 (2006).

51. Huet, S., Lavelle, C., Ranchon, H., Carrivain, P., Victor, J.-M., Bancaud, A., Hancock, R., and Jeon, K. W. Relevance and limitations of crowding, fractal, and polymer models to describe nuclear architecture. *Int. Rev. Cell Mol. Biol.* 307, 443–479 (2014).

52. Hajjoul, H., Mathon, J., Ranchon, H., Goiffon, I., Mozziconacci, J., Albert, B., Carrivain, P., Victor, J.-M., Gadal, O., Bystricky, K., and Bancaud, A. High-throughput chromatin motion tracking in living yeast reveals the flexibility of the fiber throughout the genome. *Genome Res.* 23, 1829–1838 (2013).

53. Tseng, Y., Kole, T. P., Lee, S.-H. J., and Wirtz, D. Local dynamics and visco-elastic properties of cell biological systems. *Curr. Opin. Colloid In.* 7(3), 210–217 (2002).

54. Hameed, F. M., Rao, M., and Shivashankar, G. Dynamics of passive and active particles in the cell nucleus. *Plos One* 7(10), e45843 (2012).

55. Carmo-Fonseca, M., Platani, M., and Swedlow, J. R. Macromolecular mobility inside the cell nucleus. *Trends Cell Biol.* 12(11), 491–495 (2002).

56. Platani, M., Goldberg, I. G., Bensaddek, D., and Swedlow, J. R. Cajal body dynamics and association with chromatin are ATP-dependent. *Nat. Cell Biol.* 4, 502–508 (2002).

57. Görisch, S. M., Wachsmuth, M., Ittrich, C., Bacher, C. P., Rippe, K., and Lichter, P. Nuclear body movement is determined by chromatin accessibility and dynamics. *Proc. Natl. Acad. Sci. USA* 101(36), 13221–13226 (2004).

58. Sinha, D. K., Banerjee, B., Maharana, S., and Shivashankar, G. V. Probing the dynamic organization of transcription compartments and gene loci within the nucleus of living cells. *Biophys. J.* 95(11), 5432–5438 (2008).

59. Conger, A. D. and Clinton, J. H. Nuclear volumes, DNA contents, and radiosensitivity in whole-body-irradiated amphibians. *Radiation Research* 54(1), 69–101 (1973).

60. http://www.genomesize.com.

61. Maul, G. and Deaven, L. Quantitative determination of nuclear pore complexes in cycling cells with differing DNA content. *J. Cell Biol.* 73(3), 748 (1977).

62. Baetcke, K. P., Sparrow, A. H., Nauman, C. H., and Schwemmer, S. S. The relationship of DNA content to nuclear and chromosome volumes and to radiosensitivity (ld50). *Proc. Natl. Acad. Sci. USA*58(2), 533–540 (1967).

63. Price, H. J., Sparrow, A. H., and Nauman, A. F. Correlations between nuclear volume, cell volume and DNA content in meristematic cells of herbaceous angiosperms. *Experientia* 29(8), 1028–1029 (1973).

64. http://data.kew.org/cvalues.

65. Sparrow, A. H., Rogers, A. F., and Schwemmer, S. S. Radiosensitivity studies with woody plants — I acute gamma irradiation survival data for 28 species and predictions for 190 species. *Radiation Botany* 8(2), 149–174 (1968).

66. Emanuel, M., Radja, N. H., Henriksson, A., and Schiessel, H. The physics behind the larger scale organization of DNA in eukaryotes. *Phys. Biol.* 6(2), 025008 (2009).

67. Mateos-Langerak, J. et al. Spatially confined folding of chromatin in the interphase nucleus. *Proc. Natl. Acad. Sci. USA* 106, 3812–3817 (2009).

68. Mirny, L. A. The fractal globule as a model of chromatin architecture in the cell. *Chromosome Res.* 19(1), 37–51 (2011).

69. Halverson, J. D., Smrek, J., Kremer, K., and Grosberg, A. Y. From a melt of rings to chromosome territories: The role of topological constraints in genome folding. *Rep. Prog. Phys.* 77(2), 022601 (2014).

70. Brackley, C. A., Johnson, J., Kelly, S., Cook, P. R., and Marenduzzo, D. Simulated binding of transcription factors to active and inactive regions folds human chromosomes into loops, rosettes and topological domains. *Nucleic Acids Res.* 44(8), 3503–3512 (2016).

71. Jost, D., Rosa, A., Vaillant, C., and Everaers, R. A polymer physics view on universal and sequence-specific aspects of chromosome folding. In *Nuclear Architecture and Dynamics*, Volume 2, Lavelle, C. and Victor, J.-M., editors. Academic Press (2017).

72. Rosa, A. and Everaers, R. Structure and dynamics of interphase chromosomes. *Plos Comput. Biol.* 4, e1000153 (2008).

73. Rosa, A. and Everaers, R. Ring polymers in the melt state: The physics of crumpling. *Phys. Rev. Lett.* 112, 118302 (2014).

74. Rubinstein, M. and Colby, R. H. *Polymer Physics*. Oxford University Press, New York, (2003).

75. Doi, M. and Edwards, S. F. *The Theory of Polymer Dynamics*. Oxford University Press, New York, (1986).

76. Bystricky, K., Heun, P., Gehlen, L., Langowski, J., and Gasser, S. M. Long-range compaction and flexibility of interphase chromatin in budding yeast analyzed by high-resolution imaging techniques. *Proc. Natl. Acad. Sci. USA* 101, 16495–16500 (2004).

77. Kavassalis, T. A. and Noolandi, J. New view of entanglements in dense polymer systems. *Phys. Rev. Lett.* 59, 2674–2677 (1987).

78. Rosa, A., Becker, N. B., and Everaers, R. Looping probabilities in model interphase chromosomes. *Biophys. J.* 98, 2410–2419 (2010).
79. Florescu, A.-M., Therizols, P., and Rosa, A. Large scale chromosome folding is stable against local changes in chromatin structure. *Plos. Comput. Biol.* 12(6), e1004987 (2016).
80. Louis, A., Bolhuis, P., Hansen, J., and Meijer, E. *Phys. Rev. Lett.* 85, 2522–2525 (2000).
81. Valet, M. and Rosa, A. Viscoelasticity of model interphase chromosomes. *J. Chem. Phys.* 141(24), 245101 (2014).
82. Cai, L.-H., Panyukov, S., and Rubinstein, M. Mobility of nonsticky nanoparticles in polymer liquids. *Macromolecules* 44, 7853–7863 (2011).
83. Cai, L.-H., Panyukov, S., and Rubinstein, M. Hopping diffusion of nanoparticles in polymer matrices. *Macromolecules* 48, 847–862 (2015).
84. Ge, T., Kalathi, J. T., Halverson, J. D., Grest, G. S., and Rubinstein, M. Nanoparticle motion in entangled melts of linear and nonconcatenated ring polymers. *Macromolecules* 50(4), 1749–1754 (2017).
85. Van Kampen, N. G. *Stochastic Processes in Physics and Chemistry*. Elsevier, Amsterdam, 3 edition, (2007).
86. Deam, R. T. and Edwards, S. F. Theory of rubber elasticity. *Philos. T. Roy. Soc. A* 280, 317–353 (1976).
87. Kalathi, J. T., Yamamoto, U., Schweizer, K. S., Grest, G. S., and Kumar, S. K. Nanoparticle diffusion in polymer nanocomposites. *Phys. Rev. Lett.* 112, 108301 (2014).
88. Cook, P. R. A model for all genomes: The role of transcription factories. *J. Mol. Biol.* 395, 1–10 (2010).
89. De Magistris, G. and Marenduzzo, D. An introduction to the physics of active matter. *Physica A* 418, 65–77 (2015).
90. Winkler, R. G., Elgeti, J., and Gompper, G. Active polymers — emergent conformational and dynamical properties: A brief review. *J. Phys. Soc. Jpn.* 86(10), 101014 (2017).
91. Ganai, N., Sengupta, S., and Menon, G. I. Chromosome positioning from activity-based segregation. *Nucleic Acids Res.* 42, 4145–4159 (2014).
92. Smrek, J. and Kremer, K. Small activity differences drive phase separation in active-passive polymer mixtures. *Phys. Rev. Lett.* 118, 098002 (2017).
93. Goloborodko, A., Marko, J. F., and Mirny, L. A. Chromosome compaction by active loop extrusion. *Biophys. J.* 110(10), 2162–2168 (2016).
94. Baù, D. et al. The three-dimensional folding of the alpha-globin gene domain reveals formation of chromatin globules. *Nat. Struct. Mol. Biol.* 18, 107–115 (2011).
95. Wong, H. et al. A predictive computational model of the dynamic 3d interphase yeast nucleus. *Curr. Biol.* 22, 1881–1890 (2012).
96. Di Stefano, M., Rosa, A., Belcastro, V., di Bernardo, D., and Micheletti, C. Colocalization of coregulated genes: A steered molecular dynamics study of human chromosome 19. *Plos Comput. Biol.* 9(3), e1003019 (2013).

97. Serra, F., Di Stefano, M., Spill, Y. G., Cuartero, Y., Goodstadt, M., Baù, D., and Marti-Renom, M. A. Restraint-based three-dimensional modeling of genomes and genomic domains. *FEBS Lett.* 589, 2987–2995 (2015).

98. Di Stefano, M., Paulsen, J., Lien, T. G., Hovig, E., and Micheletti, C. Hi-C-constrained physical models of human chromosomes recover functionally-related properties of genome organization. *Sci. Rep.-UK* 6, 35985, 10 (2016).

99. Tiana, G. and Giorgetti, L. Integrating experiment, theory and simulation to determine the structure and dynamics of mammalian chromosomes. *Curr. Opin. Struc. Biol.* 49(Supplement C), 11–17 (2018).

100. Guigas, G., Kalla, C., and Weiss, M. Probing the nanoscale viscoelasticity of intracellular fluids in living cells. *Biophys. J.* 93(1), 316–323 (2007).

101. Puertas, A. M. and Voigtmann, T. Microrheology of colloidal systems. *J. Phys.-Condens. Matter* 26(24), 243101 (2014).

102. Mason, T. G. and Weitz, D. A. Optical measurements of frequency-dependent linear viscoelastic moduli of complex fluids. *Phys. Rev. Lett.* 74, 1250–1253 (1995).

103. Cates, M. E. and Deutsch, J. M. Conjectures on the statistics of ring polymers. *J. Phys. (Paris)* 47, 2121–2128 (1986).

104. Vettorel, T., Grosberg, A. Y., and Kremer, K. Territorial polymers. *Phys. Today* 62, 72 (2009).

105. Khokhlov, A. R. and Nechaev, S. K. Polymer chain in an array of obstacles. *Phys. Lett.* 112A, 156–160 (1985).

106. Grosberg, A. Y., Nechaev, S. K., and Shakhnovich, E. I. The role of topological constraints in the kinetics of collapse of macromolecules. *J. Phys. France* 49, 2095–2100 (1988).

107. Grosberg, A., Rabin, Y., Havlin, S., and Neer, A. Crumpled globule model of the three-dimensional structure of DNA. *Europhys. Lett.* 23, 373–378 (1993).

108. Grosberg, A. Y. Annealed lattice animal model and Flory theory for the melt of non-concatenated rings: Towards the physics of crumpling. *Soft Matter* 10, 560–565 (2014).

109. Suzuki, J., Takano, A., Deguchi, T., and Matsushita, Y. Dimension of ring polymers in bulk studied by Monte-Carlo simulation and self-consistent theory. *J. Chem. Phys.* 131, 144902 (2009).

110. Halverson, J. D., Lee, W. B., Grest, G. S., Grosberg, A. Y., and Kremer, K. Molecular dynamics simulation study of nonconcatenated ring polymers in a melt: I. Statics. *J. Chem. Phys.* 134, 204904 (2011).

111. Uchida, N., Grest, G. S., and Everaers, R. Viscoelasticity and primitive-path analysis of entangled polymer liquids: from F-actin to polyethylene. *J. Chem. Phys.* 128, 044902 (2008).

112. Halverson, J. D., Kremer, K., and Grosberg, A. Y. Comparing the results of lattice and off-lattice simulations for the melt of nonconcatenated rings. *J. Phys. A: Math. Theor.* 46, 065002 (2013).

113. Smrek, J. and Grosberg, A. Y. Minimal surfaces on unconcatenated polymer rings in melt. *ACS Macro Lett.* 5, 750–754 (2016).

114. Lo, W.-C. and Turner, M. S. The topological glass in ring polymers. *EPL* 102(5), 58005 (2013).

115. Michieletto, D., Marenduzzo, D., Orlandini, E., Alexander, G. P., and Turner, M. S. Threading dynamics of ring polymers in a gel. *ACS Macro Lett.* 3(3), 255–259 (2014).

116. Ge, T., Panyukov, S., and Rubinstein, M. Self-similar conformations and dynamics in entangled melts and solutions of nonconcatenated ring polymers. *Macromolecules* 49(2), 708–722 (2016).

117. Michieletto, D. On the tree-like structure of rings in dense solutions. *Soft Matter* 12, 9485–9500 (2016).

118. Obukhov, S. P., Rubinstein, M., and Duke, T. Dynamics of a ring polymer in a gel. *Phys. Rev. Lett.* 73, 1263–1266 (1994).

119. Smrek, J. and Grosberg, A. Y. Understanding the dynamics of rings in the melt in terms of the annealed tree model. *J. Phys.-Condens. Matter* 27(6), 064117 (2015).

Analysis of Chromatin Dynamics and Search Processes in the Nucleus

ASSAF AMITAI AND DAVID HOLCMAN

8.1 INTRODUCTION

In this chapter, we describe chromatin dynamics using stochastic modeling and polymer physics. Brownian motion is the classical description of dynamics driven by thermal noise first observed by Robert Brown in 1827 when studying pollen grains moving in water. Brownian motion results from millions of fast collisions of the grain with solvent molecules, as explained in 1905 by Albert Einstein [1]. Stochastic processes are now routinely used to describe molecular behavior in cellular biology, reviewed recently in [2].

In a similar context, the motion of proteins, RNA and chromatin is spanning a few to hundred nanometers, is inherently stochastic. However, the Brownian description is often insufficient to describe their motion because they are embedded in a complex field of forces imposed by the nuclear environment. Our goal in this chapter is to summarize briefly modeling and analysis approaches that goes beyond classical Brownian motion and reveal how the cell imposes biological constraints on molecular motion.

In parallel to the theoretical approaches of modeling, analysis and simulations, live cell microscopy techniques generate large quantities of data of in vivo molecular dynamics that need to be interpreted. For example, single-particle tracking of chromatin [3, 4] provided many short/long, confined/unconfined trajectories, but establishing the connection between basic cellular function and nuclear organization remains a major challenge. Converting the information contained in these trajectories to molecular processes requires a physical model, which is often speculative.

Chromatin dynamics has been studied by fluorescently tagging a specific locus and tracking its motion, leading to the conclusion that chromatin movement is restricted in subregions [5] smaller than the total nuclear volume. Yet, the origin of this restriction remains unclear. Can it be due to nuclear crowding, self-avoiding interactions or simply internal forces acting on the chromatin? We present here modeling and analysis approaches to questions.

Locus trajectories were initially analyzed using classical statistical quantities derived from the study of a single Brownian motion [6–8], such as the Mean-Square-Displacement (MSD) computed along single-particle trajectories. Trajectories of an individual particle inside a fluid, on the chromatin or a membrane, are usually described and analyzed in the physical literature using the Smoluchowski's limit of the Langevin equation [9] (see equation 8.1). Under some conditions, it is often possible to recover the field of force and the diffusion tensor from a large number of trajectories. This approach has been used in a massive amount of Single-Particle Trajectory (SPT) super-resolutions [10, 11], but it remains difficult to apply to complex polymers such as chromatin. Thus, alternative methods have been found. In the following section, we show how single-particle trajectory data can be used to recover the chromatin organization and the ensemble of interactions it experiences.

8.2 SINGLE-PARTICLE TRAJECTORIES RECORDED FROM A CHROMATIN LOCUS

In this section, we describe the statistical analysis of SPTs of a chromatin locus. Several cell functions such as transcription, DNA repair or cell division can now be monitored in vivo, but interpreting trajectories in correlation with these events remains difficult due to all the mechanisms involved in generating motion. There is also a large unexplained variability in the different statistics of trajectories between cells. We first describe how SPT data are acquired and, in the second part, we present recent polymer models and the statistical analysis used to interpret data and extract biophysical parameters.

8.2.1 Live imaging of nuclear elements

Routine molecular construction (see Figure 8.1a) enables the tagging of a chromatin locus so that it becomes fluorescent and thus can be tracked using live cell microscopy. This molecular construction is based on a Lac operon, which is integrated at a specific position. For example, in the yeast genome [3], a LacI protein bound to a GFP (making a LacI-GFP complex) (Figure 8.1b) allows us to see chromatin moves. The nuclear membrane can also be visualized by tagging the elements of the nuclear pore protein. Recently, with the development of the Cas9 system [12], it became much easier to insert fluorescent tags within any chromatin site of bacteria, yeast or mammalian cells [13]. By tracking a chromatin locus for a certain time length, trajectories can be generated. Fluorescent dyes cannot be measured for too long, as they eventually bleach. Hence, imaging at a high frequency would result in a shorter trajectory duration.

Ignoring the polymer nature of the chromatin, using a Brownian description, an effective diffusion coefficient can be estimated. For example, while proteins have a fast diffusion of $D \approx 10 \, \mu m^2/s$ [14, 15], chromatin motion is characterized by an apparent diffusion coefficient of about three to four orders of magnitude slower [4]. Thus, a possible timescale to study sub-nuclear processes is seconds and below. Sometimes deconvolution procedures are necessary to recover the precise location from noisy measurement, but they can introduce various localization errors that must be differentiated from physical motions [16].

8.2.2 Statistics analysis of SPTs

The ability to follow a single locus located at different positions on chromosomes [6] revealed the heterogeneity of the nuclear organization. The yeast nucleus has a radius of the order of 1 μm. Interestingly, trajectories can be restricted to a small region (Figure 8.1c–i, where the confined ball has a radius of 221 nm). Other trajectories can be contained in an even smaller ball (radius 164 nm, Figure 8.1c–ii). The large heterogeneity of loci behavior across cell populations suggests that the loci can experience different interactions over a time scale of minutes. Thus, chromatin loci appear to be localized over an extended time. In the diluted yeast nucleus, locus

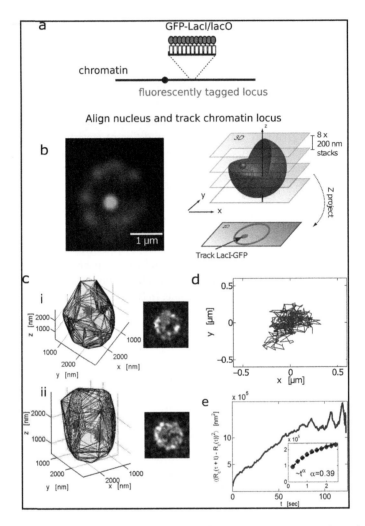

Figure 8.1 Imaging a chromatin locus. (a) A fluorescent tag inserted on chromosome III of the yeast [17]. (b) MAT locus in yeast (green) is observed using the LacI-lacO system while the nuclear membrane (red) was marked with the nup49-mCherry fusion protein (left). The nucleus is imaged in 8 stacks along the Z-axis and trajectories are projected on the XY-plane (right). (c) Two trajectories of a chromatin locus and with time resolution of 330 ms, recorded for 100 seconds. At the beginning, the locus trajectory is red and gradually becomes green, while the spindle pole body (SPB) starts red and gradually becomes blue (upper and lower left). 2d projection: the trajectory of the locus (red) and the displacement of the spindle pole body trajectory (blue) inside the nucleus is projected on a plane (upper and lower right). The nuclear membrane is reconstructed based on the nup49-mCherry fusion protein. (d) A trajectory of the MAT locus taken at a time resolution of 300 msec during 300 frames. (e) Correlation function $C(t)$ (Equation 8.19) of the position depicted (from c). At short times, the sub-diffusion regime is characterized by an anomalous exponent $\alpha = 0.39$ (inset magnification at short times). (Reproduced from [18]).

localization can be studied with respect to their distance from the centrosome [19]. In higher Eukaryotes, chromosomes reside in distinct territories [20], maintaining the cell identity. Chromatin loci appear well localized in mammalian cells. In a mouse embryonic stem cell, a locus diffuses in an area with a radius of 200 nm [21]. These constraints could be maintained via specific interaction with anchoring points on the membrane such as Lamin-A proteins [22], or via self-avoiding interactions.

ATP also proved to be important to maintaining the locus mobility. After treating E. Coli with sodium azide, which inhibits ATP synthesis, the diffusion coefficient of a DNA locus was reduced [23]. In yeast, glucose starvation abolishes large range chromatin movements [4]. Thus, chromatin movement is sensitive to various ATP energy levels. Indeed, chromatin remodeler molecules rather than RNA or DNA polymerases [24] underlie the ATP-driven motions. However, the biophysical origin of all these modifications is still not completely clear.

To extract biophysical parameters from SPTs of a chromatin locus, a physical model for the locus motion is needed. This model can be coarse-grained to account for the local organization of the medium, paved with obstacles and local forces [25]. At a molecular level, the Langevin equation is the gold standard. In that case, a stochastic particle modeled by the Langevin equation in the overdamped limit has a velocity $\dot{\mathbf{X}}$ proportional to the applied force $F(\mathbf{X})$ plus an additional white noise, leading to

$$\gamma \dot{\mathbf{X}} = \mathbf{f} + \gamma \sqrt{2D}\dot{\omega}, \tag{8.1}$$

where γ is the friction coefficient [26], D the diffusion coefficient and ω is the unit Wiener process. The source of the driving noise is the thermal agitation [26, 27]. By averaging over the ensemble of velocity realizations, it is possible to compute the first moment, which contains information about the force fields applied to the particle. However, for a polymer chain, there are also internal forces between monomers and thus, as the SPTs are measured at a single monomer location, the internal forces acting on the measured monomer should be separated from the external forces acting possibly on all monomers.

The model and analysis presented here assume that the local environment is stationary so that the statistical properties of trajectories do not change over time. In practice, the assumption of stationary is usually satisfied as trajectories are acquired for short time, where few changes of chromatin organization are expected to occur. The coarse-grained model 8.1 is recovered from the conditional moments of the trajectory increments $\Delta X = X(t + \Delta t) - X(t)$ (see [27]),

$$\frac{F(X)}{\gamma} = \lim_{\Delta t \to 0} \frac{\mathbb{E}[\Delta X(t) \mid X(t) = x]}{\Delta t}, \quad D = \lim_{\Delta t \to 0} \frac{\mathbb{E}[\Delta X(t)^T \Delta X(t) \mid X(t) = x]}{2\Delta t}. \tag{8.2}$$

Here the notation $\mathbb{E}\left[\cdot \mid X(t) = x\right]$ means averaging over all trajectories that are at point X at time t. We shall now show how this statistic can be computed for a Rouse polymer model and used to extract the forces acting on the chromatin.

8.2.3 The Rouse Polymer Model

We recall that the Rouse model is defined as a collection of beads connected by springs [28]. Monomers are positioned at $(R_n \ (n = 1,2,...N))$, subject to Brownian motions and the spring forces are due to the coupling between the nearest neighboring beads. The potential energy is defined by

$$\phi(R) = \frac{\kappa}{2} \sum_{n=1}^{N} (R_n - R_{n-1})^2. \tag{8.3}$$

In the Rouse model, only neighboring monomers interact [28]. In the Smoluchowski's limit of the Langevin equation, the dynamics of the monomer R_n is driven by the potential $\phi(R_1,..,R_N)$, which generates the force $-\nabla_{R_n}\phi(R_1,..,R_N)$. The ensemble of stochastic equations is

$$\frac{dR_n}{dt} = -D\kappa(2R_n - R_{n-1} - R_{n+1}) + \sqrt{2D}\frac{dw_n}{dt}, \tag{8.4}$$

for $n = 1,..N$. In this model, the variance of the distances between neighboring monomers averaged over different polymer realization is given by

$$\left\langle (R_{n+1} - R_n)^2 \right\rangle = b^2, \tag{8.5}$$

where b is the standard deviation of the bond length, $\kappa = dk_BT/b^2$ is the spring constant with d the spatial dimension, k_B is the Boltzmann coefficient and t the temperature.

Starting with a given configuration, the relaxation of a Rouse polymer to a steady state in a free space can be analyzed using the Fourier space

$$u_p = \sum_{n=1}^{N} R_n \alpha^n_p, \tag{8.6}$$

where the change of coordinates is encoded in the matrix

$$\alpha^n_p = \begin{cases} \sqrt{\dfrac{1}{N}}, & p = 0 \\[2ex] \sqrt{\dfrac{2}{N}}\cos\left((n-1/2)\dfrac{p\pi}{N}\right), & \text{otherwise.} \end{cases} \tag{8.7}$$

u_0 represents the motion of the center of mass. The potential ϕ is defined in equation 8.3 as

$$\phi(u_1,..,u_{N-1}) = \frac{1}{2}\sum_{p=1}^{N-1}\kappa_p u_p^2, \tag{8.8}$$

where

$$\kappa_p = 4\kappa \sin\left(\frac{p\pi}{2N}\right)^2.$$

(8.9)

Equation 8.4 is now decoupled, resulting in $(N-1)d$ independent Ornstein-Uhlenbeck (OU) processes

$$\frac{d\boldsymbol{u}_p}{dt} = -D_p\kappa_p\boldsymbol{u}_p + \sqrt{2D_p}\frac{d\tilde{\omega}_p}{dt},$$

(8.10)

where $\tilde{\omega}_p$ are independent d-dimensional Brownian motions with mean zero and variance 1 and $D_p = D$ for $p = 1..N-1$, while $D_0 = D/N$ and the relaxation times are defined by $\tau_p = 1/D\kappa_p$. The center of mass behaves as a freely diffusing particle. Starting from a stretched configuration, the relaxation time for a Rouse polymer is dominated by the slowest time constant

$$\tau_1 = \frac{1}{D\kappa_1} = \frac{1}{4D\kappa \sin\left(\frac{1\pi}{2N}\right)^2}.$$

(8.11)

8.2.4 A Rouse Polymer Driven By An External Force Applied To A Single Monomer

Relation 8.3 has been generalized to a Rouse polymer when an external force is applied on a single monomer. We define the potential $U_{\text{ext}}(\boldsymbol{R})$ to describe the total energy

$$\phi(\boldsymbol{R}) = \frac{\kappa}{2}\sum_{j=2}^{N}(\boldsymbol{R}_j - \boldsymbol{R}_{j-1})^2 + U_{\text{ext}}(\boldsymbol{R}).$$

(8.12)

The monomers dynamics follows from equation 8.4. When the monomer $\boldsymbol{R}_c$ is tracked to generate a trajectory, the first moment of the monomer increment position $\boldsymbol{R}_c(t+\Delta t) - \boldsymbol{R}_c(t)$, which is proportional to the velocity can be computed. This computation requires to average overall polymer realizations and it thus accounts for the entire polymer configuration [29]. We recall the following relation between the force and the increment [29]

$$\lim_{\Delta t \to 0} \mathbb{E}\left\{\frac{\boldsymbol{R}_c(t+\Delta t) - \boldsymbol{R}_c(t)}{\Delta t}\,\Big|\,\boldsymbol{R}_c = \boldsymbol{x}\right\} = -D\int_\Omega d\boldsymbol{R}_1..\int_\Omega d\boldsymbol{R}_N \left(\nabla_{\boldsymbol{R}_c}\phi\right)P(\boldsymbol{R}\,|\,\boldsymbol{R}_c = \boldsymbol{x}),$$

(8.13)

where $\mathbb{E}\{.\,|\,\boldsymbol{R}_c = \boldsymbol{x}\}$ denotes averaging over all polymer configuration under the condition that the tagged monomer is at the position $\boldsymbol{R}_c = \boldsymbol{x}$. Formula 8.13 is generic as it generalizes relation 8.1 and does not depend on the particular expression of the external forces acting on the polymer. No restrictions are imposed on the domain Ω where the polymer evolves, but it is reflected at the boundary $\partial\Omega$ of the domain. The conditional probability $P(R_1, R_2, \ldots, R_N \mid R_c = x)$ to observe a polymer configuration $R_1, R_2, \ldots, R_N$ conditioned that $R_c = x$ is computed from the equilibrium joint probability distribution function (pdf) $P(R_1, R_2, \ldots, R_N)$, which satisfies the Fokker-Planck equation in the phase-space $\Omega..\Omega \subset \mathbb{R}^{3N}$:

$$0 = \Delta P(R) + \nabla \cdot (\nabla \phi \, P(R)), \tag{8.14}$$

with boundary condition

$$\phi \frac{\partial P}{\partial n_i} + P \frac{\partial \phi}{\partial n_i} = 0 \text{ for } R_i \in \partial\Omega \text{ for } i = 1..N,$$

where n_i is the normal vector to the boundary $\partial\Omega$ at position R_i. The model for the external force acting on monomer n and located at position R_n is the gradient of a harmonic potential

$$U_{ext}(R_n) = \frac{1}{2} k (\mu - R_n)^2, \tag{8.15}$$

where k is the force constant and we choose $n < c$. This external potential will affect the dynamics of the entire polymer although it is applied specifically on R_n. What will be the influence on the observed locus c?

In this scenario, the mean velocity of monomer c is given by [29]

$$\lim_{\Delta t \to 0} \mathbb{E}\left\{ \frac{R_c(t + \Delta t) - R_c(t)}{\Delta t} \mid R_c(t) = x \right\} = -Dk_{cn}x, \tag{8.16}$$

$$\text{where} \qquad k_{cn} = \frac{k\kappa}{\kappa + (c - n)k}.$$

Expression 8.16 links the average velocity of the observed monomer c to the force applied on monomer n. The coefficient k_{cn} depends on the harmonic well strength k, the inter-monomer spring constant K and it is inversely proportional to the distance $|n - c|$ between monomers n and c along the chain. When more monomers are interacting with external potentials, the expression for the velocity retains its structure, but in general k_{cn} will be the sum of all tethering forces [29].

Inversion formula 8.16 assumes the Boltzmann distribution for the single monomer and that the entire polymer has reached equilibrium at the time scale of the experiment. Finally, formula 8.16 reveals how internal and external polymer forces are mixing together and are influencing the monomer velocity. It also shows the explicit decay of the force amplitude with the distance between the observed and the forced monomer.

The effective spring coefficient k_{cn} can be estimated from Brownian simulations of a tethered polymer with self-avoiding interactions. The comparison with the experimental radius of confinement measured for chromatin loci [6, 8, 30, 31], reveals that the distance from the centromere is inversely correlated with the effective spring constant and this suggests that at very long times (time between cell divisions), this distance determines the loci localization in yeast. At shorter times, transient interactions could influence significantly the locus dynamics and localization.

8.2.5 Constraint Length of a Locus

We now discuss another parameter to characterize the locus dynamics which is the standard deviation (SD) of the position with respect to its mean averaged over time. This SD is a characteristic length, called the constraint length L_C, estimated by the empirical sum

$$L_C = \sqrt{\text{Var}(R_c)} = \sqrt{\frac{1}{T}\sum_{h=1}^{T}\left(R_c(h\Delta t) - \langle R_c \rangle\right)^2}, \tag{8.17}$$

where $\langle R_c \rangle$ is the average position of the locus along the trajectory. When the chromatin is tethered, the steady-state variance $\text{Var}(R_c)_{\lim_{t\to\infty} R_c(t)}$ of the monomer's position is given by $L_C^2 = \frac{d}{K_{cn}}$. This relation is reminiscent of long-time asymptotic behavior of classical Ornstein-Uhlenbeck processes.

8.2.6 Other Empirical Estimators and Statistical Properties

The MSD, the cross-correlation function and any derived quantities are the most common statistical estimators to study stochastic trajectories. These are defined for a time series $R_c(t)$ (locus position at time t) by [9, 32]

$$C_\tau(t) = \left\langle \left(R_c(\tau + t) - R_c(\tau)\right)^2 \right\rangle, \tag{8.18}$$

where $\langle . \rangle$ denotes the average over ensemble realization. When the data contains enough recursion, the ergodicity assumption says that the function $C(t)$ can be also computed along a trajectory from the empirical estimator

$$C_\tau^T(t) = \frac{1}{T-t}\sum_{h=1}^{T-t} R_c(h\Delta t) - R_c\left((h+t)\Delta t\right)^2, \tag{8.19}$$

for $t = 1$, $T - 1$, where T is the duration of the entire trajectory. Figure 8.1d–e show the function $C(t)$ computed for a chromatin locus. Similarly, the MSD of a trajectory is defined by

$$\text{MSD}(t) = \left\langle \left(R_c(t) - R_c(0)\right)^2 \right\rangle. \tag{8.20}$$

For a Brownian particle moving in $\mathbb{R}^d$ space, characterized by the diffusion coefficient D, the MSD is $\text{MSD}(t) = 2dDt$ [9]. While $C(t)$ is linear in time for a free

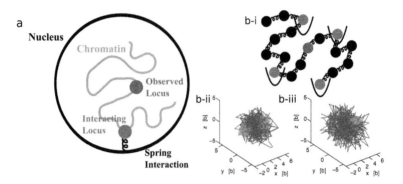

Figure 8.2 Dynamics of an observed locus when another locus is interacting with a nuclear element. (a) Schematic representation of the nucleus, where one locus is observed and followed with a fluorescent label while another (non-visible) chromatin locus is interacting with some nuclear element. (b: i) Schematic representation of a polymer, where some monomers (red) interact with fixed harmonic potential wells. Monomer c (blue) is observed. (b: ii, iii) Stochastic trajectories of three monomers, part of a polymer where the two extremities interact with potential wells fixed at the origin and at position $\mu = (5b,0,0)$ respectively. The middle monomer trajectory (blue) is more extended than the two others, shown for a polymer of length $N = 21$ (ii) and $N = 41$ (iii).

Brownian particle, it reaches an asymptotic value when the particle is moving in a confined space Ω, with an exponential rate [33].

For a particle confined in a harmonic potential by a spring force with constant k [9] and friction coefficient γ, the stochastic description is an OU-process

$$dR_c = -\frac{k}{\gamma}R_c dt + \sqrt{2D}dw,\tag{8.21}$$

and the cross-correlation function is

$$C(t) = \frac{2dk_B T}{k}\left(1 - e^{-t/\tau}\right),\tag{8.22}$$

where $\tau = k_B T / D\kappa$. We note that the auto-correlation function for an OU process is

$$\left\langle \left(R_c(t) - \left\langle R_c(t)\right\rangle\right).\left(R_c(t') - \left\langle R_c(t')\right\rangle\right)\right\rangle = \frac{dk_B T}{k}\left(e^{-(t-t')/\tau} - e^{-(t+t')/\tau}\right).\tag{8.23}$$

In summary, when the correlation function $C(t)$ converges to an asymptotic value, the underlying motion can either be restricted by a tethering force or by obstacles. However, this ambigutity can be resolved by computing the first moment (Equation 8.16), which is nonzero where there is a field of force.

8.2.7 The MSD of a Tagged Monomer

We now discuss the time evolution of the MSD for a tagged monomer, which belongs to a long polymer chain. For a Rouse polymer, the MSD of monomer R_c is a sum of independent OU-variables (see equation 8.6):

$$var\left(R_c\left(t\right)\right) = \langle\left(R_c\left(t\right) - R_c\left(0\right)\right)^2\rangle = \frac{d}{\kappa N}\sum_{p=1}^{N-1}\frac{\cos^2\left(\frac{\left(2c-1\right)p\pi}{2N}\right)}{\sin^2\left(\frac{p\pi}{2N}\right)}\left(1 - e^{-2t/\tau_p}\right) + 2dD_{cm}t,$$

(8.24)

where d is the spatial dimension. Formula 8.24 shows the deviation compared to the MSD of a Brownian motion, for which the correlation function increases linearly in time. There are three distinct regimes:

1. For a short time $t \ll \tau_{N-1}$, the variance of the normal modes can be approximated as $\sigma_p^2(t) \approx Dt$, independent of p, the sum in equation 8.24 leads to

$$var\left(R_c\right) \approx 2dDt,$$

(8.25)

which is the classical diffusion regime.

2. For a long time, $t \gg \tau_1$, the exponential terms in relation 8.24 become independent of t and only the second term in the equation corresponding to the diffusion of the center of mass gives the time-dependent behavior. This regime is dominated by normal diffusion, with a diffusion coefficient D/N.

3. For intermediate times $\tau_{N-1} \ll t \ll \tau_1$, such that $2t / \tau_p > 1$, the sum of exponentials contributes to equation 8.24. The variance 8.24 is

$$var\left(R_c\right) \approx 2\int_{p_{min}}^{N-1}\frac{\cos^2\left(\frac{\left(2c-1\right)p\pi}{2N}\right)}{\sin^2\left(\frac{p\pi}{2N}\right)}dp,$$

(8.26)

where p_{min} is such that $\tau_{p_{min}} = 2t$. We have $var\left(R_c\right) \sim t^{1/2}$. A Rouse monomer exhibits anomalous diffusion. The time interval can be arbitrarily long with the size N of the polymer.

8.2.8 Anomalous Diffusion Of A Chromatin Locus

At an intermediate time regime, a tagged monomer belonging to a Rouse polymer diffuses with MSD $\sim t^{1/2}$. This motion belongs to a class of dynamical behavior called *anomalous diffusion*. The random motion of a small molecular probe located at a position $R(t)$ at time t is characterized by the statistics of its second moment time series, given for small time t by

$$\left\langle \left(R(t) - R(0) \right)^2 \right\rangle \approx A t^{\alpha}, \tag{8.27}$$

where $\langle \cdot \rangle$ means averaging over realization and A is a constant. The exponent α characterizes the deviation from normal diffusion: when $\alpha < 1$ (resp. $\alpha > 1$) it is the subdiffusion (resp. superdiffusion) regime. The correlation function (Equation 8.19) of a DNA locus (Figure 8.1e) shows the anomalous diffusion behavior, characterized by an exponent $\alpha < 1$, [34–36].

The dynamics of the locus reflects the local chromatin organization and the ensemble of interactions it experiences. Thus, the anomalous exponent could contain physical information about the underlying polymer model describing the chromatin [37]. However, it is still unclear what the precise significance of the anomalous exponent is, computed from empirical data. Single cell analysis reveals that there is a large distribution of the anomalous exponent between cells. That could suggest that chromatin structure changes over time and across cell populations [38, 39].

Interestingly, the anomalous exponent of a moving locus was reported in many experiments to be smaller than 1/2 [18]. This result has called for a different polymer models to describe chromatin dynamics.

When topological constraints are important, the polymer is threaded around obstacles and has to move back and forth to wiggle around them. Such movement is known as reptation and the polymer moves in what is know as a reptation tube [40]. At an intermediate time regime, such motion results in subdiffusion with $\alpha = 1/4$. However, the yeast nucleus is not particularly crowded, and loci do not appear to move in a reptation tube (see Figure 8.1c), but have approximately spherically shaped trajectories. We shall now present three models that are used to explain the observed dynamics of chromatin. It is still unclear which model better represents in vivo behavior, but in both cases, the dynamics of chromatin is approximated by an unentangled polymer systems.

We recall that chromatin is not structured as a simple linear chain. Using Hi-C experiments to study chromatin organization, it was shown [41] that chromatin is packed more compactly than a flexible chain. This observation leads to a model where the distance between monomers l and m scales like

$$\left\langle \left(R_l - R_m \right)^2 \right\rangle \sim \left| l - m \right|^{2/d_f} \tag{8.28}$$

where $d_f > 0$ is called the fractal dimension of the polymer. For a Rouse polymer $d_f = 2$, while for a fractal globule $d_f = 3$ [42]. Experimentally, the distance between chromatin sites of distance s along the chain scales as s^{γ}, where γ is empirically found in the range between 1 and 1.5. Thus, it was suggested that chromatin is organized as a fractal globule polymer [42]. This scaling probably originates from interlinking between far away chromatin sites, resulting in unique topological domains along the strand [43–45]. The fractal description of the chromatin remains quite speculative. However, an important question remains: How can we relate chromatin dynamics to its structure? We now introduce a generalization of the Rouse model to account for homogeneous long-range connections.

8.2.9 The β-Polymer Is A Generalized Rouse Model

How is it possible to construct a polymer model with a prescribed anomalous exponent α? We will now demonstrate the construction and the resulting polymer model, called the β-polymer model introduced in [46]. We shall see that prescribing the anomalous exponent α imposes intrinsic long-range interactions between monomers. The construction of this polymer is based on extending monomer interactions beyond the closest neighbors starting with the Rouse model.

We recall that the dynamics of a polymer is determined by the scaling relaxation time $\tau_p = 1/D\kappa_p$. The construction is based on modifying the spring coefficient to

$$\tilde{\kappa}_p = 4\kappa \sin^\beta\left(\frac{p\pi}{2N}\right),$$

(8.29)

where we will specify later on the value of β. This new power will change the scaling time constants $\tilde{\tau}_p = 1/D\tilde{\kappa}_p \sim p^{-\beta}$, where $1 < \beta \leq 2$. Note that $\beta = 2$ corresponds to a Rouse polymer [46]. The new Hamiltonian of the polymer is

$$\tilde{\phi}(u_1,..u_N) = \frac{1}{2}\sum_p \tilde{\kappa}_p u_p^2 = \frac{1}{2}\sum_{l,m} R_l R_m A_{l,m},$$

(8.30)

where the coefficients are given by

$$A_{l,m} = \sum_{p=1}^{N-1}\tilde{\kappa}_p \alpha_p^l \alpha_p^m = 4\kappa\frac{2}{N}\sum_{p=1}^{N-1}\sin^\beta\frac{p\pi}{2N}\cos\left(\left(l-\frac{1}{2}\right)\frac{p\pi}{N}\right)\cos\left(\left(m-\frac{1}{2}\right)\frac{p\pi}{N}\right).$$

(8.31)

Such a construction defines a unique ensemble of long-range interactions, and the associated potential energy differs from that of a flexible chain (Equation 8.3). For $\beta < 2$, all monomers become coupled and the strength of the interaction decays with the distance along the chain, as shown in Figure 8.3a–b.

The modified Hamiltonian can be diagonalized, and the β-polymer dynamics is described by the Smoluchowski's limit of the Langevin equation (8.10) with the modified spring coefficient $\tilde{\kappa}_p$. At an intermediate time scale where $\tilde{\tau}_{N-1} \ll t \ll \tilde{\tau}_1 \approx N^\beta \tilde{\tau}_{N-1}$ (anomalous regime), the asymptotics of the time cross-correlation function is given by [46]

$$\langle(R_c(t_0+t) - R_c(t_0))^2\rangle \approx \frac{d}{\kappa N}\frac{2N}{\pi}\left[\frac{1}{\beta-1} + \left(\frac{1}{2} + \frac{\beta}{12}\right)(1 - e^{-t/\tau_1})\right]$$

(8.32)

$$+ \frac{\pi}{\Gamma(2-1/\beta)\sin(\pi/\beta)\beta}\left(\frac{t}{\tau_1}\right)^{2-\frac{1}{\beta}} + \frac{2dDt}{N}.$$

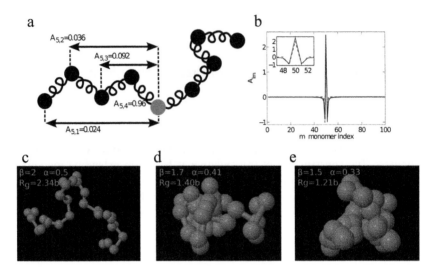

Figure 8.3 The β-polymer model. (a) Representation of a β-polymer, where all monomers are connected together with springs whose strength decay depending on their distance along the chain. The central monomer (blue) interacts with all other monomers in a chain of length $N = 9$ for $\beta = 1.5$ (interaction unit $\kappa = 3/b^2$). (b) Monomer–monomer interactions in the modified Rouse polymer model (β-model). The coefficients A_{lm} (in units of κ) measure the strength of the interaction between two monomers. Shown are the coefficients A_{lm} for the polymer with $\beta = 1.1$, where $l = 50$ and $N = 100$. All monomers interact with each other and the strength of the interaction decays with the distance along the chain (adapted from [46]). (c–e) Examples of β-polymer configuration for different values of β. The anomalous exponent at an intermediate regime is given by $\alpha = 1 - 1/\beta$. $\beta = 2$ corresponds to a Rouse polymer. The radius of gyration is R_g. Adapted from [18].

Hence,

$$\operatorname{var}(\mathbf{R}_c) \sim t^\alpha, \tag{8.33}$$

with $\alpha = 1 - 1/\beta$. This model maps the dynamics of the chain to a structure parameter β representing the interaction between different parts of the chain. Consequently, the monomer of a denser polymer (smaller β) will move slower, characterized by a smaller anomalous exponent (Figure 8.3c–e).

8.2.9.1 ANOMALOUS DIFFUSION IN FRACTAL GLOBULES

The relation between chromatin structure/organization can be described using the dynamics of a tagged monomer when the chromatin is described as a static fractal globule. Assuming that a piece of the strain moves together with an effective diffusion coefficient that scales as $D_{\delta s} \sim D\tau^{-\alpha}$, and moves a distance that scales with $\delta s^{2/d_f}$, then the anomalous exponent is $\alpha_f = 2/(2 + d_f) = 0.4$ [47, 48], similar to the one experimentally observed. The fractal globule model remains speculative and is not sufficient to explain the large heterogeneity observed in both chromatin structure and its dynamics.

8.2.10 Fractional Brownian Motion Of A Locus

In the previous subsection, we assumed that the solvent surrounding the polymer chain is a Newtonian liquid, where friction acting on the beads constituting polymer chain is inversely proportional to their velocity. This is not necessarily the case, and the motion of a probe particle in non-Newtonian/viscoelastic solvents can result in long-memory effects. This modeling approach offers a different explanation of correlated motion inside the nucleus.

The probability density function of a stochastic motion with long-memory is described by the fractional Fokker-Planck equation [49] or fractional Brownian motion (FBM). Phenomenological models [35, 50] based on the fractional Langevin equation leads to a MSD that exhibits a power law. The construction of the associated polymer model relies on the Langevin equation with a memory kernel with an algebraical decay. This kernel reflects the properties of the viscoelastic fluid, which slows down the loci dynamics [36].

A subdiffusive process $B_H(t)$ is generated by the FBM [51, 52] and has the following properties

$$\langle B_H(t)\rangle = 0,$$

$$\langle B_H(t)B_H(s)\rangle = \frac{1}{2}(t^{2H} + s^{2H} - |t-s|^{2H}),$$

(8.34)

where the parameter H is called the Hurst exponent ($0 < H < 1$). The generalized Langevin's equation [53] is

$$m\frac{dv(t)}{dt} = -\gamma\int_{-\infty}^{t} v(t')K(t-t')dt' + \sqrt{2D}dw^f(t),$$

(8.35)

where m is the mass, γ is the friction coefficient and D the associated diffusion coefficient. The memory kernel $K(t-t')$ is associated with the fractional noise $dw^f(t)$ so that the autocorrelation function satisfies the relation

$$\langle dw^f(t)dw^f(t')\rangle = k_B T\gamma K(t-t').$$

(8.36)

When the kernel is a δ-function, we recover the classical Langevin's equation. However, for a kernel with a long-time decay (such as a power law), the motion is described as sub-diffusion. For example, with the kernel [54, 55]

$$K(t) = 2H(2H-1)\frac{1}{|t|^{-2H+2}},$$

(8.37)

the anomalous diffusion exponent is $\alpha = 2\,H$. The variance of the position of a tagged monomer governed by equation 8.35 is given by

$$var\left(R_c\right) \sim \frac{k_B T}{\gamma}\, \frac{\sin\left(2H\pi\right)}{\pi H\left(1-2H\right)\left(2-2H\right)} t^{2-2H}. \tag{8.38}$$

To conclude, in that framework, a particle performs an anomalous diffusion with exponent $\alpha = 2-2\,H$. The FBM was used in [36] to describe chromatin locus dynamics. The power-law decaying kernel accounts for the motion of the loci in a viscoelastic fluid. In that medium, the motion of a locus as well as the chromosome dynamics are slowed down, resulting in a sub-diffusion regime. Relating the exponent H to the local chromatin properties or the nuclear environment remains a challenge and we refer to the references mentioned in this section for further information.

8.3 DIRECTED MOTION OF CHROMATIN: ACTIVE MOTION AND ARTIFACTS

Some trajectories revealed by a chromatin locus appear long and directed (Figure 8.4a). This behavior can result from the motion of the entire nucleus (or reference frame) or can be due to internal forces. Can these two cases be discriminated? In the first case, such motion will be classified as an artifact because it is due to the movement of the entire domain where the locus is embedded. Actually, the motion of the locus with respect to other nuclear bodies can be studied by considering the projection to the perpendicular direction of the trajectory (Figure 8.4a).

To detect a directed motion, a fixed and marked reference point is chosen in the nucleus. In yeast, it is the spindle pole body (SPB). Interestingly, when the SPB is marked, it does jitter around its averaged position (Figure 8.4b), which leads to the conclusion that the yeast nucleus performs random precessions (oscillation around an axis, that can change over time) [18]. Correcting for such precession is not simple and it actually requires tagging more than one point on the nucleus to detect the precession in three dimensions. This precession motion can be abolished using the sponge toxin Latrunculin A that depolymerize the actin mesh wrapped around the nucleus in the yeast cytoplasm. When Latrunculin A was applied in yeast, it stopped the precession motion [18], possibly by decoupling the nucleus from the cytoplasm actin and microtubule network. Following this procedure (addition of Latrunculin A), the anomalous exponent drops from a value of $\alpha = 0.48 \pm 0.14$ to $\alpha = 0.3 \pm 0.14$ [18], confirming that a directed component of the motion is removed.

A direction motion couples the motion of monomers and can be introduced in the polymer model, starting from the Rouse equations, by adding in the Hamiltonian of equation 8.30 a noise component

$$\langle A_{ni}\left(t\right)\rangle = 0,$$
$$\langle A_{ni}\left(t\right)A_{mj}\left(t\right)\rangle = \delta_{ij}B\left(t-t'\right)C\left(n-m\right), \tag{8.39}$$

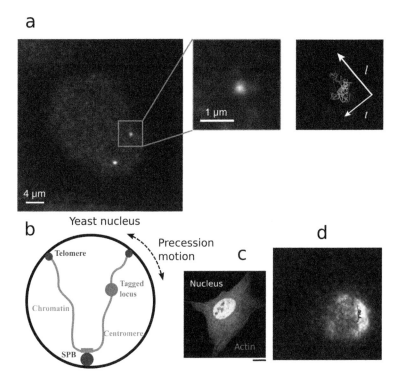

Figure 8.4 Directed motion of a chromatin locus. (a) Two tagged loci on the X chromosomes of a mouse ESC expressing Tet repressor-EGFP, visualized in live-cell fluorescence microscopy. The 3D position of each chromosomal locus is determined over a period of 180 s at 900 ms intervals, and principal component analysis is used to extract the longitudinal (*l*) and transversal (*t*) component of the trajectory (identified as the directions of maximum (longitudinal) and minimum (transversal) variance across the trajectory). (adapted from [21]). (b) Schematic representation of the nucleus of the yeast *Saccharomyces cerevisiae*. Each chromosome is connected at the centromere to filaments connecting to the spindle pole body (SPB), which is a fixed point on the nucleus membrane. (c) The nucleus is embedded in a network of actin filaments that cause it to move in precession motion. (d) The trajectory of a tagged locus (red) and that of the SPB (blue). The SPB is performing angular motion suggesting the nucleus is undergoing random precession.

where n,m are the monomers and i,j are spatial directions. Here $B(t-t')$ is a function correlated in time. Thus, the model can account for an anomalous diffusion larger than the Rouse one.

For example, an exponentially correlated colored noise $B(t-t') \sim e^{-t/t_A}$ can result in a larger anomalous exponent or even ballistic motion for the tagged monomer at an intermediate time regime [56]. However, at later times, the motion is again subdiffusive, before reaching a steady-state behavior for long periods. Hence, a noise with a characteristic time similar to the relaxation mode of the polymer can affect the dynamics during an observation window of the same time order. When the active noise component is coupled between different monomers reflected in the correlation function $C(n-m)$, we expect an increase of the anomalous exponent.

By correlating the motion of monomers in time or space along the polymer chain, it is possible to model and analyze active noise. However, it cannot be used to directly infer the properties of an active component such as the directed motion of the media and rotation. Directed motion of physiological importance has also been shown to affect the anomalous exponent [57] by following telomere motion in ALT (alternative lengthening of telomere) cancer cells. (Telomeres are repetitive elements at the end of chromosomes that shorten upon cell division). In stem cells and cancer cells, telomeres are elongated to allow the cell to divide many times. In ALT cells, telomeres maintain their length via homologous recombination (HR), when they encounter and use each other as a template for elongation: Telomeres perform long-range movement to actively search for recombination partners. This motion is characterized by anomalous diffusion with $\alpha = 0.8$ [57] on average. Interestingly, the motion is aided by Rad51 filaments, allowing the telomeres to move on the order of microns, and perform ballistic motion at long timescales. These filaments mediate the search and repair of breaks.

8.3.1 In Vivo Modification Of Chromatin Mobility

The formation of a double-stranded DNA break (DSB) at a molecular level causes a massive perturbation of the entire nucleus. Many DSBs can occur continuously as our genetic material is constantly exposed to radiation and attacked by free radicals. A break is potentially very harmful, as it can lead to cell death or to cancer. Hence, from bacteria to high Eukaryotes, cells have developed highly sophisticated mechanisms to repair these damages [58].

The choice of repair mechanism often depends on the cell phase or on the location of the break. During the growth phase (G1), yeast prefers to repair DSBs using the *non-homologous end joining* pathway [59]. In this pathway, the two broken ends of the DSB are quickly re-ligated. However, a piece of the DNA sequence material can be lost, leading to a loss of genetic information after the break is resolved. When the yeast is in duplication mode (S phase), it often prefers the *homologous recombination* (HR) pathway, in which the break searches for a homologous sequence on a different chromosome and uses it as a template for repair [60]. In S phase, the presence of a nearby newly synthesized duplicate of the chromosome allows the break to find the target rapidly, but cells which are diploids will also use the other chromosome as a template.

DSBs at random locations in the genome can be induced using radiation [61] or toxins such as Zeocin [62]. These actions cause massive chromatin modifications. Indeed, following treatment with Zeocin chromatin starts to move rapidly and can be seen by the formation of repair proteins foci at break sites [62]. When measuring different dynamical parameters of the foci, they show a significant increase (diffusion coefficient, anomalous exponent, and localization). To interpret these modifications, the polymer models introduced in the previous sections are used [18, 63].

Another method to study the consequence of DSBs is to induce them locally using a restriction enzyme, cutting DNA in proximity to a fluorescent label (Figure 8.5a). Interestingly, once a break is induced, in S phase, the tagged locus increase its motion, possibly to scan a larger sub-nuclear domain (Figure 8.5b) [7, 8, 64]. This increased motion could facilitate repair via homologous recombination by allowing the break to find its homologous partner on another chromosome. This change is reflected in the biophysical properties of the chromatin strand. Indeed, different repair proteins evict histones in the proximity of the break site, causing a decrease in parameters such as the spring constant k_c [18], suggesting a reduction in tethering forces (Figure 8.5c–d). This is accompanied by an increase of the anomalous exponent (Figure 8.5e).

The FBM model of chromatin dynamics interprets the subdiffusion of chromatin loci to be a consequence of the viscoelastic behavior of the nuclear plasma. Thus, the FBM model of chromatin cannot readily explain such a modification in

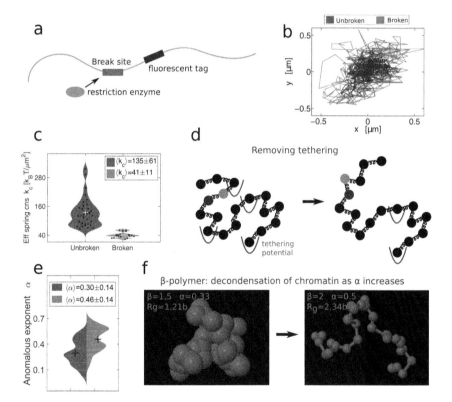

Figure 8.5 Dynamics of a double-stranded DNA break. (a) Double-stranded DNA break (DSB) induced in yeast using a restriction enzyme and monitored by a fluorescent tag. (b) Trajectory of an unbroken locus (blue), which following break induction, increases its motion (red). (c–d) The effective spring constant k_c (computed from equation 8.16) of the locus decreases following a break, reducing the tethering interactions. (e–f) Increase of the anomalous exponent α following break induction. (Adapted from [18]).

the anomalous exponent α, as the DSB is a local event, and any local modification to chromatin will not affect the overall viscoelastic properties of the nuclear solution. However, according to the β-polymer model, an increase in the exponent α corresponds to an open or a local de-condensation of chromatin (Figure 8.5f). Following break induction, such de-condensation was observed using super-resolution microscopy [18, 61, 62]. The de-condensation seems to result from nucleosome eviction and their degradation.

Another possible explanation for the increased mobility is the release of the centrosome from its tight grip on the centromere [65], which is connected to the SPB (see Figure 8.4b). The release of tethering forces would allow the locus to scan a larger domain. From a modeling perspective, this scenario agrees with a decrease in the spring constant k_c, which represents the sum of all tethering forces acting on the chromatin around the observed locus.

Finally, changes in chromatin mobility could be attributed to a modification in the persistence length l_p [64, 66, 67]. We recall that the DNA has a persistence length of 50 nm, which corresponds to 150 base pairs. However, it is still unclear what the persistence length l_p of chromatin is. It is probably not constant along the strand, as chromatin is epigenetically modified locally. In yeast, it was initially estimated to be of the order $l_p = 197 \pm 62$nm [6], which corresponds to about 28kbs. However, recent Hi-C experiments [68], give an estimate less than 5kbp, much smaller than previous estimates for the 30 nm fiber. Another approach is to estimate l_p from SPTs, but it is limited due to the motion artifact described above.

We recall that a rigid rod can result in an anomalous exponent of 0.75 [69]. Thus, partial nucleosome removal associated with a break will certainly increase l_p, which could be directly measured from a locus dynamics. Such experiments would require a fast sampling time, at a rate below the flexible chain relaxation time. While in the Rouse model, the relaxation modes depend on $\sim p^{-2}$, where p is the mode (introduced previously), the relaxation modes of a rigid rod scales with the fourth power $\sim p^{-4}$, as predicted from classical elasticity theory. Neither scaling was yet demonstrated for chromatin. Another possibility to study chromatin rigidity is the longitudinal and transversal oscillations of the strand. If rigidity plays a significant role in loci dynamics, we predict that the stand should move much more in the transversal direction.

8.3.2 Is The Directed Motion Of A Break During HR An Active Or A Passive Mechanism?

Following the induction of DSBs in the yeast nucleolus, which is the location of ribosomal DNA, breaks are expelled *outside* this dense domain [70] (see Figure 8.6a [18]). Later on, proteins associated with the repair process assemble at the break location, by a mechanism that remains unclear. Similar behavior is shown in heterochromatin (in *Drosophila* [71]), where breaks slowly relocate outside of a dense sub-nuclear domain.

Since these domains contain many repetitive DNA sequences, repair by NHEJ (simple re-ligation) could result in mismatch repair and chromosomal translocations (incorrect repair of two strands). These translocations could also be

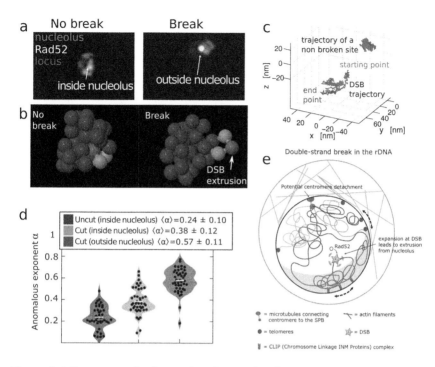

Figure 8.6 Dynamics of a dsDNA break outside of a dense domain. (a) Once a chromatin locus in the yeast nucleolus is cleaved by a restriction enzyme, the DSB moves outside the nucleolus, and the repair protein Rad52 can be recruited to form a focus. Rad52 is largely excluded from the nucleolus. (b) Steady-state configuration of a β-polymer model with Lennard–Jones interactions (left). Once a break is induced (right), the Hamiltonian changes (Equation 8.40) and the broken site (red) does not participate in long-range interaction with far away monomers. Polymer simulations predict a break extrusion as is seen experimentally in (a). (c) Two trajectories following a monomer unaffected by the extrusion, and the broken monomer shown in (b). The time during the 0.05 s simulation, is represented by the color change of the trajectory from red to green. (d) Anomalous exponent for the locus shown in (a). Latrunculin A was added to remove nuclear precession. (adapted from [18]). (e) Schematics of a DSB repair via HR in the yeast nucleolus.

produced if a break was repaired inside the dense domain via HR. Thus, it has been suggested that DSBs relocate to prevent abnormal recombination. Such a major reorganization of chromatin is not very common in higher Eukaryotes [72]. It could potentially be the result of an active process pulling the DSBs outside of the domains, as discussed in the context of the ALT cell, where directed motion was mediated by Rad51 filaments. However, Rad51 is only recruited to the break once it is outside of the dense domain and cannot assist in such motion.

The physical mechanism of DSB relocation remains unclear but could simply be *passive*. Indeed, chromatin local decondensation at the break site can be modeled using the β polymer model [18]. The local chromatin modification at the DSB is modeled at the "broken" monomer by decreasing the connections to the ones located far away but remains connected only to its nearest neighbors. In such a scenario, for a break at monomer n, the associated Hamiltonian $\tilde{\phi}$ (Equation 8.30) is

$$\tilde{\phi}_{break}(R_1,..R_N) = \frac{1}{2}\sum_{l,m\neq n} R_l R_m A_{l,m} + \frac{\kappa}{2}\left[(R_{n+1}-R_n)^2 + (R_n - R_{n-1})^2\right], \quad (8.40)$$

It was shown [18], that with such modifications, in a stochastic polymer simulations with Hamiltonian 8.40, after a transient regime, the polymer is reorganized where the break is extruded and relocated to the surface of the polymer globule (Figure 8.6b). This redistribution is a passive process. It is the result of the energy minimization of the polymer configuration and does not require an active mechanism of transport (Figure 8.6c). In addition, this relocalization is associated with an increase of the anomalous exponent α of the break due to chromatin decondensation and relocation to the periphery of the domain. This prediction was observed experimentally following break initiation, reported as an increase of the anomalous exponent while the DSB is gradually moving outside the nucleolus (Figure 8.6d). Such a model can potentially explain large-scale rearrangement of chromatin, that would require a lot of energy and would be difficult to achieve in the dense environment of the nucleolus.

8.3.3 Long-Range Correlations Of Chromatin Mobility

A directed or persistent loci motion can be due to local or global histone modifications, arising from DNA breaks. However, long-range correlations are often observed without a precise perturbation of the strand. For example, the strand can have a micron range correlation in its velocity [73], leading to displacement correlation with a magnitude of 4–5 μm (see Figure 8.7).

The origin of these correlations remains unclear, yet, ATP depletion abolishes them completely. Thus, ATP-active motion could be the physical source although the exact physical mechanism is unclear. Since hydrodynamical interactions decay slowly in a solution (inversely proportional to the spatial distance), they could used as a model to explain the observed correlation, when the dense chromatin is approximated by a fluid medium propagating the velocity fields. This model is known as chromatin hydrodynamics [74].

Fluctuations of the nuclear membrane can lead to long-range mobility correlation of chromatin and perhaps some of these correlations could generate a flow inside the nucleus, which could be significant (see Figure 8.4a).

As a possible modeling perspective, it would be interesting to see how the result of such a continuum model of chromatin mobility coincides with the description of chromatin as a linear chain discussed in the previous sections. Recent experiments suggest that some bodies/regions in the nucleus behave as if undergoing liquid–liquid phase separation from the rest of the nuclear cytoplasm [75]. The nucleolus and heterochromatin are some examples [76]. A complete theory of chromatin should be able to explain the fast relaxation times associated

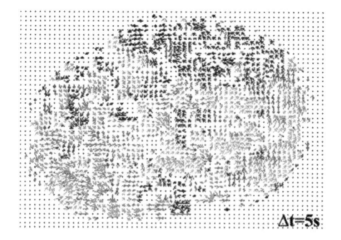

Figure 8.7 Long range correlations of chromatin mobility. (a) Velocity correlation of chromatin. Sections with the same colors move together in the same direction. (Adapted from [73]).

with such domains, as well as the localized behavior of chromatin loci that would not be expected for a particle in a liquid drop.

8.4 CONCLUSION: WHAT HAVE WE LEARNED SO FAR FROM ANALYSING SPTS OF CHROMATIN?

SPTs of a chromatin locus behavior has revealed the heterogeneity of molecular dynamics, driven by diffusion, and short and long range drift motion. Yet, relating the anomalous exponent and the diffusion coefficient extract from SPT of chromatin loci to the local chromatin organization has remained a major challenge. The relationship between the anomalous exponent and interactions between different segments of the strand (mediated by cohesin molecules for example) has been partially addressed in [44, 45].

It is now clear that polymer modeling plays a key role in the reconstruction of chromatin structure inside the nucleus, but also by defining parameters that can be extracted from SPTs to reveal the physical forces acting on the strand. Such models also reveal how the local chromatin organization can confine a locus motion in a small domain. Several estimators are now being used routinely to extract biophysical parameters, such as the effective spring coefficient k_{cn}, the local drift due to external forces, the apparent spring coefficient, the anomalous exponents or the size of a confinement domain. The predictions of polymer models are now tested experimentally. Because classical processes such as transcription repair, differentiation, and different pathologies correlate with chromatin motion, studying their dynamics reveal information about the physics of key steps.

To conclude, the field of polymer models is now mature enough for the nucleus to be studied as a physical system and at a given scale. More high throughput data of both structural (Hi-C for example) and dynamical nature are being generated. A future challenge will be to integrate and develop sophisticated statistical methods to interpret these data. Since imaging is becoming more and more accurate in the optical resolution, the analysis of multiple chromatin loci will certainly reveal the manner and frequency in which different parts of our genome interact with one another. This should allow us to understand both the dynamics and organization of the nucleus.

REFERENCES

1. S. Chandrasekhar. Stochastic problems in physics and astronomy. *Reviews of modern physics*, 15(1):1, 1943.
2. D. Holcman and Z. Schuss. 100 years after Smoluchowski: Stochastic processes in cell biology. *Journal of Physics A: Mathematical and Theoretical*, 50(9):093002, 2017.
3. C. C. Robinett, A. Straight, G. Li, C. Willhelm, G. Sudlow, A. Murray, and A. S. Belmont. In vivo localization of DNA sequences and visualization of large-scale chromatin organization using lac operator/repressor recognition. *J. Cell Biol.*, 135:1685–1700, 1996.
4. W. F. Marshall, A. Straight, J. F. Marko, J. Swedlow, A. Dernburg, A. Belmont, A. W. Murray, D. A. Agard, and J. W. Sedat. Interphase chromosomes undergo constrained diffusional motion in living cells. *Curr. Biol.*, 7(12):930–939, 1997.
5. V. Dion and S. M. Gasser. Chromatin movement in the maintenance of genome stability. *Cell*, 152:1355–1364, 2013.
6. K. Bystricky, P. Heun, L. Gehlen, J. Langowski, and S. M. Gasser. Long-range compaction and flexibility of interphase chromatin in budding yeast analyzed by high-resolution imaging techniques. *Proc. Natl. Acad. Sci. USA*, 47:16495–16500, 2004.
7. V. Dion, V. Kalck, C. Horigome, B. D. Towbin, and S. M. Gasser. Increased mobility of double-strand breaks requires mec1, rad9 and the homologous recombination machinery. *Nat. Cell Biol.*, 14:502–509, 2012.
8. J. Miné-Hattab and R. Rothstein. Increased chromosome mobility facilitates homology search during recombination. *Nat. Cell Biol.*, 14:510–517, 2012.
9. Z. Schuss. *Diffusion and Stochastic Processes. An Analytical Approach*. Springer-Verlag, New York, NY, 2009.
10. N. Hoze, D. Nair, E. Hosy, C. Sieben, S. Manley, A. Herrmann, J-B. Sibarita, D. Choquet, and D. Holcman. Heterogeneity of AMPA receptor trafficking and molecular interactions revealed by superresolution analysis of live cell imaging. *Proc. Natl. Acad. Sci.*, 109(42):17052–17057, 2012.
11. Nathanaël Hozé and David Holcman. Statistical methods for large ensembles of super-resolution stochastic single particle trajectories in cell biology. *Annual Review of Statistics and Its Application*, 4:189–223, 2017.
12. L. Cong, F. A. Ran, D. Cox, S. Lin, R. Barretto, P. D. Hsu, X. Wu, W. Jiang, and L. A. Marraffini. Multiplex genome engineering using CRISPR/VCas systems. *Science*, 339(6121):819–823, 2013.

13. B. Chen, L A. Gilbert, B A. Cimini, Joerg Schnitzbauer, Wei Zhang, Gene-wei Li, Jason Park, E H. Blackburn, J S. Weissman, S. Lei, and B. Huang. Dynamic imaging of genomic loci in living human cells by an optimized CRISPR/Cas system. *Cell*, 155(7):1479–1491, 2013.

14. Michael R. Hübner and David L. Spector. Chromatin dynamics. *Annual Review of Biophysics*, 39(1):471–489, 2010.

15. I. Izeddin, V. Récamier, L. Bosanac, I. Cissé, L. Boudarene, C. Dugast-Darzacq, F. Proux, O. Bénichou, R. Voituriez, O. Bensaude, M. Dahan, and X. Darzacq. Single-molecule tracking in live cells reveals distinct target-search strategies of transcription factors in the nucleus. *eLife*, 3, 2014.

16. D. Holcman, N. Hoze, and Z. Schuss. Analysis of single particle trajectories: When things go wrong. *arXiv*, 1502.00286, 2015.

17. S. Nagai, K. Dubrana, M. Tsai-Pflugfelder, M. B. Davidson, T. M. Roberts, G. W. Brown, E. Varela, F. Hediger, S. M. Gasser, and N. J. Krogan. Functional targeting of DNA damage to a nuclear pore-associated sumo-dependent ubiquitin ligase. *Science*, 322:597–602, 2008.

18. A. Amitai, A. Seeber, C. Horigome, S. M. Gasser, and D. Holcman. Changes in local chromatin organization during the homology search: Effects of local contacts on search time. *Cell Reports*, 18:1200–1214, 2017.

19. P. Therizols, T. Duong, B. Dujon, C. Zimmer, and E. Fabre. Chromosome arm length and nuclear constraints determine the dynamic relationship of yeast subtelomeres. *Proc. Natl. Acad. Sci. USA*, 107:2025–2030, 2010.

20. T. Cremer and C. Cremer. Chromosome territories, nuclear architecture and gene regulation in mammalian cells. *Nat. Rev. Genet.*, 2:292–301, 2001.

21. G. Tiana, A. Amitai, T. Pollex, T. Piolot, D. Holcman, E. Heard, and L. Giorgetti. Structural fluctuations of the chromatin fiber within topologically associating domains. *Biophysical Journal*, 110(6):1234–1245, 2016.

22. I. Bronshtein, E. Kepten, I. Kanter, S. Berezin, M. Lindner, Abena B. Redwood, S Mai, S. Gonzalo, R. Foisner, Y. Shav-Tal, and Y. Garini. Loss of lamin a function increases chromatin dynamics in the nuclear interior. *Nature Communications*, 6:8044, 2015.

23. S. C. Weber, A. J. Spakowitz, and J. A. Theriot. Nonthermal ATP-dependent fluctuations contribute to the in vivo motion of chromosomal loci. *Proc. Natl. Acad. Sci. USA*, 109(19):7338–7343, 2012.

24. A. Taddei, H. Schober, and S. M. Gasser. The budding yeast nucleus. *Cold Spring Harbor Perspect. Biol.*, 2:a000612, 2010.

25. D. Holcman, N. Hoze, and Z. Schuss. Narrow escape through a funnel and effective diffusion on a crowded membrane. *Phys. Rev. E.*, 84:021906, 2011.

26. Z. Schuss. *Brownian Dynamics at Boundaries and Interfaces In Physics, Chemistry*, and *Biology Applied Mathematical Sciences*. Springer-Verlag, New York, NY, 2013.

27. Z. Schuss. *Theory and Applications of Stochastic Differential Equations. Wiley Series in Probability and Statistics - Applied Probability and Statistics Section*. Wiley, Hoboken, NJ, 1980.

28. M. Doi and S. F. Edwards. *The Theory of Polymer Dynamics*. Clarendon Press, Oxford, 1986.

29. A. Amitai, M. Toulouze, K. Dubrana, and D. Holcman. *Extracting In Vivo Interactions Acting On The Chromatin From A Statistical Analysis Of Single Locus Trajectories. Personal Communication*, 2015.

30. F. R. Neumann, V. Dion, L. R. Gehlen, M. Tsai-Pflugfelder, R. Schmid, A. Taddei, S. M. Gasser. Targeted INO80 enhances subnuclear chromatin movement and ectopic homologous recombination. *Genes & Development*, 26(4):369–83, 2012.

31. P. Heun, T. Laroche, K. Shimada, P. Furrer, and S. M. Gasser. Chromosome dynamics in the yeast interphase nucleus. *Science*, 294(5549):2181–6, 2001.

32. H. Qian, M. P. Sheetz, and E. L. Elson. Single particle tracking. analysis of diffusion and flow in two-dimensional systems. *Biophys. J.*, 60(4):910–21, 1991.

33. A. Amitai D. Holcman. Polymer physics of nuclear organization and function. *Physics Reports*, 678:1–83, 2017.

34. R. Metzler and J. Klafter. The random walk's guide to anomalous diffusion: a fractional dynamics approach. *Phys. Rep.*, 339(1):1–77, 2000.

35. E. Kepten, I. Bronshtein, and Y. Garini. Improved estimation of anomalous diffusion exponents in single-particle tracking experiments. *Phys. Rev. E.*, 87:052713, 2013.

36. S. C. Weber, J. A. Theriot, and A. J. Spakowitz. Subdiffusive motion of a polymer composed of subdiffusive monomers. *Phys. Rev. E.*, 82:011913, 2010.

37. H. Hajjoul, J. Mathon, H. Ranchon, I. Goiffon, J. Mozziconacci, B. Albert, P. Carrivain, J.-M. Victor, O. Gadal, K. Bystricky, and A. Bancaud. High-throughput chromatin motion tracking in living yeast reveals the flexibility of the fiber throughout the genome. *Genome Research*, 23(11):1829–1838, 2013.

38. B. Albert, J. Mathon, A. Shukla, H. Saad, C. Normand, I. Léger-Silvestre, D. Villa, A. Kamgoue, J. Mozziconacci, H. Wong, C. Zimmer, P. Bhargava, A. Bancaud, and O. Gadal. Systematic characterization of the conformation and dynamics of budding yeast chromosome XII. *Journal of Cell Biology*, 202(2):201–210, 2013.

39. A. Amitai, M. Toulouze, K. Dubrana, and D. Holcman. *Heterogeneity Of DNA-locus dynamics revealed by polymer physics analysis. Personal Communication*, 2015.

40. P. G. de Gennes. *Scaling Concepts in Polymer Physics*. Ithaca, New York, Cornell Univ. Press, 1979.

41. E. Lieberman-Aiden, N. L. van Berkum, L. Williams, M. Imakaev, T. Ragoczy, A. Telling, I. Amit, B. R. Lajoie, P. J. Sabo, M. O. Dorschner, R. Sandstrom, B. Bernstein, M. A. Bender, M. Groudine, A. Gnirke, J. Stamatoyannopoulos, L. A. Mirny, E. S. Lander, and J. Dekker. Comprehensive mapping of long-range interactions reveals folding principles of the human genome. *Science*, 326:289–293, 2009.

42. A. Y. Grosberg, Y. Rabin, S. Havlin, and A. Neer. Crumpled globule model of the three-dimensional structure of DNA. *Europhys. Lett.*, 23:373, 1993.

43. L. Giorgetti, R. Galupa, E. P. Nora, T. Piolot, F. Lam, J. Dekker, G. Tiana, and E. Heard. Predictive polymer modeling reveals coupled fluctuations in chromosome conformation and transcription. *Cell*, 157:950–963, 2014.

44. O. Shukron and D. Holcman. Transient chromatin properties revealed by polymer models and stochastic simulations constructed from chromosomal capture data. *Plos Computational Biology*, 13(4):e1005469, 2017.

45. O. Shukron and D. Holcman. Statistics of randomly cross-linked polymer models to interpret chromatin conformation capture data. *Phys. Rev. E.*, 96(1):012503, 2017.

46. A. Amitai and D. Holcman. Polymer model with long-range interactions: Analysis and applications to the chromatin structure. *Phys. Rev. E.*, 88:052604, 2013.

47. M. V. Tamm, L. I. Nazarov, A. A. Gavrilov, and A. V. Chertovich. Anomalous diffusion in fractal globules. *Phys. Rev. Lett.*, 114:178102, 2015.

48. K. E. Polovnikov, M. Gherardi, M. Cosentino-Lagomarsino, and M. V. Tamm. Fractal folding and medium viscoelasticity contribute jointly to chromosome dynamics. *Phys. Rev. Lett.*, 120:088101, 2018.

49. R. Metzler, E. Barkai, and J. Klafter. Anomalous diffusion and relaxation close to thermal equilibrium: A fractional fokker-planck equation approach. *Phys. Rev. Lett.*, 82:3563–3567, 1999.

50. E. Kepten, I. Bronshtein, and Y. Garini. Ergodicity convergence test suggests telomere motion obeys fractional dynamics. *Phys. Rev. E.*, 83:041919, 2011.

51. G. Samorodnitsky and M. Taqqu. *Stable Non-Gaussian Random Processes*. Chapman and Hall, New York, 1994.

52. J. Jeon and R. Metzler. Fractional Brownian motion and motion governed by the fractional Langevin equation in confined geometries. *Phys. Rev. E.*, 81:021103, 2010.

53. S. C. Kou. Stochastic modeling in nanoscale biophysics: Subdiffusion within proteins. *The Annals of Applied Statistics*, 2(2):501–535, 2008.

54. S. C. Kou and X. S. Xie. Generalized Langevin equation with fractional Gaussian noise: Subdiffusion within a single protein molecule. *Phys. Rev. Lett.*, 93:180603, 2004.

55. W. H. Deng and E. Barkai. Ergodic properties of fractional Brownian–Langevin motion. *Phys. Rev. E.*, 79:011112, 2009.

56. D. Osmanovic and Y. Rabin. Dynamics of active Rouse chains. *Soft Matter*, 13(5):963–968, 2017.

57. N. W. Cho, R. L. Dilley, M. A. Lampson, and R. A. Greenberg. Interchromosomal homology searches drive directional ALT telomere movement and synapsis. *Cell*, 159(1):108–21, 2014.

58. F Pâques and J E Haber. Multiple pathways of recombination induced by double-strand breaks in Saccharomyces cerevisiae. *Microbiology and Molecular Biology Reviews : MMBR*, 63(2):349–404, 1999.

59. J K Moore and J E Haber. Cell cycle and genetic requirements of two pathways of nonhomologous end-joining repair of double-strand breaks in Saccharomyces cerevisiae. *Molecular and Cellular Biology*, 16(5):2164–2173, 1996.

60. W. M. Hicks, M. Yamaguchi, and J. E. Haber. Real-time analysis of double-strand dna break repair by homologous recombination. *Proc. Natl. Acad. Sci. USA*, 108:3108–3115, 2011.

61. S. Adam, J. Dabin, O. Chevallier, O. Leroy, C. Baldeyron, A. Corpet, P. Lomonte, O. Renaud, G. Almouzni, and S. E. Polo. Real-time tracking of parental histones reveals their contribution to chromatin integrity following DNA damage. *Mol. Cell.*, 64(1):65–78, 2016.

62. M. Hauer, A. Seeber, V. Singh, R. Thierry, R. Sack, A. Amitai, J. Kryzhanovska, M. Eglinger, D. Holcman, T. Owen-Hughes, and S. M. Gasser. Histone degradation in response to dna damage triggers general chromatin decompaction. *Nat. Struc. and Mol. Bio.*, 24:99107, 2017.

63. A Amitai and D Holcman. Encounter times of chromatin loci influenced by polymer decondensation. *Physical Review E.*, 97(3):032417, 2018.

64. A. Seeber, V. Dion, and S. M. Gasser. Checkpoint kinases and the INO80 nucleosome remodeling complex enhance global chromatin mobility in response to DNA damage. *Genes & Development*, 27(18):1999–2008, 2013.

65. J. Strecker, G. D. Gupta, W. Zhang, M. Bashkurov, M.-C. Landry, L. Pelletier, and D. Durocher. DNA damage signalling targets the kinetochore to promote chromatin mobility. *Nat. Cell Biol.*, 18(3):281–90, 2016.

66. S. Herbert, A. Brion, J.-M. Arbona, M. Lelek, A. Veillet, B. Lelandais, J. Parmar, F. G. Fernández, E. Almayrac, Y. Khalil, E. Birgy, E. Fabre, and C. Zimmer. Chromatin stiffening underlies enhanced locus mobility after DNA damage in budding yeast. *The EMBO Journal*, 36(17):2595–608, 2017.

67. J. Miné-Hattab, V. Recamier, I. Izeddin, R. Rothstein, and X. Darzacq. Multi-scale tracking reveals scale-dependent chromatin dynamics after DNA damage. *Molecular Biology of the Cell*, 28(23):3323–3332, 2017.

68. A. L. Sanborn, S. S. P. Rao, S.-C. Huang, N. C. Durand, M. H. Huntley, A. I. Jewett, I. D. Bochkov, D. Chinnappan, A. Cutkosky, J. Li, K. P. Geeting, A. Gnirke, A. Melnikov, D. McKenna, E. K. Stamenova, E. S. Lander, and E. Lieberman Aiden. Chromatin extrusion explains key features of loop and domain formation in wild-type and engineered genomes. *Proc. Natl. Acad. Sci.*, 112(47):201518552, 2015.

69. A. Caspi, R. Granek, A. Lachish, D. Zbaida, and M. Elbaum. Semiflexible polymer network: A view from inside. *Physical Review Letters*, 80(5): 1106–1109, 1998.

70. J. Torres-Rosell, I. Sunjevaric, G. De Piccoli, M. Sacher, N. Eckert-Boulet, R. Reid, S. Jentsch, R. Rothstein, L. Aragón, and M. Lisby. The Smc5-Smc6 complex and SUMO modification of Rad52 regulates recombinational repair at the ribosomal gene locus. *Nature Cell Biology*, 9(8):923–931, 2007.

71. I. Chiolo, A. Minoda, S. U. Colmenares, A. Polyzos, S. V. Costes, and G. H. Karpen. Double-strand breaks in heterochromatin move outside of a dynamic HP1a domain to complete recombinational repair. *Cell*, 144: 732–744, 2011.

72. E. Soutoglou and T. Misteli. Mobility and immobility of chromatin in transcription and genome stability. *J. Natl Cancer Inst. Monogr.*, 39:16–19, 2008.

73. A. Zidovska, D. A. Weitz, and T. J. Mitchison. Micron-scale coherence in interphase chromatin dynamics. *Proc. Natl. Acad. Sci. USA*, 110(39):15555–15560, 2013.

74. R. Bruinsma, A. Y. Grosberg, Y. Rabin, and A. Zidovska. Chromatin hydrodynamics. *Biophysical Journal*, 106(9):1871–1881, 2014.

75. S. F. Banani, H. O. Lee, A. A. Hyman, and M. K. Rosen. Biomolecular condensates: Organizers of cellular biochemistry. *Nat. Rev. Mol. Cell Biol.*, 18(5):285–298, 2017.

76. A. R. Strom, A. V. Emelyanov, M. Mir, D. V. Fyodorov, X. Darzacq, and G. H. Karpen. Phase separation drives heterochromatin domain formation. *Nature*, 547(7662):241–245, 2017.

9

Chromosome Structure and Dynamics in Bacteria: Theory and Experiments

MARCO GHERARDI, VITTORE SCOLARI, REMUS THEI
DAME, AND MARCO COSENTINO LAGOMARSINO

9.1 INTRODUCTION

This chapter is dedicated to chromosome organization and dynamics in bacteria. We will review some of the main discoveries and open questions on the bacterial chromosome that emerged from recent studies, through the experimental, theoretical and computational work performed around them. Our aim is to highlight research directions in this area where theoretical and experimental work performed on bacteria can lead the way to formulating new questions concerning eukaryotic chromosomes, and, vice versa, attempt to identify where studies performed on eukaryotes can inform the field of the bacterial chromosome.

An important distinction in terms of cellular structure is that eukaryotes have a dedicated organelle for storage of the genetic material, the nucleus, which is absent in prokaryotes. The genetic material of prokaryotes is immersed in the surrounding cytoplasm, but occupies a spatially defined region, inside a nucleoprotein complex called the nucleoid, where gene transcription and DNA replication take place. In this region of the cell, DNA is organized in one or more chromosomes carrying the genetic information. In bacteria, each chromosome has typically a single replication origin, which plays an important role in its structure and function [1]. Why is it interesting to study chromosome organization and dynamics in bacteria? An important motivation to study the bacterial chromosome is that it is one of the main players in mediating physiological shifts in response to rapidly changing environments and stimuli [2, 3], as the physiological changes in the course of each cell cycle. Chromosomal changes impact on genome replication and gene transcription, both at the local and at the global level, via the interplay of protein binding, self-tethering, and tethering to other cellular structures. In this kind of response, mechanical and physical aspects of the genome are deeply interconnected with biological processes affecting genes. Recently, we have described this response using the technological metaphor of "smart polymers" [4] a polymer with sensors embedded and responding to environmental changes with dynamic structural transitions.

As many excellent reviews on the biology of bacterial chromosome organization are available (see, e.g., refs [2, 5–11]), our aim is not to be exhaustive. Rather, we select some topics that are under debate today, and where the cross-fertilization between the bacterial chromosome community and that focusing on eukaryotes seems plausible. Our account is divided into four different but inter-related areas of (i) chromosome folding, (ii) DNA-organizing proteins (iii) chromosome dynamics, and (iv) role of molecular crowding, presented in separate sections (Figure 9.1).

9.2 CHROMOSOME FOLDING

The comprehensive assessment of the folding of bacterial chromosomes has benefited from progress in several high-throughput techniques, including chromosomal conformation capture (3C) [13, 14], single-cell omics [15], cryo-electron tomography [16], and super-resolution microscopy techniques, such as *in vivo* photo-activation localization microscopy (PALM) [17].

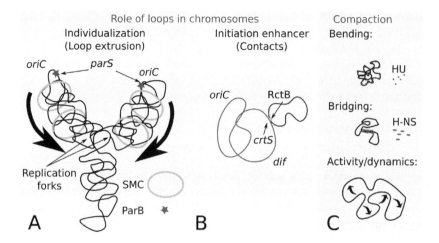

Figure 9.1 Examples of the influence of structuring effects of DNA-organizing proteins on the folded structure and function of the bacterial chromosome. (A) Individualization of newly replicated chromatin. For example, SMC is loaded by ParB at parS sites and moves along the DNA chain in the Ori-Ter axis in association with loop extrusion. (B) Chromosome folding and DNA-organizing proteins. For example, the loop formed by the RctB protein on chromosome two in *V. cholerae* and the crtS locus on chromosome one initiates chromosome two replication. Specifically, chromosome two replication initiates when the crtS locus is duplicated, allowing for a perfect synchronization between the replication processes of the two chromosomes. (C) Compaction can be induced by DNA-organizing proteins, such as HU through bending and H-NS through bridging, affecting mobility and key functions (e.g., transcription). Recent results show a significant correlation between the short-time dynamics of chromosomal loci and estimates of compaction levels by 3C data [12].

The presence of interactions between distant loci results in the formation of patterns in chromosome conformation at different length scales: an exhaustive classification of these patterns, as well as the identification of the proteins and the sequence elements that cause their emergence, has been the subject of chromosome conformation capture studies of prokaryotes as well as eukaryotes. 3C techniques allow the measurement of the mean contact frequencies between genomic loci in a cell population through stabilization using formaldehyde crosslinking and genome fragmentation, followed by identification of interacting segments via re-ligation and sequencing. Results are usually represented in the form of a matrix, quantifying the relative contact frequencies between pairs of loci – the "contact matrix". As in eukaryotes, the three-dimensional (3D) architecture of bacterial genomes was found to both reflect and regulate its functional state.

The main pattern observed in any 3C experiment is a thick diagonal signal along the main matrix axis, reflecting the DNA polymer backbone. Indeed, since the frequency of interaction reflects the spatial proximity of genomic loci, sequences that are closer to each other along the genome coordinate show a higher contact frequency than sequences that are farther away [13]. For the same reason, in presence of multiple chromosomes, these chromosomes are detected as well-defined blocks on the contact matrix. Finally, intra-chromosomal patterns may also show up as blocks in the contact matrix of the same chromosome. The formation of structural patterns of functional importance correlates with DNA-structuring proteins (also called nucleoid-associated proteins, NAPs), replication, and transcription (reviewed in ref. [3, 5]).

9.2.1 Replication and segregation are the main determinants of genome folding at large scales

Each organism, in order to survive, needs to ensure the proper segregation of duplicated chromosomes between mother and daughter cells. Different bacteria adopt different strategies in order to compact and spatially resolve the replicated chromosomes. In *Bacillus subtilis* and *Caulobacter crescentus*, SMC condensins and the ParB partitioning protein have been observed to play a central role in the interplay of chromosome folding and segregation [18–20]. SMC condensins are ring-shaped complexes composed of two subunits that constrain and bridge DNA segments [21, 22]. Cells lacking condensin subunits are viable, but they are inefficient in spatially segregating their replication origins, producing anucleated cells [23, 24]. In both *C. crescentus* and *B. subtilis*, SMC is associated with the presence of a secondary diagonal in the interaction matrix, called the main hairpin, perpendicular to the main polymer diagonal, spanning from the terminus of replication to an origin-proximal region and holding the left and right replichores together.

The generation of this pattern occurs in association with the so-called "loop-extrusion" process [25], whereby the condensin is recruited on the chromosome by ParB which is bound to the centromeric *parS* site near the origin of replication. During chromosome replication, the SMC rings translocate from *parS* (by diffusion or active transport) toward Ter along the two chromosomal arms joining at the replication fork. This activity helps in the resolution of knots and entanglements in the DNA backbone, and allows the separation of the individual chromosomes. The continuous loading of additional SMC at *parS* sustains the formation of the hairpin structure, and the SMC rings move along with the replication complex toward the replication terminus, until segregation is fully accomplished. A single copy of *parS* is sufficient to load SMC on the chromosome. The introduction of multiple ectopic *parS* sites generates multiple independent hairpins, associated with SMC loading, along the chromosomal arms [19].

Curiously, the *B. subtilis* genome harbors several *parS* binding sites in the origin-proximal region. The presence of multiple copies leads to the internal restructuring of the Ori (replication origin) genomic segment. Additionally, a second bow-shaped, intra-arm hairpin, has been detected originating at the *parS* loci positioned ~350 kb away from Ori on the left arm. The Ori-region structuring and the second intra-arm hairpin are not related to segregation. Instead, it was proposed that they exist in order to bring distant binding sites for DnaA, the protein responsible for initiation of DNA replication, into the vicinity of OriC, thus playing a role in the regulation of this process [20]. The role of SMC-like proteins in the coordination of replication and segregation has been also established in eukaryotic cells [26–31].

9.2.2 Genome folding role in replication initiation

Little is known about the possible role of chromosome structure changes as a regulatory mechanism for initiation, but recent evidence implicates chromosome structure as a possible player, working jointly with the DnaA initiator protein circuit [32]. The mechanisms behind the proposed role of the structuring of the *B. subtilis* Ori region in the control of replication initiation still need to be fully elucidated. Even less is known about the possible role of chromosome structuring in the initiation of DNA replication in *E. coli*, but a recent genomic study in *Vibrio cholerae* revealed a new control mechanism for the replication initiation of the secondary chromosome in this bacterium [33]. *V. cholerae*, like 10% of all bacterial species, carries a second chromosome 1Mb in size, in addition to its main (3Mb) chromosome. Initiation of replication in the first chromosome is controlled by DnaA; the second chromosome has a plasmid-like origin controlled by a *Vibrio*-specific factor, called RctB. This protein binds to the origin of replication of the second chromosome, called the "iteron", as well as to a 39-mer regulatory site on the second chromosome, strongly inhibiting initiation [34].

RctB was also found to bind to a locus on the first chromosome, *crtS*, acting upon chromosome 2 as an enhancer of initiation [35]. The initiation of both chromosomes occurs only once per cycle and terminates at the same time, implying a tight concerted control between replication of the first chromosome and the initiation of the replication of the second chromosome. Marker frequency analysis, a method that allows the measurement of the relative mean copy number of different chromosomal loci, indicates that the timing of replication at the RctB binding site on the first chromosome tightly controls the timing of replication initiation of the second chromosome. 3C analysis reveals that the two chromosomes are individual entities in terms of interactions, apart from the two Ter regions exhibiting strong contact. Their interaction is possibly due to structuring by the Ter-condensing protein MatP [36]. The origin of replication of the second chromosome and the *crtS* locus interact, pointing toward a mechanical process driving origin initiation in the second chromosome.

9.2.3 Transcription defines chromatin-interaction domains

Chromosome conformation capture analyses also led to the discovery of local domains of increased interaction frequency, called chromosomal interaction domains (CIDs). For example, experiments performed at high genomic resolution (obtained by using frequent-cutter restriction enzymes and deep sequencing) show short-range domains of interactions in *C. crescentus* [25]. 23 CIDs ranging in length from 30 to 420 kb have been identified, generally flanked by highly expressed genes. Inhibition of transcription using rifampicin causes disruption of CID boundaries, suggesting that high transcription induces domain formation. CIDs persist throughout the cell cycle and are independent of the

SMC protein (see below). Similar patterns of genome organization have been detected in *V. cholerae* (see previous paragraph) and *B. subtilis*. In *B. subtilis*, 20 CID barriers were identified, ranging in size from 50 to 300 kb. As in *V. cholerae*, the barriers coincided with highly expressed genes and were sensitive to rifampicin treatment [20]. Finally, in *Mycoplasma pneumoniae* similar patterns were also found. For this organism, it was demonstrated that CIDs contain genes that are preferentially coexpressed and coregulated [37]. This specific point has not been addressed for the other organisms. At larger scales, domains of different natures can emerge. In *B. subtilis* the Ori and Ter (replication origin and terminus) regions show a higher level of compaction than the rest of the chromosome. In *E. coli*, the NAP MatP appears to structure the Ter region [12, 36, 38, 39], and the NAP HU (discussed in the next section) increases short-distance interactions in *C. crescentus* [25] and *E. coli* [12].

The 3C technique detects a population average of the folded state and the local compaction of chromatin. Interestingly, these measurements correlate with the local dynamics of chromatin, measured in single cells. Indeed, recent 3C measurements in *E. coli* show that local compaction correlates with the mobility and fluctuations in time of chromosomal loci [12]. In this study, local compaction was inferred from a locus-dependent estimate of the interaction probability $P(s)$ from 3C data, while short-time subdiffusive dynamics of loci around the chromosome had been previously measured through fluorescent-tag microscopy of labeled chromosomal loci [40] (see below).

9.3 DNA-ORGANIZING PROTEINS

A prominent role in organizing chromosome folding at different length scales is played by a set of DNA-binding proteins with diverse properties. Both prokaryotes and eukaryotes abundantly express DNA-binding proteins involved in the organization and compaction of their genomes [10, 41, 42]. Generically, these proteins are referred to as "chromatin proteins", proteins that shape the prokaryotic or eukaryotic chromatin.

9.3.1 Genome compaction

Typical of eukaryotes are the histone proteins, which assemble as octamers on DNA to yield nucleosomes; DNA with a unit length of about 150 bp is wrapped in a left-handed manner around an octameric histone core. DNA decorated with nucleosomes is referred to as the "10 nm fiber". Depending on the separation distance (referred to as linker length) and its regularity, 10 nm fibers can assemble into thicker "30 nm fibers", formed due to nucleosome-nucleosome stacking interaction [5]. Among prokaryotes, a distinction needs to be made between bacteria and archaea. Both bacteria and archaea express many different chromatin proteins, or NAPs [41, 43]. Bacteria do not express homologues of eukaryotic histones. Some archaeal phyla, which are considered to share a common predecessor with eukaryotes, do express histone homologues [44, 45]. Therefore, studies of archaeal chromosome organization might provide useful information on the evolution of eukaryotic chromosome organization.

In bacteria and archaea, the functional equivalents of eukaryotic histones achieve compaction by their ability to bend DNA. The best-characterized (and cellularly most abundant) example of a DNA-bending protein is HU [46]. This protein lacks sequence specificity; it effectively compacts DNA by binding and bending at multiple apparently random sites along its contours *in vitro* [47]. HU is the only protein for which the ability to compact DNA has also been corroborated *in vivo* [48]. Interestingly, the binding of HU *in vitro* is cooperative and leads to the formation of stiff filamentous structures at elevated protein densities [47, 49]. While the physiological relevance of this mode of binding is currently unclear, it highlights architectural diversity due to protein–protein interaction as an important feature of nucleoid-associated proteins. Among archaea, examples of simple DNA-bending proteins include the Sul7, Cren7, MC1, and Sso10a proteins [50]. All these proteins compact DNA by bending, but apart from Sso10a they do not share with HU the ability to assemble into stiff filaments [51]. Archaeal histones have been recently proposed to be associated with the genome *in vivo* in different multimeric states, as single dimers or multimers [52], a property reminiscent of that described previously for simple DNA-bending proteins. The X-ray crystallography structure of archaeal-histone DNA complexes [53] reveals that histones associate side-by-side, giving rise to a hypernucleosome structure, in which DNA is wrapped in a left-handed manner around an "endless" histone core [44, 53]. Distinct from the other characterized DNA-bending proteins, side-by-side association of histones, in this case, yields a tenfold linear compaction. It is tempting to consider the archaeal hypernucleosome and the eukaryotic nucleosome as the product of simple DNA-bending proteins evolved to self-associate.

9.3.2 Genome organization

While the proteins described above contribute to genome compaction, a plastic genome responsive to ambient signals, as well as robust to the consequences of endogenous and exogenous DNA damage, requires functional organization. Such functional organization is achieved by proteins capable of mediating DNA–DNA interactions [54], which due to their global structural impact also affect gene activity. The prototypical example of a protein with such functionality is the H-NS protein, conserved among gram-negative bacteria such as *E. coli*. H-NS was originally described as a major constituent of the bacterial nucleoid [55] and later identified as a protein involved in regulation (usually repression) of numerous genes [56]. Characteristic of the protein is its modular build-up and resulting bi/multivalency. Despite its small size (15.6 kDa) the protein harbors at least three functional domains, a dimerization domain, a multimerization domain, and a DNA-binding domain [57]. Recently, a fourth functional domain, involved in sensing physico-chemical signals was identified [58]. At low concentrations in solution, the protein exists as a dimer and is thus bivalent. Self-association along DNA leads to stiffening of the DNA [58–60]. This multivalent DNA-bound multimer is capable of interaction with DNA provided in *trans* [61, 62]. This property of the protein is referred to as "DNA bridging".

H-NS can bind along an extended region of the chromosome, and represses 5–10% of genes in *E. coli*, with a preference for genes that are recently gained from horizontal transfers [63–67]. Many of these genes can be then de-repressed, leading to gene expression, in response to environmental signals [68, 69]. The mechanism by which H-NS exerts gene repression has been under scrutiny and discussion for over three decades. Rather than a single mechanism operating at all H-NS repressed genes, repression can, in fact, be achieved through a number of mechanisms: 1) occlusion of the transcription machinery from the transcription start site and regulatory elements; 2) trapping of the transcription machinery at the transcription start site, not allowing transcription to take off; and 3) permitting transcription to take off, but hindering the progression of the transcription machinery by binding intragenically along the transcribed gene [10, 70]. Note that both modes of H-NS binding, the lateral filament, as well as the bridged complex, are expected to be able to occlude the transcription machinery from the transcription start site. There is possibly a difference in the strength of repression for the two modes of binding *in vivo*, separating different "classes" of H-NS repressed genes. Trapping relies on DNA bridging of upstream and downstream elements. Progression could theoretically be hindered by both lateral filaments and bridged complexes, but experimentally it has been shown that the lateral filaments do not provide a barrier to progression of the transcription machinery [71].

Additionally, the bi/multivalency of the protein allows it to form repressive loops on short scales, on the scale of genes and operons, described above, but the protein has also been implied in mediating long-distance loops [72]. This model is based on the observation that the genome is divided into so-called topologically isolated domains [73–75], which could arise due to the formation of loops, as seen in classical images of isolated bacterial chromosomes [76]. The size of topologically isolated domains, as well as that of the observed loops, is on average 10 kbp, which corresponds with the average spacing between H-NS bound regions along the genome *in vivo* [77]. According to this model global organization of the genome in loops coincides with clustering of H-NS repressed genes. This model has not been verified *in vivo* yet, but an approach combining chromosome conformation capture and specific pull-down of H-NS-bound-DNA could shed light on this issue.

In addition to H-NS, there are other proteins capable of mediating long-range genomic interactions, SMC-like proteins [24, 25, 31]. These proteins are much larger in size than H-NS, exist as ring-like structures, and are conserved throughout all domains of life. Finally, an emerging theme in the bacterial field is that many nucleoid-associated proteins exhibit different structural properties (combining bending and stiffening, or bending, stiffening, and bridging), depending on the DNA-binding density [5].

9.3.3 "Smart polymer" behavior

The combination of targeted binding, sensory behavior, regulatory capabilities and role in the global folding of the genome of nucleoid-associated proteins like H-NS suggests the existence of a complex but highly coordinated interplay of global and local factors orchestrating the physiological response of a cell from both the biological and the physical viewpoint. This has been described by the technological metaphor of a "smart polymer" [4] for the behavior of the bacterial nucleoid. In soft-matter physics and engineering, smart polymers are designed to respond to a wide range of external stimuli and perform a wide range of mechanical and chemical tasks. The bacterial nucleoid is similar, in that it modulates its compaction and conformation according to growth conditions, cell-cycle stage, and environmental and internal cues. We are just beginning to understand this behavior [7]. From a theoretical standpoint, capturing the mechanisms coupling the physical properties of the chromosome and its surrounding medium to the dynamic changes in cell composition and chromosome state in response to conditions, and the internal/external environment is the main challenge for the coming years.

9.4 DYNAMICS OF CHROMOSOMES AND THEIR EMBEDDING MEDIUM

At faster time scales than those of the orchestrated movements, such as chromosome segregation, macromolecules immersed in the cytoplasm perform random jiggly movements, and such fluctuations carry a wealth of information on the physics of the intra-cellular environment [78]. In bacteria, while active forces and processes may be very relevant, active directed transport is less prevalent than in eukaryotes, and molecular interactions are mainly limited by cytoplasmic diffusion. Yet, it is becoming more and more evident that short-time fluctuations play an important role in eukaryotes as well [79, 80]. In spite of the complexity and diversity of the cytosolic environment, the observed diffusion of tracer particles across cells, conditions, and species presents certain common statistical features, which has spawned research on a common set of analytical tools. Such robustness also encourages a modeling approach where the most relevant aspects of the medium are taken into account, and more specific details can be left out. This lies at the core of a framework based on statistical physics, which is guiding important discoveries on the physical nature of the bacterial cytoplasm and chromosome.

9.4.1 Subdiffusion of chromosomal loci and particles

The classic theory of a particle suspended in a simple fluid is that of Brownian motion. The hallmark of Brownian behavior is the linear scaling of fluctuations with time. A commonly used measure of the fluctuations, computed from the individual trajectories $r(t)$, is the mean squared displacement, defined as

$$\text{MSD}(\tau) = \left\langle \left[r(t+\tau) - r(t) \right]^2 \right\rangle_t, \tag{9.1}$$

where the average is performed by using all time points t available and all trajectories. For Brownian motion, $\text{MSD}(\tau) = D\tau$, where the diffusivity $D = \mu k_B T$

depends on the mobility μ and the temperature T. Additionally, the step-size distribution is Gaussian at all time scales τ. Both the power-law scaling of the MSD and this scale invariance are essentially consequences of the Central Limit Theorem. However, experiments in many areas have shown that such a description is inadequate for more complex fluids and materials, where the relation between the MSD and time deviates from linearity, often taking the form of a power law

$$\mathrm{MSD}(\tau) = D_{app}\tau^{\alpha}. \tag{9.2}$$

Here D_{app} is an apparent diffusion coefficient, and the exponent α measures the deviation from normal diffusion [81]. Sub-diffusion, signaled by a less-than-unity α, has been observed *in vivo* both for bacterial chromosomal loci [78, 82, 83] and for cytoplasmic RNA-protein particles [83–85] as well as for complexes such as replisomes [86]. The definition of MSD is useful for single-molecule tracking experiments, e.g., in time-lapse fluorescence microscopy, giving access to the full traces $r(t)$. The relevant quantities α and D_{app} can also be obtained with other experimental techniques, such as fluorescence recovery after photobleaching or fluorescence correlation spectroscopy [87].

9.4.2 Challenges to the interpretation of data

The precise physical picture emerging from the experiments focusing on bacterial nucleoid dynamics is unclear. Roughly, we should picture it as a complex heterogeneous fluid where the folded supercoiled chromosome structure, the crowding from cytosolic macromolecules, and the presence of diffusing elastic elements and nucleoid-associated proteins contribute to the overall sub-diffusion of both chromosomal loci and intracellular particles [40, 83, 84,]. Recent studies found systematic dependencies of short-time mobility indicators with nucleoid structure and cell physiology. For instance, it was observed that the short-time diffusivity of genomic loci in *E. coli* depends on their chromosomal coordinate and subcellular localization [40], and is affected by sublethal doses of clinically relevant antibiotics [88]. Therefore, an important challenge is to disentangle the different factors contributing to the short-time dynamics, and to characterize their signatures in the data. Much effort is being put into theoretical analyses and computer simulations of models probing different physical scenarios, with the aim of establishing a corpus of model-guided data-analysis techniques [78].

The classic Rouse model [8, 89] may provide a first simplified description of the chromosome as an ideal polymer chain fluctuating in a simple fluid. This model predicts $\alpha = 1/2$ and normally distributed fluctuations of the steps $r(t + \delta t) - r(t)$. However, the exponents measured in most experiments are sensibly smaller than this prediction, and the motion of cytoplasmic particles, not attached to the chromosome, appears to be itself subdiffusive. These observations led to a more sophisticated model where a non-interacting polymer chain is immersed in a viscoelastic medium, i.e., a fluid responding both viscously and elastically to stress. This model gives predictions for the values that the anomalous exponent takes for chromosome-bound tracers (α_{chr}) and for cytosolic particles (α_{cyt}). In particular, it predicts that the ratio $\alpha_{cyt} / \alpha_{chr}$ is 2 in an ideal case [89].

The picture has been further expanded by the recent theoretical discovery that a compacted chromosome, arranged in a fractally organized globule, but immersed in a simple viscous fluid, would yield subdiffusion exponents in agreement with those observed in bacteria (around 0.4) [8, 90]. Experimentally, it is still very unclear to what extent the observed anomalous diffusion is due to the dynamics of the folded chromosome or to the complexity of the cytosolic medium. A related question regards the degree of viscoelasticity of the intracellular medium. A clear elastic response is observed by measuring the velocity autocorrelation function of tracer trajectories [91], which robustly presents anticorrelation peaks (see Figure 9.2). However, it is not clear whether such behavior is a signature of the presence of elastic elements (such as the chromosome) or could arise from the complex interaction with a "disordered" environment (possibly as a consequence of high molecular crowding, see Section 9.4.3). Further information on the rheology of the medium could be extracted by considering the full space-time dependence of correlation functions.

On the theoretical side, attempts have been made to develop a polymer dynamics framework predicting the correlation in the motion of distant chromosomal loci, taking into account their connection through the DNA backbone and the subdiffusive properties of the embedding medium [91, 92]. This approach predicts that the time scale of stress propagation through the polymer segment between a pair of loci can be obtained by analyzing time-delayed correlations between the two loci. The original theory by Lampo and coworkers [91] considers the effect of cytoplasm viscoelasticity on an unorganized polymer (a Gaussian chain) representing the chromosome. More recent work extends this framework to the case of a polymer with a complex folded state, describing the chromosome, immersed in a viscoelastic medium [92]. This model can disentangle the intrinsic contribution to the dynamics induced by the chromosome folding from the specific properties of the cytoplasm. These theoretical studies indicate that dynamic measurements from multiple tagged chromosomal loci and cytoplasmic particles are a way to resolve the question of the relative contributions of chromosome and surrounding medium to the physical properties of the cytoplasm as a complex fluid. In particular, correlations in the fluctuations of two tagged loci with controllable arc-length distance along the chromosome are expected to carry rich information on stress propagation through the polymer backbone. Such measurements will allow a quantitative and independent assessment of both the local degree of chromosomal compaction and the rheology of the cytoplasm or nucleoplasm [91, 92] (see Figure 9.2).

9.4.3 New concepts and models

Particle dynamics in the bacterial chromosome-cytosol environment were initially approached theoretically mostly as an equilibrium thermal process. However, recent advances are depicting a much more heterogeneous situation, where non-equilibrium, possibly athermal, processes superimpose on the

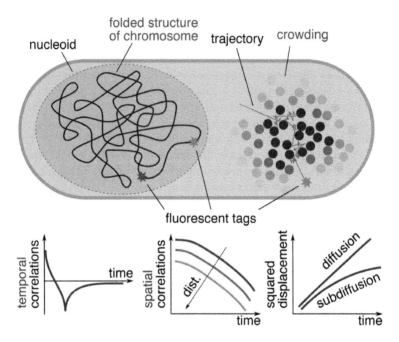

Figure 9.2 Sketch of the main determinants of chromosome and cytoplasmic particle dynamics in bacteria. The folded structure of the chromosome in the nucleoid is connected to its local (short-time) dynamics, and to the crowding levels in the cytoplasm [8]. Crowding affects biochemical processes and mobility of particles, as well as causing polymer collapse. Dynamic tracking of single tags gives access to temporal correlations and displacements. Tracking multiple intracellular tags can give access to spatio-temporal correlations. Such correlations starting to be explored theoretically, and are expected to be an important tool for investigating the physical properties of both chromosome and cytoplasm.

subdiffusive background, yielding a rich — and largely unexplored — spectrum of crossovers. ATP-dependent enzymatic activity has been shown to affect the motion of chromosomal loci in *E. coli*, giving rise to a "super-thermal" response of fluctuations at increasing temperature [80]. Cellular metabolism was shown to also have a fluidizing effect on the cytoplasm alone (see the following section), allowing larger particles to become mobile, whereas they are trapped in ATP-depleted cells (an observation that was interpreted as a sign of glassy behavior) [84, 85].

In *E. coli*, rapid fluctuations and relocations of nucleoid mass [93] were detected, as well as sporadic quasi-ballistic motions of chromosomal loci, qualitatively different from the subdiffusive background [94]. Phenomenological modeling analysis shows that the intensity and typical time scales of the forces related to the fast relocations vary across the chromosome and are associated with increased noise levels [95]. More generally, these rapid movements may emerge from global or local chromosomal rearrangements, stress relocation due to ruptures of a self-tethered structure, the action of molecular motors, etc. [7]. Their precise nature is not known and they may emerge from several of these sources at the same time. Additionally, such active movements have been proposed to have some periodicity or characteristic time scale and may play an important role in cell cycle progression [7].

The methods and concepts developed in this context could be useful in studies specifically targeted at eukaryotes, where they can be applied to disentangle the contributions of active movements and basal (sub)diffusion. For instance, a crossover between transient subdiffusion at short time spans and nearly normal diffusion at long time spans was found by looking at telomere-bound proteins in mammalian cells [96]. Very recently, the origin of such behavior has been connected to the action of stochastic forces of cytoskeletal origin [79].

9.5 ROLE OF MOLECULAR CROWDING

The problems discussed in the previous section are deeply interconnected with the open question of the physical consequences of molecular crowding in the bacterial cell, and in particular its impact on chromosome dynamics and organization. The high concentration of charged and polydisperse macromolecules in a cell affects both cytoplasm and chromosome. Such macromolecular crowding is responsible for a number of physical effects of entropic and energetic nature, which are still incompletely understood (see Figure 9.2). Crowding agents can favor molecular association events and accelerate molecular reactions, but also dramatically decrease molecular mobility in a strongly size-dependent way. Depletion interactions, a consequence of crowding, create an effective short-range attraction between larger macromolecules. Such interactions are felt by large molecular assemblies in the presence of smaller particles, and are due to a reduction in the total excluded volume by the latter.

9.5.1 Compaction by molecular crowding

DNA condensation by crowding agents such as PEG or dextran can be observed *in vitro* under controlled conditions. In bacteria such as *E. coli*, the chromosome occupies roughly one sixth of the available cell volume, and upon cell lysis it expands to about five times the cell radius, while many of the nucleoid-associated proteins remain bound [97]. These observations already suggest that cytoplasmic crowding plays a major role in chromosome compaction.

This was hypothesized early on by Odijk [98], and observed in purified nucleoids from both wild-type and mutants lacking nucleoid-associated proteins [99–101], leading to the hypothesis that the effects of crowding on compaction are independent of the NAP composite background. In a more recent study by Pelletier and coworkers [97], PEG was added to purified nucleoids in microfluidic channels and the compaction transition was quantified, as well as the response of nucleoids to the applied force. The results of this work suggest that the *in vivo* levels of crowding make the chromosome very close to the transition to the compact state. This speculation is intriguing, as this condition would be close to the optimal compromise between accessibility and size.

After the pioneering work by Odijk, simulation and theoretical studies have approached the question only in recent years [102]. A combined theoretical and computational study considered the collapse of a polymer due to a two-body short-range attraction describing crowding and discrete bridging representing the action of bridging proteins such as H-NS [103]. The authors find that bridging leads to a sudden collapse transition, and to a spontaneous tendency to fold into separate compartmentalized domains without the need of intra-specific interactions. These domains are stable under compaction due to depletion interactions, provided that such interactions do not become stronger than bridging.

However, crowding is not a two-body force. Molecular dynamics simulations with simple but explicit descriptions of crowders are available, and allow, for example, the study of the role of size and polydispersity of crowding agents on chromosome compaction [104, 105]. These studies indicate that the compaction transition is continuous, and occurs at lower volume fractions, but larger concentrations with smaller crowding agents. A recent study [106] followed the same approach to describe the compression experiments by Pelletier and coworkers. They found that the osmotic compression force by crowders dominates over the entropic expansion force exerted by the compressed DNA, even at low density.

Depletion interactions are expected to occur between any large objects, such as ribosomes [107], and may also be important in modulating bridging interactions [103, 108]. More in general, depletion provides a likely explanation for the formation of several macromolecular assemblies such as transcription factories [107, 109], which are also observed in bacteria [110, 111]. A recent study combining experiments with theoretical arguments has looked at the combined collapse of purified nucleoids by the bridging protein H-NS and a crowding agent (PEG) [112]. While H-NS alone had little impact on nucleoid size, the addition of this bridging protein strongly affected nucleoid collapse by PEG. The authors interpret this effect as an enhancement of depletion due to the increased effective diameter of the DNA helix, causing an increase of the self-attraction induced by PEG. They also conclude that in presence of H-NS, the free energy of the nucleoid depends so weakly on H-NS concentration, that the latter is essentially not relevant for the transition. Hence, while the cooperation between crowders and the bridging protein is evident, the predicted effect is binary, and not gradual or tunable by protein amounts.

9.5.2 Crowding and cytoplasmic mobility

Crowding also affects cytosol dynamics, and, as discussed in the previous section, this has consequences on the mobility of chromosomal loci. The physical origin of the anomalous diffusion behavior of both chromosomal loci and cytoplasmic particles in bacteria is subject of an intense debate. As mentioned above, tracked cytoplasmic particles of different size and origin on the minute time scale show heterogeneous (and bimodal) mobility distributions [84], as well as highly probe-size-dependent motion, typical of colloidal glasses. ATP-dependent processes appear to be the main driver of the mobility of such cytoplasmic particles, as they drastically reduce their motion under ATP depletion. More recent work [85] extracted further details of this complex picture for cytoplasmic objects. These tracked objects show velocity–velocity correlation functions with an anticorrelated part, pointing to viscoelastic behavior. Additionally, the ensemble of particle displacements reveals a non-Gaussian Laplace-like distribution at a wide range of time scales, rather than the Gaussian distribution predicted by the central limit theorem. The non-Gaussian behavior may emerge from both heterogeneity across trajectories of different particles (e.g. due to spatially heterogeneous confinement) and dynamic heterogeneity (disorder) along single trajectories. The emergent picture is that disorder due to extreme crowding may be one of the main drivers of mobility of any cytosolic object, but on the theoretical side, we lack a more specific description of the processes at play.

9.5.3 Crowding as a player in cell physiology

A final, more biological comment goes to the wider context of the role of crowding (and hence of chromosome organization). It is well known from classic work in bacterial cell physiology that bacteria perform strong buffering of total molecular concentration [113]. More recent work has linked this behavior to several key cellular properties, leading to the hypothesis that the level of crowding in the cytoplasm must be under tight homeostatic control [114], and cells actively maintain the overall concentration of macromolecules within a narrow range.

Intriguingly, crowding and chromosome compaction can also be an integral *effector* of physiological changes. At the theoretical level, this factor has been addressed in the framework of the quantitative approach to bacterial physiology [115]. Experimentally, the joint physiological switch affecting chromosome and total cytoplasmic density remains relatively unexplored, but recent intriguing evidence appears to come from work on eukaryotes. For example, a recent study in yeast [116] has shown that glucose starvation triggers cell and vacuole size changes that slow down the physiology making the cells quiescent. Such a chain of events clearly involves crowding (measured as chromosomal and cytoplasmic mobility using tagged particles). Another recent (unpublished) study [117] in eukaryotes has shown that the well known mTORC1 pathway, the primary pathway for cell growth, tunes macromolecular crowding and therefore the size-dependent mobility of intracellular macromolecules and intracellular phase separation through the regulation of ribosome concentration.

In conclusion, the emergent picture is that crowding may be a constraint for several core cellular processes, but is also a regulatory tool to drive chromosome organization, and this question is still under-explored in eukaryotes. Conversely, recent work on eukaryotes focuses more on the physiological consequences of crowding, which appear relatively less explored in bacteria. The physical nature of the crowded cytoplasm/nucleoid system remains a largely open problem, to address both experimentally and theoretically in bacteria.

9.6 CONCLUSIONS

Overall, the organization and dynamics of prokaryotic and eukaryotic chromosomes bear a set of common features, ranging from a compartmentalized structure, to their short-time dynamics, and to their dynamics on the scale of cell cycle transitions, such as replication initiation, and chromosome segregation [7]. Our knowledge is still very limited and many central questions appear to be open, such as the physical nature of the cytoplasm and nucleoplasm, and their interplay with cell decisions, homeostasis, and chromosome state and dynamics. The "smart-polymer" paradigm for the chromosome appears to be a promising guiding principle for both kingdoms, as increasing evidence supports the picture of a joint role of physical organization and biological degrees of freedom to perform key physiological changes in the cell.

BIBLIOGRAPHY

1. E. P. C. Rocha. The organization of the bacterial genome. *Annu. Rev. Genet.*, 42:211–233, 2008.
2. C. J. Dorman. Nucleoid-associated proteins and bacterial physiology. *Adv. Appl. Microbiol.*, 67:47–64, 2009.
3. M. Cosentino Lagomarsino, O. Espeli, and I. Junier. From structure to function of bacterial chromosomes: Evolutionary perspectives and ideas for new experiments. *FEBS Lett*, 589(20 Pt A):2996–3004, 2015.
4. V. F. Scolari, B. Sclavi, and M. C. Lagomarsino. The nucleoid as a smart polymer. *Front. Microbiol.*, 6: 424, 2015.
5. R. T. Dame and M. Tark-Dame. Bacterial chromatin: Converging views at different scales. *Curr. Opin. Cell Biol.*, 40:60–65, 2016.
6. A. Badrinarayanan, T. B. Le, and M. T. Laub. Bacterial chromosome organization and segregation. *Annu. Rev. Cell Dev. Biol.*, 31:171–199, 2015.
7. N. Kleckner, J. K. Fisher, M. Stouf, M. A. White, D. Bates, and G. Witz. The bacterial nucleoid: Nature, dynamics and sister segregation. *Curr. Opin. Microbiol.*, 22:127–137, 2014.
8. V. G. Benza, B. Bassetti, K. D. Dorfman, V. F. Scolari, K. Bromek, P. Cicuta, and M. C. Lagomarsino. Physical descriptions of the bacterial nucleoid at large scales, and their biological implications. *Rep. Prog. Phys.*, 75(7):076602, 2012.
9. S. Rimsky and A. Travers. Pervasive regulation of nucleoid structure and function by nucleoid-associated proteins. *Curr. Opin. Microbiol.*, 14(2):136–141, 2011.
10. S. C. Dillon and C. J. Dorman. Bacterial nucleoid-associated proteins, nucleoid structure and gene expression. *Nat. Rev. Microbiol.*, 8(3):185–195, 2010.
11. E. Toro and L. Shapiro. Bacterial chromosome organization and segregation. *Cold Spring Harb. Perspect. Biol.*, 2(2):a000349, 2010.

12. V. S. Lioy, A. Cournac, M. Marbouty, S. Duigou, J. Mozziconacci, O. Espeli, F. Boccard, and R. Koszul. Chromosome replication and the division cycle of *Escherichia coli* b/r. *Cell* 172(4):771–783, 2018.

13. J. Dekker, K. Rippe, M. Dekker, and N. Kleckner. Capturing chromosome conformation. *Science*, 295(5558):1306–1311, 2002.

14. J. Dekker. Gene regulation in the third dimension. *Science*, 319(5871):1793–1794, 2008.

15. E. Shapiro, T. Biezuner, and S. Linnarsson. Single-cell sequencing-based technologies will revolutionize whole-organism science. *Nat. Rev. Genet.*, 14(9):618–630, 2013.

16. O. Medalia, I. Weber, A. S. Frangakis, D. Nicastro, G. Gerisch, and W. Baumeister. Macromolecular architecture in eukaryotic cells visualized by cryoelectron tomography. *Science*, 298(5596):1209–1213, 2002.

17. W.-K. Cho, N. Jayanth, B. P. English, T. Inoue, J. O. Andrews, W. Conway, J. B. Grimm, J.-H. Spille, L. D. Lavis, T. Lionnet, and L.L. Cisse.RNA polymerase II cluster dynamics predict mRNA output in living cells. *Elife*, 5:e13617, 2016.

18. M. A. Umbarger, E. Toro, M. A. Wright, G. J. Porreca, D. Ba, S.-H. Hong, M. J. Fero, L. J. Zhu, M. A. Marti-Renom, H. H. McAdams, L. Shapiro, J. Dekker, and G. M. Church. The three-dimensional architecture of a bacterial genome and its alteration by genetic perturbation. *Mol. Cell*, 44(2):252–264, 2011.

19. X. Wang, T. B. Le, B. R. Lajoie, J. Dekker, M. T. Laub, and D. Z. Rudner. Condensin promotes the juxtaposition of DNA flanking its loading site in *Bacillus subtilis*. *Genes Dev.*, 29(15):1661–1675, 2015.

20. M. Marbouty, A. Le Gall, D. I. Cattoni, A. Cournac, A. Koh, J.-B. Fiche, J. Mozzi- conacci, H. Murray, R. Koszul, and M. Nollmann. Condensin-and replication-mediated bacterial chromosome folding and origin condensation revealed by HI-C and super- resolution imaging. *Mol. Cell*, 59(4):588–602, 2015.

21. K. Kimura and T. Hirano. ATP-dependent positive supercoiling of DNA by 13s condensin: A biochemical implication for chromosome condensation. *Cell*, 90(4):625–634, 1997.

22. Z. M. Petrushenko, Y. Cui, W. She, and V. V. Rybenkov. Mechanics of DNA bridging by bacterial condensin mukbef in vitro and in singulo. *The EMBO journal*, 29(6):1126–1135, 2010.

23. S. Gruber, J.-W. Veening, J. Bach, M. Blettinger, M. Bramkamp, and J. Errington. Interlinked sister chromosomes arise in the absence of condensin during fast replication in b. subtilis. *Curr. Biol*, 24(3):293–298, 2014.

24. X. Wang, O. W. Tang, E. P. Riley, and D. Z. Rudner. The SMC condensin complex is required for origin segregation in *Bacillus subtilis*. *Curr. Biol.*, 24(3):287–292, 2014.

25. T. B. K. Le, M. V. Imakaev, L. A. Mirny, and M. T. Laub. High-resolution mapping of the spatial organization of a bacterial chromosome. *Science*, 342(6159):731–734, 2013.

26. J. H. Gibcus, K. Samejima, A. Goloborodko, I. Samejima, N. Naumova, M. Kanemaki, L. Xie, J. R. Paulson, W. C. Earnshaw, L. A. Mirny, et al. Mitotic chromosomes fold by condensin-dependent helical winding of chromatin loop arrays. *bioRxiv*, 174649, 2017.

27. Y. Kakui, A. Rabinowitz, D. J. Barry, and F. Uhlmann. Condensin-mediated remodeling of the mitotic chromatin landscape in fission yeast. *Nature Genetics*, 49(10):1553–1557, 2017.

28. L. Lazar-Stefanita, V. F. Scolari, G. Mercy, H. Muller, T. M. Guerin, A. Thierry, J. Mozziconacci, and R. Koszul. Cohesins and condensins orchestrate the 4d dynamics of yeast chromosomes during the cell cycle. *The EMBO Journal*, 36(18):2684-2697, 2017.

29. T. Nagano, Y. Lubling, C. Varnai, C. Dudley, W. Leung, Y. Baran, N. M. Cohen, S. Wingett, P. Fraser, and A. Tanay. Cell cycle dynamics of chromosomal organisation at single-cell resolution. *Nature* 5;547(7661):61–67, 2017. doi: 10.1038/nature23001.nature23001 page 094466, 2016.

30. N. Naumova, M. Imakaev, G. Fudenberg, Y. Zhan, B. R. Lajoie, L. A. Mirny, and J. Dekker. Organization of the mitotic chromosome. *Science*, 342(6161):948–953, 2013.

31. S. A. Schalbetter, A. Goloborodko, G. Fudenberg, J.-M. Belton, C. Miles, M. Yu, J. Dekker, L. Mirny, and J. Baxter. Smc complexes differentially compact mitotic chromosomes according to genomic context. *Nat. Cell Biol.*, 19:1071, 2017.

32. D. Magnan and D. Bates. Regulation of DNA replication initiation by chromosome structure. *J. Bacteriol.*, 197(21):3370–3377, 2015.

33. M.-E. Val, M. Marbouty, F. de Lemos Martins, S. P. Kennedy, H. Kemble, M. J. Bland, C. Possoz, R. Koszul, O. Skovgaard, and D. Mazel. A checkpoint control orchestrates the replication of the two chromosomes of *Vibrio cholerae*. *Sci. Adv.*, 2(4):e1501914, 2016.

34. T. Venkova-Canova and D. K. Chattoraj. Transition from a plasmid to a chromosomal mode of replication entails additional regulators. *Proc. Natl. Acad. Sci.*, 108(15):6199–6204, 2011.

35. J. H. Baek and D. K. Chattoraj. Chromosome i controls chromosome ii replication in vibrio cholerae. *PLoS Genet.*, 10(2):e1004184, 2014.

36. O. Espeli, R. Borne, P. Dupaigne, A. Thiel, E. Gigant, R. Mercier, and F. Boccard. A matp-divisome interaction coordinates chromosome segregation with cell division in e. coli. *EMBO J.*, 31(14):3198–3211, 2012.

37. M. Trussart, E. Yus, S. Martinez, D. Bau, Y. O. Tahara, T. Pengo, M. Widjaja, S. Kretschmer, J. Swoger, S. Djordjevic, et al. Defined chromosome structure in the genome-reduced bacterium *Mycoplasma pneumoniae*. *Nature Communications*, 8: 14665, 2017.

38. M. Valens, S. Penaud, M. Rossignol, F. Cornet, and F. Boccard. Macrodomain organization of the *Escherichia coli* chromosome. *EMBO J.*, 23(21):4330–4341, 2004.

39. R. Mercier, M.-A. Petit, S. Schbath, S. Robin, M. E. Karoui, F. Boccard, and O. Espeli. The matp/mats site-specific system organizes the terminus region of the e. coli chromosome into a macrodomain. *Cell*, 135(3):475–485, 2008.

40. A. Javer, Z. Long, E. Nugent, M. Grisi, K. Siriwatwetchakul, K. D. Dorfman, P. Cicuta, and M. Cosentino Lagomarsino. Short-time movement of e. coli chromosomal loci depends on coordinate and subcellular localization. *Nat. Commun.*, 4:3003, 2013.

41. R. T. Dame. The role of nucleoid-associated proteins in the organization and compaction of bacterial chromatin. *Mol. Microbiol.*, 56(4):858–870, 2005.

42. M. S. Luijsterburg, M. F. White, R. van Driel, and R. T. Dame. The major architects of chromatin: architectural proteins in bacteria, archaea and eukaryotes. *Crit. Rev. Biochem. Mol. Biol.*, 43(6):393–418, 2008.

43. E. Peeters, R. P. Driessen, F. Werner, and R. T. Dame. The interplay between nucleoid organization and transcription in archaeal genomes. *Na.t Rev. Microbiol.*, 13(6):333–341, 2015.

44. B. Henneman, M. van Emmerik, H. van Ingen, and R. T. Dame. Structure and function of archaeal histones. *Biophys. J.*, 114(Special Issue 3):446a-446a, 2018.

45. K. Sandman and J. N. Reeve. Archaeal histones and the origin of the histone fold. *Curr. Opin. Microbiol*, 9(5):520–525, 2006.

46. K. K. Swinger and P. A. Rice. Ihf and hu: Flexible architects of bent DNA. *Curr. Opin. Struct. Biol.*, 14(1):28–35, 2004.

47. J. van Noort, S. Verbrugge, N. Goosen, C. Dekker, and R. T. Dame. Dual architectural roles of hu: formation of flexible hinges and rigid filaments. *Proc. Natl. Acad. Sci. USA*, 101(18):6969–6974, 2004.

48. R. Hurme and M. Rhen. Temperature sensing in bacterial gene regulation-what it all boils down to. *Mol. Microbiol.*, 30(1):1–6, 1998.

49. D. Skoko, B. Wong, R. C. Johnson, and J. F. Marko. Micromechanical analysis of the binding of DNA-bending proteins hmgb1, nhp6a, and hu reveals their ability to form highly stable DNA-protein complexes. *Biochemistry*, 43(43):13867–13874, 2004.

50. R. P. Driessen and R. T. Dame. Structure and dynamics of the crenarchaeal nucleoid. *Biochem. Soc. Trans.*, 41(1):321–325, 2013.

51. R. P. Driessen, S. N. Lin, W. J. Waterreus, A. L. van der Meulen, R. A. van der Valk, N. Laurens, G. F. Moolenaar, N. S. Pannu, G. J. Wuite, N. Goosen, and R. T. Dame. Diverse architectural properties of sso10a proteins: Evidence for a role in chromatin compaction and organization. *Sci. Rep.*, 6:29422, 2016.

52. H. Maruyama, J. C. Harwood, K. M. Moore, K. Paszkiewicz, S. C. Durley, H. Fukushima, H. Atomi, K. Takeyasu, and N. A. Kent. An alternative beads-on-a- string chromatin architecture in thermococcus kodakarensis. *EMBO Rep.*, 14(8):711–717, 2013.

53. F. Mattiroli, S. Bhattacharyya, P. N. Dyer, A. E. White, K. Sandman, B. W. Burkhart, K. R. Byrne, T. Lee, N. G. Ahn, T. J. Santangelo, J. N. Reeve, and K. Luger. Structure of histone-based chromatin in archaea. *Science*, 357(6351):609–612, 2017.

54. R. A. van der Valk, J. Vreede, F. Cremazy, and R. T. Dame. Genomic looping: a key principle of chromatin organization. *J. Mol. Microbiol. Biotechnol*, 24:344–359, 2014.

55. A. Spassky, S. Rimsky, H. Garreau, and H. Buc. H1a, an e. coli DNA-binding protein which accumulates in stationary phase, strongly compacts DNA in vitro. *Nucleic Acids Res.*, 12(13):5321–5340, 1984.

56. S. H. Hong and H. H. McAdams. Compaction and transport properties of newly replicated caulobacter crescentus DNA. *Mol. Microbiol.*, 82(6):1349–1358, 2011.

57. C. J. Dorman, J. C. Hinton, and A. Free. Domain organization and oligomerization among H-NS-like nucleoid-associated proteins in bacteria. *Trends Microbiol.*, 7(3):124–8, 1999.

58. R. A. van der Valk, J. Vreede, L. Qin, G. F. Moolenaar, A. Hofmann, N. Goosen, and R. T. Dame. Mechanism of environmentally driven conformational changes that modulate H-NS DNA-bridging activity. *Elife*, 6: e27369, 2017.

59. R. Amit, A. B. Oppenheim, and J. Stavans. Increased bending rigidity of single DNA molecules by H-NS, a temperature and osmolarity sensor. *Biophys. J.*, 84(4):2467–2473, 2003.

60. Y. Liu, H. Chen, L. J. Kenney, and J. Yan. A divalent switch drives H-NS/DNA-binding conformations between stiffening and bridging modes. *Genes Dev.*, 24(4):339–344, 2010.

61. R. T. Dame, C. Wyman, and N. Goosen. H-NS mediated compaction of DNA visualised by atomic force microscopy. *Nucleic Acids Res*, 28(18):3504–3510, 2000.

62. R. Schneider, R. Lurz, G. Luder, C. Tolksdorf, A. Travers, and G. Muskhelishvili. An architectural role of the *Escherichia coli* chromatin protein fis in organising DNA. *Nucleic Acids Res.*, 29(24):5107–5114, 2001.

63. C. J. Dorman. H-NS: A universal regulator for a dynamic genome. *Nat. Rev. Microbiol.*, 2(5):391–400, 2004.

64. C. J. Dorman. Genome architecture and global gene regulation in bacteria: making progress towards a unified model? *Nat. Rev. Microbiol.*, 11(5):349–355, 2013.

65. C. Kahramanoglou, A. S. N. Seshasayee, A. I. Prieto, D. Ibberson, S. Schmidt, J. Zim- mermann, V. Benes, G. M. Fraser, and N. M. Luscombe. Direct and indirect effects of H-NS and FIS on global gene expression control in *Escherichia coli*. *Nucleic Acids Res*, 39(6):2073–2091, 2011.

66. M. Zarei, B. Sclavi, and M. C. Lagomarsino. Gene silencing and large-scale domain structure of the e. coli genome. *Mol. BioSyst.*, 9(4):758–767, 2013.

67. R. Srinivasan, V. F. Scolari, M. C. Lagomarsino, and A. S. N. Seshasayee. The genome-scale interplay amongst xenogene silencing, stress response and chromosome architecture in *Escherichia coli*. *Nucleic Acids Res.* 43(1):295–308, 2015.

68. T. Atlung and H. Ingmer. H-NS: A modulator of environmentally regulated gene expression. *Molecular Microbiology*, 24(1):7–17, 1997.

69. F. Hfling and T. Franosch. Anomalous transport in the crowded world of biological cells. *Rep. Prog. Phys.*, 76(4):046602, 2013.

70. R. Landick, J. T. Wade, and D. C. Grainger. H-NS and RNA polymerase: a love-hate relationship? *Curr. Opin. Microbiol.*, 24C:53–59, 2015.

71. M. V. Kotlajich, D. R. Hron, B. A. Boudreau, Z. Sun, Y. Lyubchenko, and R. Landick. Bridged filaments of histone-like nucleoid structuring protein pause RNA polymerase and aid termination in bacteria. *Elife*, 4, 2015.

72. M. C. Noom, W. W. Navarre, T. Oshima, G. J. Wuite, and R. T. Dame. H-NS promotes looped domain formation in the bacterial chromosome. *Curr. Biol.*, 17(21):R913–R914, 2007.

73. C. D. Hardy and N. R. Cozzarelli. A genetic selection for supercoiling mutants of *Escherichia coli* reveals proteins implicated in chromosome structure. *Mol. Microbiol.*, 57(6):1636–1652, 2005.

74. F. Hommais, E. Krin, C. Laurent-Winter, O. Soutourina, A. Malpertuy, J. P. Le Caer, A. Danchin, and P. Bertin. Large-scale monitoring of pleiotropic regulation of gene expression by the prokaryotic nucleoid-associated protein, H-NS. *Mol. Microbiol.*, 40(1):20–36, 2001.

75. L. Postow, C. Hardy, J. Arsuaga, and N. Cozzarelli. Topological domain structure of the *Escherichia coli* chromosome. *Genes Dev*, 18(14):1766–1779, 2004.

76. R. Kavenoff and B. C. Bowen. Electron microscopy of membrane-free folded chromosomes from *Escherichia coli*. *Chromosoma*, 59(2):89–101, 1976.

77. W. W. Navarre, S. Porwollik, Y. Wang, M. McClelland, H. Rosen, S. J. Libby, and F. C. Fang. Selective silencing of foreign DNA with low gc content by the H-NS protein in salmonella. *Science*, 313(5784):236–238, 2006.

78. M. Gherardi and M. Cosentino Lagomarsino. *Procedures for Model-Guided Data Analysis of Chromosomal Loci Dynamics at Short Time Scales*, pages 291–307. Springer New York, New York, NY, 2017.

79. L. Stadler and M. Weiss. Non-equilibrium forces drive the anomalous diffusion of telomeres in the nucleus of mammalian cells. *New J. Phys.*, 19(11):113048, 2017.

80. S. C. Weber, A. J. Spakowitz, and J. A. Theriot. Nonthermal ATP-dependent fluctuations contribute to the in vivo motion of chromosomal loci. *Proc. Natl. Acad. Sci. USA*, 109(19):7338–7343, 2012.

81. R. Metzler. and J. Klafter. The random walk's guide to anomalous diffusion: a fractional dynamics approach. *Physics Reports*, 339:1–77, 2000.

82. O. Espeli, R. Mercier, and F. Boccard. DNA dynamics vary according to macrodomain topography in the e. coli chromosome. *Mol. Microbiol.*, 68(6):1418–1427, 2008.

83. S. C. Weber, A. J. Spakowitz, and J. A. Theriot. Bacterial chromosomal loci move subdiffusively through a viscoelastic cytoplasm. *Phys. Rev. Lett.*, 104(23):238102, 2010.

84. B. R. Parry, I. V. Surovtsev, M. T. Cabeen, C. S. O'Hern, E. R. Dufresne, and C. Jacobs-Wagner. The bacterial cytoplasm has glass-like properties and is fluidized by metabolic activity. *Cell*, 156(1):183–194, 2014.

85. T. J. Lampo, S. Stylianidou, M. P. Backlund, P. A. Wiggins, and A. J. Spakowitz. Cy- toplasmic RNA-protein particles exhibit non-gaussian subdiffusive behavior. *Biophys. J.*, 112(3):532–542, 2017.

86. R. Reyes-Lamothe, C. Possoz, O. Danilova, and D. J. Sherratt. Independent positioning and action of *Escherichia coli* replisomes in live cells. *Cell*, 133(1):90–102, 2008.

87. N. P. Higgins, X. Yang, Q. Fu, and J. R. Roth. Surveying a supercoil domain by using the gamma delta resolution system in *Salmonella typhimurium*. *J. Bacteriol.*, 178(10):2825–2835, 1996.

88. M. Wlodarski, B. Raciti, J. Kotar, M. C. Lagomarsino, G. M. Fraser, and P. Cicuta. Both genome and cytosol dynamics change in e. coli challenged with sublethal rifampicin. *Phys. Biol.*, 14(1):015005, 2017.

89. S. C. Weber, J. A. Theriot, and A. J. Spakowitz. Subdiffusive motion of a polymer composed of subdiffusive monomers. *Phys. Rev. E*, 82:011913, 2010.

90. M. V. Tamm, L. I. Nazarov, A. A. Gavrilov, and A. V. Chertovich. Anomalous diffusion in fractal globules. *Phys. Rev. Lett.*, 114:178102, 2015.

91. T. Lampo, A. Kennard, and A. Spakowitz. Physical modeling of dynamic coupling between chromosomal loci. *Biophys. J.*, 110(2):338–347, 2016.

92. K. Polovnikov, M. Gherardi, M. Cosentino-Lagomarsino, and M. Tamm. Folding and cytoplasm viscoelasticity contribute jointly to chromosome dynamics. *arXiv preprint arXiv:1703.10841*, 2017. *Phys. Rev. Lett.* 120, 088101, 2018.

93. J. K. Fisher, A. Bourniquel, G. Witz, B. Weiner, M. Prentiss, and N. Kleckner. Four-dimensional imaging of e. coli nucleoid organization and dynamics in living cells. *Cell*, 153(4):882–895, 2013.

94. A. Javer, N. J. Kuwada, Z. Long, V. G. Benza, K. D. Dorfman, P. A. Wiggins, P. Cicuta, and M. C. Lagomarsino. Persistent super-diffusive motion of *Escherichia coli* chromosomal loci. *Nat. Commun.*, 5:3854, 2014.

95. M. Gherardi, L. Calabrese, M. Tamm, and M. Cosentino Lagomarsino. Model of chromosomal loci dynamics in bacteria as fractional diffusion with intermittent transport. *Phys. Rev. E*, 96:042402, 2017.

96. I. Bronstein, Y. Israel, E. Kepten, S. Mai, Y. Shav-Tal, E. Barkai, and Y. Garini. Transient anomalous diffusion of telomeres in the nucleus of mammalian cells. *Phys. Rev. Lett.*, 103(1):018102, 2009.

97. J. Pelletier, K. Halvorsen, B.-Y. Ha, R. Paparcone, S. J. Sandler, C. L. Woldringh, W. P. Wong, and S. Jun. Physical manipulation of the *Escherichia coli* chromosome reveals its soft nature. *Proc. Natl. Acad. Sci. USA*, 109(40):E2649–E2656, 2012.

98. T. Odijk. Osmotic compaction of supercoiled DNA into a bacterial nucle- oid. *Biophys. Chem.*, 73(1–2):23–9, 1998.

99. S. B. Zimmerman. Studies on the compaction of isolated nucleoids from *Escherichia coli. J. Struct. Biol.*, 147(2):146–58, 2004.

100. S. B. Zimmerman. Shape and compaction of *Escherichia coli* nucleoids. *J. Struct. Biol.*, 156(2):255–261, 2006.

101. S. B. Zimmerman. Cooperative transitions of isolated *Escherichia coli* nucleoids: implications for the nucleoid as a cellular phase. *J. Struct. Biol.*, 153(2):160–175, 2006.

102. R. de Vries. DNA condensation in bacteria: Interplay between macro-molecular crowding and nucleoid proteins. *Biochimie*, 92(12):1715–1721, 2010.

103. V. F. Scolari and M. Cosentino Lagomarsino. Combined collapse by bridging and self-adhesion in a prototypical polymer model inspired by the bacterial nucleoid. *Soft Matter*, 11(9): 1677–1687, 2015.

104. J. Kim, C. Jeon, H. Jeong, Y. Jung, and B.-Y. Ha. A polymer in a crowded and confined space: effects of crowder size and poly-dispersity. *Soft Matter*, 11:1877–1888, 2015.

105. T. Shendruk, M. Bertrand, H. deHaan, J. Harden, and G. Slater. Simulating the entropic collapse of coarse-grained chromosomes. *Biophys. J.*, 108(4):810–820, 2015.

106. M. C. F. Pereira, C. A. Brackley, J. S. Lintuvuori, D. Marenduzzo, and E. Orlandini. Entropic elasticity and dynamics of the bacterial chromosome: A simulation study. *The J. Chem. Phys.*, 147(4):044908, 2017.

107. D. Marenduzzo, C. Micheletti, and P. R. Cook. Entropy-driven genome organization. *Biophys. J.*, 90(10):3712–3721, 2006.

108. A. S. Wegner, S. Alexeeva, T. Odijk, and C. L. Woldringh. Characterization of Escherichia coli nucleoids released by osmotic shock. *J. Struct. Biol.*, 178(3):260–269, 2012.

109. D. Marenduzzo, K. Finan, and P. R. Cook. The depletion attraction: An underappreciated force driving cellular organization. *J. Cell Biol.*, 175(5):681–686, 2006.

110. J. E. Cabrera, C. Cagliero, S. Quan, C. L. Squires, and D. J. Jin. Active transcription of rRNA operons condenses the nucleoid in *Escherichia coli*: Examining the effect of transcription on nucleoid structure in the absence of transertion. *JBacteriol.*, 191(13):4180–4185, 2009.

111. D. J. Jin and J. E. Cabrera. Coupling the distribution of RNA polymerase to global gene regulation and the dynamic structure of the bacterial nucleoid in *Escherichia coli*. *J. Struct. Biol.*, 156(2):284–291, 2006.

112. A. S. Wegner, K. Wintraecken, R. Spurio, C. L. Woldringh, R. de Vries, and T. Odijk. Compaction of isolated *Escherichia coli* nucleoids: Polymer and H-NS protein synergetics. *J. Struct. Biol.*, 194(1):129–137, 2016.

113. H. Bremer and P. P. Dennis. Modulation of chemical composition and other parameters of the cell by growth rate. In F. C. Neidhardt, editor, *Escherichia coli and Salmonella*, pages 1553–1569. ASM Press Washington, DC, 1996.

114. J. van den Berg, A. J. Boersma, and B. Poolman. Microorganisms maintain crowding homeostasis. *Nat. Rev. Microbiol.*, 15:309–318, 2017.

115. S. Klumpp, M. Scott, S. Pedersen, and T. Hwa. Molecular crowding limits translation and cell growth. *Proc. Natl. Acad. Sci. USA*, 110(42):16754–16759, 2013.

116. R. P. Joyner, J. H. Tang, J. Helenius, E. Dultz, C. Brune, L. J. Holt, S. Huet, D. J. Mller, and K. Weis. A glucose-starvation response regulates the diffusion of macromolecules. *eLife*, 5:e09376, 2016.

117. M. Delarue, G. Brittingham, I. Surovtsev, K. J. Kennedy, I. Gutierrez, J. Chung, J. T. Groves, C. Jacobs-Wagner, and L. J. Holt. Mtorc1 controls phase-separation and the biophysical properties of the cytoplasm by tuning crowding. *bioRxiv*, 073866, 2017.

PART 2

Data-Driven Models

10

Restraint-Based Modeling of Genomes and Genomic Domains

MARCO DI STEFANO AND MARC A. MARTI-RENOM

10.1 INTRODUCTION

The function of the genome in living cells depends on various layers of regulation [1]. The first one is the nucleotide sequence of the DNA molecules. The sequencing of the entire human genome has been a milestone of biology in the last decades [2], and opened the way to the annotation of the genes and regulatory sequences on the 1D DNA sequence (promoters, enhancers, etc...). The second and still largely unexplored level of organization is how the genome is folded in 3D space [3]. In fact, similar to the words in a book, genes are meaningful when they can be read and put in the correct context [4]. In the genome, this happens only if the DNA fiber is accessible to the transcription machinery, and if specific, and eventually distant, sequences (e.g., enhancers, promoters) are brought close in 3D space [5, 6].

To explore this aspect of genome function, several experimental techniques have been introduced to investigate the 3D organization of the DNA in cells at different scales. On the one hand, imaging (e.g., 3D FISH) has revealed the large-scale organization of the chromosomes in separated areas of the nucleus, called chromosome territories [7, 8], which typically occupy different nuclear radial positions depending on the gene content [9–11]. On the other hand, atomic resolution techniques, such as X-ray crystallography, provided a detailed description of the stacked organization of the DNA base-pairs [12] and of the structures of the double-helix DNA molecule and its wrapping around histone proteins [13].

Only recently it has been possible to start filling the gaps between the entire chromosome scale and the atomistic one [14]. This effort has been led by Chromosome Conformation Capture (3C) technologies (e.g., 3C [15], 4C (3C-on-chip or Circular 3C) [16–18], 5C [19], Hi-C [20], in situ-Hi-C [21], TCC [22] or Capture C [23, 24], among others), which are here referred as 3C-based technologies. These experimental setups can differ in the biochemical reactions involved, but they have at least two aspects in common. Firstly, all these approaches are normally applied to a population of cells and, so, study the cumulated contacts over millions of cells. These cells can be heterogeneous in terms of cell stage or cell development, or synchronized at a given stage [25]. Secondly, the read from these techniques are the numbers of captured interactions between pairs of DNA fragments that were close in 3D space. This quantitative information is typically visualized as heat maps (Figure 10.1A).

Since the very first introduction of the 3C-based technologies, they have been complemented by computational techniques that take as input the measured interaction frequencies to determine the structural folding of genomes and genomic domains [15]. These computational methods, which have been

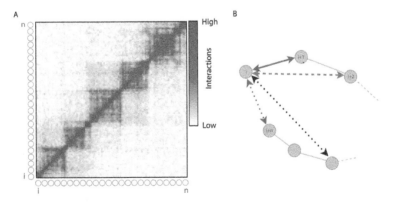

Figure 10.1 Representation. (A) Binned 3-C-based interaction matrices are represented in TADbit using individual particles for each bin in the matrix. (B) Each particle in the model is given spatial restraints to place them with respect to each other based on the data from the interaction matrix.

further developed over recent years [22, 26–28], have a common workflow made of three main stages: the *representation* of the chromatin as a set of points, the *scoring* of the conformations that these points can assume in 3D space, based on the translation of the 3C interaction frequencies into spatial restraints, and the *sampling* of the model conformations which optimally satisfy the imposed restraints, and as a consequence optimally represent the input 3C interactions. In this chapter, we will discuss the modeling strategy behind TADbit (https://github.com/3DGenomes/tadbit), a computational tool developed in our group, which is designed with the aim of providing a single tool for the complete analysis and 3D modeling of 3C-based experiments [29].

10.2 3C INTERACTIONS PROCESSING IN TADBIT

TADbit is a Python library (https://github.com/3DGenomes/tadbit) to manage 3C-based data from their production, to the analysis of the obtained interaction maps and the 3D modeling based on the interaction data. Specifically, TADbit is capable of evaluating the quality of the sequenced reads using the Phred score [30], mapping the read in the FASTQ files to the reference genome using GEM [31], obtaining raw interaction 3C heat maps binned at the desired resolution, normalizing and correcting the obtained interaction matrices to remove experimental biases by applying several normalization algorithms [32, 33], filtering the heat maps removing rows with low coverage and studying the hierarchical organization of the genome in A/B compartments [20] and in Topological Associating Domains (TADs) [34]. It is also possible to compare Hi-C maps using Spearman correlation analysis, or eigenvectors decomposition analysis [32], and to compare the hierarchical organization inferred from 3C matrices to describe the differences and similarities between cell types. Once the 3C-based heat maps have been obtained, TADbit also provides all the tools necessary to determine 3D models that optimally reproduce the interaction patterns of the input 3C matrix and allow analysis of the various structural properties from the models. In the following, we will focus on the technical details of the 3D modeling strategy.

10.3 RESTRAINT-BASED MODELING STRATEGY FOR STRUCTURAL DETERMINATION

Once a 3C-based interaction map is obtained, a simple strategy to reconstruct a 3D structure of chromatin that best reproduces the data is to force regions with a high number of interactions to stay close together, while regions with few interactions are kept far apart. In simple words, one can use interaction frequencies as a proxy of spatial proximity. The TADbit strategy implements this simple yet useful way of reconstructing 3D genomic domains by using the points-and-restraints strategy implemented in the Integrative Modelling Platform [35–37] (IMP, http://integrativemodeing.org), a general framework for restraint-based modeling of 3D bio-molecular structures. Inheriting from IMP, the TADbit modeling strategy is designed in three steps, which are discussed in details in the following sections.

10.3.1 Representation

The modeling strategy starts by describing each bin of the 3C interaction matrix as a spherical particle of a given radius (Figure 10.1B). Spheres have the advantage of being isotropic objects described by only four numbers: the Cartesian coordinates (x,y,z) of the center, and the magnitude of the radius. While the sphere position is decided to get the optimal structure (see Section 10.3.2), the radius is set to be proportional to the amount of DNA contained in the described bin (b_w). The conversion from the DNA amount (in base pairs) to the radius length (in nanometers) is done via a parameter called *scale* (*s*), which is equal to the linear space in nanometers occupied by a single base pair. That is, the radius of a particle is equal to

$$R = 0.5 \, s \, b_w \qquad (10.1)$$

The *scale* parameter describes the typical linear compaction of the chromatin fiber expected *a priori* for the studied region. The most used value is by default equal to *s = 0.01 nm/bp*, which corresponds to the expected physical compaction of the so-called 30-nm fiber [38, 39], which typically spans 30 nm every 3 kilobases of DNA. However, in TADbit, it is possible to change and to optimize the value of the scale parameter adapting the produced models to the desired level of nominal compaction.

10.3.2 Scoring: The distance restraints

One of the crucial steps in the TADbit modeling approach is converting the 3C interaction frequencies (c_{ij}) into harmonic distance restraints. A harmonic distance restraint is an energetic term expressed by a harmonic function that enforces the distance between two particles in the system to be in a certain range. The harmonic form is particularly convenient because it is defined by only two free parameters: the spring constant (*k*) and the equilibrium distance (d_{eq}). Given any pair of particles (*i* and *j*) in the system at distance d_{ij}, the harmonic energy (penalty) associated with the distance restraints between them is:

$$H\left(d_{ij}\right) = 0.5 \, k_{ij} \left(d_{ij} - d_{eq}^{(ij)}\right)^2 \qquad (10.2)$$

where k_{ij} and $d_{eq}^{(ij)}$ are respectively the specific spring constant and equilibrium distance for the particle pair (*i* and *j*). Usually, in restraint-based modeling approaches, there are three possible different types of harmonic restraints that could be applied between any two particles depending on the range of distances one wants to enforce: (a) *Harmonic distance restraint* is used to keep two particles around the equilibrium distance $d_{eq}^{(ij)}$, and is implemented using Equation 10.2

only for any possible value of distance between particles, d_{ij}; (b) *harmonic lower-bound distance restraint* is used to keep two particles further than the equilibrium distance, and is implemented using Equation 10.2 only if this condition is not accomplished (that is, $d_{ij} < d_{eq}$); and (c) *harmonic upper-bound distance restraint* is used to keep two particles closer than the equilibrium distance, and is implemented using Equation 10.2 only if $d_{ij} > d_{eq}$.

The sum of all the harmonic distance restraints between all the pairs of particles in the system involved in spatial restraints is the so-called objective function. By definition, the objective function is exactly equal to zero when all the distance restraints are simultaneously satisfied, and is high when many of them are violated. As a consequence, the aim of the restraint-based modeling strategy is to obtain a 3D conformation of points that closely satisfy as many restraints as possible, which invariably would result in a low objective function.

However, it has to be noted that since the objective function may contain upper- and lower-bound restraints and since the restraints act on the pairwise distance between particles, a value of zero could be associated with different 3D conformations, which are all legitimate, because they all satisfy the entire set of restraints. In 3D coordinate space, it is usually found that the differences between models are largely the result of the mirroring of portions of the structures. The presence of mirrors is expected, because perfect mirror images are indistinguishable in the distance space [40].

The next step in the modeling procedure is converting the 3C interaction frequencies into restraints that will score the objective function. To map the 3C interaction frequency onto pairwise harmonic distance restraints, one has to define the strengths k_{ij} and the equilibrium distances $d_{eq}^{(ij)}$ for all the pairs of particles (i and j) as a function of c_{ij}. In this procedure, the contact frequencies are re-normalized by log_{10} transformation and Z_{score} computation:

$$Z_{score\,ij} = \frac{\mu - log_{10}\left(c_{ij}\right)}{\sigma} \tag{10.3}$$

where μ and σ are the average and standard deviation of the log_{10} interaction counts for the entire 3C matrix. The log_{10} transformation is meant to smooth the values of the 3C interaction frequencies which can have very high values next to zeros, and the Z_{score} is used to weigh each interaction value by its statistical significance, because it increases with the number of standard deviations that the 3C interaction is far from the average value.

The strength, k_{ij}, is set equal to the square root of the absolute value of the 3C Z_{score} observed between a pair of bins: $k_{ij} = \sqrt{|Z_{score\,ij}|}$. In this way, the larger the Z_{score} (i.e., the more significant the 3C interaction), the stronger the harmonic force applied to the restraint. To obtain the equilibrium distance, we hypothesize an inverse linear relationship between the d_{eq} and the *absolute* Z_{score}. Specifically, a straight interpolation line is drawn to connect the point with the min(Z_{score}) and the maximum distance (*maxdist*) admitted between two particles, and the point with the max(Z_{score}) and the minimum target distance (*mindist*). The equation of the resulting linear interpolation is:

$$d_{eq} = mindist + \frac{maxdist - mindist}{\max(Z_{score}) - \min(Z_{score})}\left(\max(Z_{score}) - Z_{score}\right) \qquad (10.4)$$

where *mindist* is equal to a particle diameter, and *maxdist* is a free parameter in the TADbit optimization strategy. *maxdist* corresponds to the maximum distance at which two particles are kept by the 3C-based (lower-bound) distance restraints. Using this formula, it is possible to assign a unique value of equilibrium distance to each value of Z_{score}.

To filter the restraints to use for modeling, which greatly decreases the computational time for modeling, the Z_{score} values close to zero corresponding to the values of 3C interaction frequency close to average are usually discarded. To do this, other two parameters are introduced, the *lowfreq* and the *upfreq* that correspond to the lower and larger values of the discarded Z_{score}. To decide which type of restraint (harmonic, harmonic lower-bound or harmonic upper-bound) for a given pair of particles, TADbit implements a strategy based on stratifying the interaction depending on the genomic distance between the involved fragments. This idea comes from the general observation that due to the polymeric structure of chromatin, regions closer in the 1D sequence than others have also a larger probability to be captured by 3C-based techniques, and to have higher interaction frequencies [20]. This implies that 3C data close to the main diagonal should be evaluated differently than data far away. The stratification is done as following (Figure 10.2): (a) When two bins are nearest neighbors along the chromosome sequence, they are connected using a harmonic restraint of strength $k = 5.0$. If the 3C interaction is higher than the parameter *upfreq* the equilibrium distance d_{eq} of the harmonic restraints are set based on the 3C Z_{score} of the interaction value, while if it is lower a physics-based harmonic restraint with d_{eq} equal to 2 particle radii is enforced. This data-driven implementation of the nearest neighbor's interaction produces useful biological measures from the models since it takes into account possible variations of the local chromatin compaction in contrast to the average value of the *scale* used by default (see Section 10.3.1). This change in compaction is a sign of close (or open) chromatin conformation [26, 40–43] and can be directly correlated with experimental data, such as ATAC-seq profiles [44]. As discussed later, it is possible to measure the level of closeness (or openness) of the chromatin fiber on the models, computing the chromatin density in TADbit (see Sections 10.5). (b) A second nearest neighbor connectivity is further enforced in TADbit by upper-bound harmonic restraints between particles at positions i and $i+2$ along the model sequence. The strength and the target distance of these restraints are constant for all the particles pairs, and are $k = 5.0$ and d_{eq} equal to 2 particle diameters. (c) For distances larger than 2 particles the imposed restraint depends on the 3C interaction frequency of the corresponding bins in the interaction heat map. If the Z_{score} is higher than the *upfreq* parameters, a harmonic restraint (Equation 10.2) is enforced to keep the particles close to each other, if lower than the *lowfreq* parameter a lower-bound harmonic restraint is imposed to keep the two particles far apart. After filtering, in the 3C interaction maps there can be high entries flanking a discarded bin. To smooth the signal for the long-range

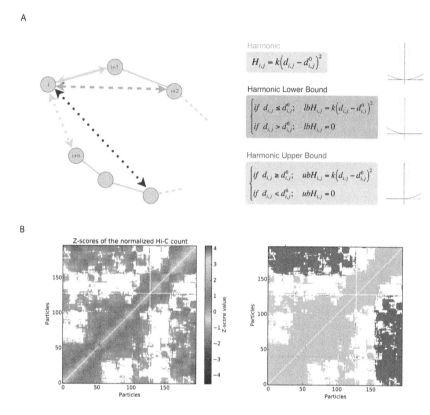

Figure 10.2 Scoring. (A) The final score of a model conformation is obtained by the weighted sum of all the scores (penalties) for each of the imposed restraints in the model. In TADbit there are three types of harmonic restraints imposed on the models. Harmonic restraints are used as "attractive forces" that aim to obtain an optimal distance between two particles, they are used for consecutive particles as well as for detected long-range (i, $i+n$ where n is 3 or more) interactions in the input matrix. Harmonic lower-Bound restraints are used as "repulsive forces" to avoid that two particles get closer than an equilibrium distance if they barely interact. Finally, harmonic upper-bound restraints are used only for keeping the connectivity of the chain by applying them between i and $i+2$ particles. B) Interaction matrices, represented by their Z_{score}, are then transformed into a series of harmonic (orange) and harmonic lower-bound (blue) restraints that will enforce the final conformation after optimization.

interactions only, a Z_{score} is assigned also to the discarded bin as the average of the nearest neighbor values. From the interpolated Z_{score}, the equilibrium distance, d_{eq}, of the corresponding harmonic oscillator is computed using the conversion curve in Equation 10.4 with a spring constant equal to half of the usual value, $k = 0.5\sqrt{|Z_{score}|}$.

Finally, in some applications, additional restraints are applied to take into account other factors affecting the genome organization, which are not explicitly related to the 3C interactions. These restraints are provided by non-3C experimental sources (biological restraints) [45], and by the general principles of (polymer) physics (physical restraints). An example of biological restraints is the phenomenological confining of the genome inside the nucleus [40]. To implement this, TADbit provides the option to define a confining environment composed of a cylinder capped with two half-spheres (capsule shape). This flexible geometrical construction can be customized to model, for example, spherical, or ellipsoidal confinements varying the height of the cylinder. Model confinement has been used to model the *Caulobacter crescentus* genome [40]. In this case, all the particles of the model have been constrained to be inside a parallelepiped box of dimensions of 2,350 nm × 600 nm × 600 nm, which is slightly larger than the dimensions of *Caulobacter* cells. These restraints were implemented using harmonics with a large spring constant of 100 to be sure that no fragments are lying outside of the box [40].

Additionally, physics-based restraints are based on the general principle that each particle in the model should not have large overlaps with any of the other particles. To this purpose, excluded volume harmonic interactions with strength $k_{excl} = 5.0$ are applied between all pairs of particles in the system using standard IMP commands [37]. We notice that excluded volume restraints are costly to implement, because their numbers grow as the number of particles in the system squared, and in some cases, they can be unnecessary. Specifically, depending on the coarse-graining of the models the physical hindrance of the chromatin fiber can play a marginal role. For example, if a model is produced at 10 kb, which following TADbit representation in Equation 10.1 has particles of diameter 100 nm, the fraction of the particle volume physically occupied by the chromatin fiber is only about 10%. The particle should be considered, in fact, mostly void. This implies that two distinct particles could largely overlap without breaking any steric hindrance physical restraint.

10.3.3 Sampling of the chromatin conformational space

Once the bins are represented as spherical particles, and the experimental 3C contacts are converted into distance restraints, we face the problem of finding the 3D arrangements of the particles that best satisfies the imposed restraints. To solve this complex optimization problem TADbit builds on the engines of the Integrative Modelling Platform (IMP) [37], and follows the steps below.

All the particles are initially positioned in a random, yet non-overlapping, manner inside a cubic box, whose typical size largely exceeds the expected size of the final modeled region [26, 41]. However, the initial cubic box doesn't act as a confining environment, and any particle of the system can escape it during the simulation protocol without any additional penalty in the objective function. Starting from these unbiased locations, the optimal positions of the particles are found by carrying out an optimization approach combining Monte Carlo sampling (MC), and molecular dynamics simulation (MD). Typically, 10 combinations of 500 MC rounds and 5 local steps of MD are carried out within a standard simulated annealing scheme [26, 41, 46]. At each MC round, a model conformation is proposed, moving all the particles in the system using MD, this

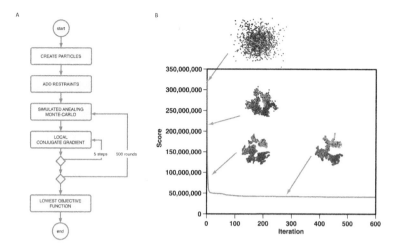

Figure 10.3 Optimization. A) Flowchart of a typical optimization protocol in TADbit based on simulated annealing and Monte-Carlo moves. B) Modeling score as a function of modeling iteration. All particles in the model are randomly placed at iteration zero and are optimally placed to globally minimize the scoring function in TADbit. A few models are shown as examples of how the model conformation changes during the optimization.

possibly new model is evaluated by computing the correspondent objective function value, and it is accepted (or rejected) according to the Metropolis criterion [47]. The 500 rounds of this hybrid optimization procedure are enough to ensure that the convergence of the objective function is to a minimum (Figure 10.3). As a consequence, the conformation associated with this minimum is retained as the representative model for the single simulation run.

Since the 3C input restraints represent an average picture of a possibly heterogeneous population of cells (see Introduction), a single model typically describes only a fraction of the proximities in the 3C interaction map. To provide a more complete description of the chromatin structures compatible with the input data, the sampling procedure is repeated thousands of times (usually 5,000), obtaining at the end an equal number of optimal model conformations (resampling procedure) [45]. By ranking the obtained models according to the associated objective function value (the lower the better), the top 20% of the obtained structures (about 1,000) are used to compose the ensemble of optimal structures which best satisfy the input restraints. Since the latter depends on the internal TADbit parameters (*maxdist, lowfreq, upfreq*), we will discuss in the following the procedure used to obtain the triplet of optimal values.

10.3.4 Optimization of TADbit parameters

The values of the free TADbit parameters (*maxdist*, lowfreq*, upfreq**) to be used for modeling the genomic domain under investigation are determined by a systematic grid search exploring several of the possible combinations. Given a triplet of values (*maxdist, lowfreq, upfreq*), this empirical search is done as follows:

1. A limited number of independent model conformations (usually 500) are generated.
2. The top 20% (usually 100) conformations, which are the ones associated with the lowest values of objective function, are used to compute the model contact map at a contact radius equal to 2 particle diameters. This contact map obtained from the structural models is the counterpart of the 3C interaction map.
3. The Spearman correlation coefficient is computed between the input 3C interaction map and the model contact map.

The optimal triplet (*maxdist*, lowfreq*, upfreq**) corresponds to the set of parameters that gives the highest Spearman correlation value. Such a procedure ensures that the ensemble of obtained models best describes the input 3C interaction frequency. The contact map computed on the best TADbit models typically has a Spearman correlation value higher than 0.6 with the input 3C interaction map, which is normally for large matrices statistically significant [43, 48]. Unfortunately, a general rule to select the triplets to explore in the grid search is not available, and a simple empirical approach is usually followed. As a rule of thumb, the *maxdist* is varied from 3 to 8 times the particle diameter, *lowfreq* between the lower value of the Z_{score} (~ −3.0) up to 0.0 (to use positive values would mean to keep far apart genomic regions with 3C interactions above average), and *upfreq* between 0.0 and the larger Z_{score} value (~3.0). The initial search along each of the 3 parameters can be done with large steps (e.g., 1.0). Next, when the sub-region of the maximum correlation is pinned down, one can do a finer search (e.g., in steps of 0.2) inside this limited parameter space.

10.4 MODEL VALIDATION

In principle, the optimal ensemble of models is consistent with the input 3C data. However, the sole comparison with the input data is obviously tautological and a proper validation is needed. Validation means controlling the consistency of the structural properties of the models with the experimental evidences obtained from techniques independent of 3C. This process ensures that the models are consistent not only with the data used to obtain them but also with other unrelated data, and further structural predictions obtained from them are significant and reliable.

The 3C-based technologies probe the pattern of contacts between chromosomal regions, and are *a priori* oblivious of other structural features. For example, these technologies don't provide direct measures of the spatial distance between pairs of genomic loci, but only in how many cells this distance is below an interaction cutoff [49, 50]. For this reason, a crucial step in TADbit modeling procedure is the conversion of the 3C interactions into special distances (see Section 10.3.2). This is why restraint-based models obtained from 3C data are usually validated using imaging data, which can aptly provide information on the spatial distances between selected pairs of genomic loci in cells [27]. Therefore, models are normally validated by comparing the measured distances in the models with the correspondent distances observed by imaging. For example, fluorescence in situ hybridization (FISH) [51–53] was used to measure the spatial distances between a pair of genomic *loci* of the ENm008 region of human chromosome 16 in two different cell lines K562, and GM12878 [41]. The loci, positioned at a genomic distance of 500 kb, were found at an average spatial distance of 318.8 ± 17.0 nm in GM12878, and of 391.9 ± 23.4 nm in K562 cells, showing a more extended conformation of the genomic region in the latter. Consistently, the distances computed on the models obtained using TADbit were 198.9 ± 0.7 nm and 434.6 ± 1.4 nm for GM12878 and K562 models, respectively, providing the first validation of the TADbit modeling approach. Similar validations based on FISH were carried out in other works, in which TADbit has been applied to study eukaryote and prokaryote organisms [40, 42, 43].

Another structural feature of the obtained TADbit models, which is not directly encoded in the 3C data, is the overall volume occupied by the chromosome region. This quantity depends on the specific conversion of 3C interaction into spatial distances (see Section 10.3.2), and on the *maxdist* of TADbit parameters, but is only partially tested using point-wise distances as in FISH-based validations. Interestingly, it has been possible to use the volume occupied by models of the entire *Mycoplasma pneumoniae* genome for model validation [43]. Specifically, they measured the volume of *M. pneumoniae* cells using a combination of transmission electron microscopy (TEM), quick-freeze deep-etch replica method [54], and computer 3D reconstruction. This volume of 0.075 mm^3 was consistent with the chromosome volume estimated form the ensemble of models obtained using TADbit of 0.074 mm^3.

10.5 MODEL ANALYSIS

It could be seen as obvious, but once the models have been produced the first thing to do is to visualize them. For this purpose, TADbit offers the possibility to save the obtained model structures in formats compatible with Chimera, a state-of-the-art software for the visualization of bio-molecules [55], as well as TADkit (http://www.3dgenomes.org/tadkit), a new 3D genome browser that allows mapping genomic tracks to the 3D models. The visual inspection of the models has, in fact, many advantages that can drive the entire modeling process. On the one hand, by looking at the obtained structures it is possible to spot errors in the modeling procedure immediately. For example, the overall size of the models could be too small, or larger than expected. In this regard, opening the best model using Chimera [55], and computing the distances between particles at the extremes of the models using the Chimera interface is a simple operation. If these distances fit with the expected spatial confinement of the structure, that in the case of genome-wide models can be, for instance, the nuclear (in eukaryotes) or the cell size (in prokaryotes), is a preliminary hint that the models are biologically sound, and a deeper quantitative analysis can be performed. Secondly, the model visual inspection can suggest further quantitative analyses to perform on the models. A part of the vast choice of model analysis provided in TADbit is discussed subsequently.

One of the most evident characteristics of the restraint-based 3D models is that the thousands of structures generated typically bear large similarities in their overall shape. To quantify this, TADbit performs model structure comparison using pair-wise rigid-body superposition to minimize the root mean square deviation (RMSD) between the conformations [56]. This optimal superposition is summarized in an all-against-all matrix storing the number of particles that align within a given distance cut-off. This cut-off (e.g., 75 nm and 100 nm have been previously used in [42] and [40], respectively) depends on the granularity of the models, and the overall size of the modeled chromatin region. Next, this scoring matrix is processed using the Markov Cluster Algorithm (MCL) program [57] to generate unsupervised sets of clusters of related structures. Two main parameters, the pre-inflation (pi) and the inflation (I), affect the clustering outcome of the MCL program. Similar to the empirical optimization of TADbit parameters (see Section 10.3.4), the values of these two parameters are chosen to maximize the Spearman correlation between the contact map computed on the models of the top cluster and the input 3C interaction map. A striking example of the biological insight that is possible to get from clustering analysis has been previously discussed [40]. Here, 10,000 models of the structural organization of the entire genome of *Caulobacter crescentus* were generated using TADbit. Interestingly, the clustering analysis revealed the presence of only 4 significantly different spatial organizations, and a visual inspection of the main representative structures of each of the clusters clarified that the 4 different spatial organizations were mainly mirror (or partial mirror) structures of each other (Figure 10.4A). This is a practical case in which the reduction of the structural variance helped to rationalize in simple terms the expectedly large variability of genomic structures.

Once the most representative structures of the models' ensemble have been identified, TADbit offers a set of quantitative tools to analyze the global properties of the modeled regions. For example, a typical quantity is the radius of gyration (R_g) of a chromatin region. This is the root mean square distance of the objects' particles from its center of mass, and measures the linear size of the spatial region occupied by the object. Another related quantity is the particle accessibility. It measures the accessible fraction of a particle in the model to a hypothetical spherical object of a given radius. The default value of this radius is 75 mn, which corresponds to the expected average size of the protein complexes typically interacting with the chromatin fiber [58]. These two quantities have been used to measure the size and the structural accessibility of topological associated domains (TADs) in human breast cancer cells [59]. It has been possible to show using TADbit models that after progesterone treatment specific TADs open (larger R_g), and get more structurally accessible (larger accessibility) than the control wild-type condition (Figure 10.4B).

In addition to the global structural arrangement, the TADbit model analysis also implements a tool to determine the spatial distances between any two loci (particles) in the models. This feature, also discussed for validation purposes in Section 10.4, has been used to characterize the proximity of the alpha-globin genes and their enhancer (HS40) in K562 cells, in which the genes are active, and

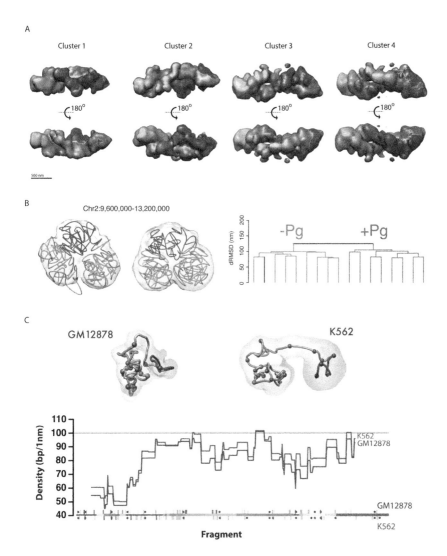

Figure 10.4 Analysis. A) The *Caulobacter crescentus* 3D modeling exercise [40] resulted in thousands of different models that could be clustered in four different conformations (clusters 1 to 4). Those clusters represented mirrors or partial mirrors of each other. B) The modeling of several Topological Associating Domains (TADs) of the human genome before and after progesterone induction [41] resulted in clear different conformations where TADs that get their resident genes activated after progesterone increase their overall radius of gyration, size and accessibility. C) The modeling of the alpha-globin domain [42] indicated that there was a change in local DNA density around the locus control region of the alpha-globin when the genes are expressed (K5562 cell line) compared to when they are silent (GM12878 cell line).

the substantially larger spatial separation in GM12878 cells, in which they are inactive [26, 41] (Figure 10.4C). In this application, the structural models and the associated analysis provided new biological insight into the mechanisms of gene expression regulation.

The local organization of the chromatin fiber can be also quantitatively studied by looking at the typical compaction of the model chromatin fiber along its contour length. This quantity on the models is measured by computing the number of DNA base-pairs contained in a single nanometer along the model contour length. Interestingly, the chromatin fiber compaction of the models can be directly compared to experimental measures probing analogous properties of the chromatin in real nuclei as ATAC-seq [44], or can be correlated to the presence of specific histone markers associated with open (e.g., H3K4me3) or closed (e.g., H3K9me3, and H3K27me3) chromatin using ChiP-Seq [60]. This analysis, called *density* in TADbit, has been used in many different systems. For example, in the human alpha-globin locus to show that transcriptionally active loci have a more open chromatin state [41], and in a genome-wide study of *C. crescentus* [40] to show that the region spanning 100–200 kb on the left of specific sites (ParS) assume a compact chromatin conformation in wild type and mutant strains, and to study the different degrees of compaction of chromatin of different types (colors) in *Drosophila melanogaster*, where inactive black chromatin has the highest density in TADbit models (median 212 bp/nm), slightly more than blue (207 bp/nm), and substantially more than green, yellow and red chromatin (182 bp/nm, 180 bp/nm, and 179 bp/nm, respectively) [29].

10.6 LIMITS OF THE TADbit MODELING APPROACH

When producing 3D models of biomolecules, it is important to be able to evaluate the limits of the approach. For the TADbit modeling strategy, this issue has been extensively addressed [48]. The TADbit benchmark strategy was based on producing datasets of simulated toy circular chromosomes embedding expected structural features of realistic chromosomes. Half of the produced toy models were organized in local globular domains to mimic the presence of TADs, while the other half was not. Per each of these two groups (hereafter named TAD-like and not-TAD-like), three groups of models were produced, varying the chromatin content per model particle or the so-called density. All models spanned 1Mb, one group of models was composed of 202 particles each containing 5 kb (density 150 bp/nm), the second had 402 particles each of 2.5 kb (density 75 bp/nm), and the third had 626 particles 1.6 kb each (40 bp/nm).

Next, for each of the six different conditions, simulations were done to reproduce *in silico* typical biases of the 3C technologies. Specifically, for each of the 6 architectural organizations, a total of 7 groups of 100 models each were produced using the Monte Carlo method [48, 61]. The models in each group were separated by a different number of sampling steps. Models in the first group were separated by 10^0 step (group 0), in the second by 10^1 steps (group 1), etc., up to group 7 whose models were separated by 10^6 steps. The increasing number of steps separating the models reproduced an increasing degree of structural variability in the ensemble of toy conformations. This procedure reproduced the expected cell-to-cell structure variability typical of a 3C sample [45, 50, 62]. Finally, the analysis of the toy models leading to the computation of the contact map reproduced *in silico* typical biases of the 3C technologies. Specifically, the noise level introduced by the 3C capturing procedure [62, 63] was reproduced in the toy models by computing the contact maps using a Monte Carlo procedure with 4 different levels of noise compared to a direct contact map generated from the models [48]. The entire procedure resulted in a total of 168 simulated 3C interaction maps with six architectures, seven levels of structural variability and four levels of noise in the data.

In the second part of the study, from each of the matrices, TADbit 3D models were generated using the standard procedure presented in Section 10.3. Finally, the comparison between the contact maps computed on the TADbit reconstructed models and the input-simulated interaction maps demonstrated at which degree each simulated dataset was accurately reconstructed by the TADbit modeling approach. The benchmark of the TADbit modeling strategy revealed that reconstructed models are robust to noise but sensitive to structural variability. This implies that experimental noise, which could originate from limitations in any of the main steps in 3C-based methods (cell fixation, DNA fragmentation, DNA ligation and read-out by sequencing), is not highly relevant for 3D reconstruction. However, 3C interaction maps obtained from a homogeneous as possible population of cells (e.g., synchronized in cell cycle, same cell state, unique cell type, etc.) are more useful for 3D reconstruction. Another experimental aspect that yields accurate models is producing high-resolution matrices, which corresponded to the low-density toy models. Those invariably resulted in a larger proportion of restraints per particle, which in turn resulted in more accurate models. Therefore, increasing the sequencing depth of a Hi-C experiment will result not only in higher resolution models (i.e., finer binning of the interaction matrix) but also in models of higher overall accuracy. Finally, the local organization in domains (such as TADs) gives more accurate models at any levels of noise and structural variability.

10.7 DATA "MODELABILITY": ARE 3C DATA GOOD ENOUGH TO OBTAIN 3D MODELS?

An important result of the benchmark analysis [48] was the introduction of the Matrix Modeling Potential (MMP) score, which evaluates *a priori* whether the 3C-based interaction matrix is suitable to generate accurate 3D models using TADbit. This score is defined, in fact, using three statistical quantities of the 3C input map: the *significant eigenvectors score* (SEV), the Skewness (SK) and the Kurtosis (KU) of the Z_{score} distribution in the input 3C matrix. The MMP score is calculated then as:

$$MMP = -0.0002 * Size + 0.0335 * SK - 0.0229 * KU + 0.0069 * SEV + 0.8126$$

where *Size* is the number of bins in the generated 3C interaction map. In general, 3C matrices with high MMP scores are likely to result in accurate 3D reconstructed models. For example, for 3C interaction maps of *M. pneumoniae* at a 10 kb resolution obtained using two different restriction enzymes (HindIII and HpaII) the MMP scores ranged from 0.71 to 0.74 [43], with a maximum predicted model accuracy of 0.70 (0.58–0.81 at 95% confidence interval), suggesting the feasibility of the TADbit modeling, which in the study provided accurate 3D genome-wide models.

10.8 CONCLUSION

In this chapter, we have discussed in detail the TADbit computational package. TADbit here represents the so-called restraint-based 3D genome reconstruction methods but many other computational tools have recently been produced [45, 64]. Such new methods aim at addressing the challenge of analyzing and visualizing chromatin interaction data. Such datasets are more common today since there has been improved protocols, as well as a significant reduction in the costs of performing such experiments [65]. We have also shown specific examples where the resulting 3D models based on interaction data have already provided significant insights into how the 3D spatial conformation of the genome impacts gene expression and gene regulation.

REFERENCES

1. Misteli, T., Beyond the sequence: Cellular organization of genome function. *Cell*, 2007. 128(4): p. 787–800.
2. Watson, J.D., The human genome project: Past, present, and future. *Science*, 1990. 248(4951): p. 44–49.
3. Misteli, T., Self-organization in the genome. *Proc Natl Acad Sci USA*, 2009. 106(17): p. 6885–6886.
4. Lanctot, C., et al., Dynamic genome architecture in the nuclear space: regulation of gene expression in three dimensions. *Nat Rev Genet*, 2007. 8(2): p. 104–115.
5. Tolhuis, B., et al., Looping and interaction between hypersensitive sites in the active beta-globin locus. *Mol Cell*, 2002. 10(6): p. 1453–65.
6. Vernimmen, D., et al., Long-range chromosomal interactions regulate the timing of the transition between poised and active gene expression. *EMBO J*, 2007. 26(8): p. 2041–2051.
7. Cremer, T. and C. Cremer, Chromosome territories, nuclear architecture and gene regulation in mammalian cells. *Nat Rev Genet*, 2001. 2(4): p. 292–301.
8. Branco, M.R. and A. Pombo, Intermingling of chromosome territories in interphase suggests role in translocations and transcription-dependent associations. *PLoS Biol*, 2006. 4(5): p. e138.
9. Bolzer, A., et al., Three-dimensional maps of all chromosomes in human male fibroblast nuclei and prometaphase rosettes. *PLoS Biol*, 2005. 3(5): p. e157.

10. Cremer, M., et al., Non-random radial higher-order chromatin arrangements in nuclei of diploid human cells. *Chromosome Res*, 2001. 9(7): p. 541–567.

11. Boyle, S., et al., *The spatial organization of human chromosomes within the nuclei of normal and emerin-mutant cells.* Hum Mol Genet, 2001. 10(3): p. 211–219.

12. Watson, J.D. and F.H. Crick, Molecular structure of nucleic acids; a structure for deoxyribose nucleic acid. Nature, 1953. 171(4356): p. 737–8.

13. Luger, K., et al., *Crystal structure of the nucleosome core particle at 2.8 A resolution. Nature*, 1997. 389(6648): p. 251–260.

14. Marti-Renom, M.A. and L.A. Mirny, Bridging the resolution gap in structural modeling of 3D genome organization. *PLoS Comput Biol*, 2011. 7(7): p. e1002125.

15. Dekker, J., et al., Capturing chromosome conformation. *Science*, 2002. 295(5558): p. 1306–1311.

16. Simonis, M., et al., Nuclear organization of active and inactive chromatin domains uncovered by chromosome conformation capture-on-chip (4C). *Nat Genet*, 2006. 38(11): p. 1348–1354.

17. Zhao, Z., et al., Circular chromosome conformation capture (4C) uncovers extensive networks of epigenetically regulated intra- and interchromosomal interactions. *Nat Genet*, 2006. 38(11): p. 1341–1347.

18. Fullwood, M.J., et al., An oestrogen-receptor-alpha-bound human chromatin interactome. *Nature*, 2009. 462(7269): p. 58–64.

19. Dostie, J., et al., Chromosome Conformation Capture Carbon Copy (5C): A massively parallel solution for mapping interactions between genomic elements. *Genome Res*, 2006. 16(10): p. 1299–1309.

20. Lieberman-Aiden, E., et al., Comprehensive mapping of long-range interactions reveals folding principles of the human genome. *Science*, 2009. 326(5950): p. 289–293.

21. Rao, S.S., et al., A 3D map of the human genome at kilobase resolution reveals principles of chromatin looping. *Cell*, 2014. 159(7): p. 1665–1680.

22. Kalhor, R., et al., Genome architectures revealed by tethered chromosome conformation capture and population-based modeling. *Nat Biotechnol*, 2011. 30(1): p. 90–98.

23. Schoenfelder, S., et al., The pluripotent regulatory circuitry connecting promoters to their long-range interacting elements. *Genome Res*, 2015. 25(4): p. 582–597.

24. Hughes, J.R., et al., Analysis of hundreds of cis-regulatory landscapes at high resolution in a single, high-throughput experiment. *Nat Genet*, 2014. 46(2): p. 205–212.

25. Naumova, N., et al., Organization of the mitotic chromosome. *Science*, 2013. 342(6161): p. 948–953.

26. Baù, D. and M.A. Marti-Renom, Genome structure determination via 3C-based data integration by the integrative modeling platform. *Methods*, 2012. 58(3): p. 300–306.

27. Giorgetti, L., et al., Predictive polymer modeling reveals coupled fluctuations in chromosome conformation and transcription. *Cell*, 2014. 157(4): p. 950–963.

28. Duan, Z., et al., A three-dimensional model of the yeast genome. *Nature*, 2010. 465(7296): p. 363.

29. Serra, F., et al., Automatic analysis and 3D-modelling of Hi-C data using TADbit reveals structural features of the fly chromatin colors. *PLoS Comput Biol*, 2017. 13(7): p. e1005665.

30. Ewing, B., et al., Base-calling of automated sequencer traces using phred. I. Accuracy assessment. *Genome Res*, 1998. 8(3): p. 175–85.

31. Marco-Sola, S., et al., The GEM mapper: fast, accurate and versatile alignment by filtration. *Nat Methods*, 2012. 9(12): p. 1185–1188.

32. Imakaev, M., et al., Iterative correction of Hi-C data reveals hallmarks of chromosome organization. *Nat Methods*, 2012. 9(10): p. 999–1003.

33. Yaffe, E. and A. Tanay, Probabilistic modeling of Hi-C contact maps eliminates systematic biases to characterize global chromosomal architecture. *Nat Genet*, 2011. 43(11): p. 1059–1065.

34. Dixon, J.R., et al., Topological domains in mammalian genomes identified by analysis of chromatin interactions. *Nature*, 2012. 485(7398): p. 376–380.

35. Alber, F., et al., Determining the architectures of macromolecular assemblies. *Nature*, 2007. 450(7170): p. 683–694.

36. Alber, F., et al., The molecular architecture of the nuclear pore complex. *Nature*, 2007. 450(7170): p. 695–701.

37. Russel, D., et al., Putting the pieces together: integrative modeling platform software for structure determination of macromolecular assemblies. *PLoS Biol*, 2012. 10(1): p. e1001244.

38. Harp, J.M., et al., Asymmetries in the nucleosome core particle at 2.5 A resolution. *Acta Crystallogr D Biol Crystallogr*, 2000. 56(Pt 12): p. 1513–1534.

39. Finch, J.T. and A. Klug, Solenoidal model for superstructure in chromatin. *Proc Natl Acad Sci USA*, 1976. 73(6): p. 1897–1901.

40. Umbarger, M.A., et al., The three-dimensional architecture of a bacterial genome and its alteration by genetic perturbation. *Mol Cell*, 2011. 44(2): p. 252–264.

41. Baù, D., et al., The three-dimensional folding of the alpha-globin gene domain reveals formation of chromatin globules. *Nat Struct Mol Biol*, 2011. 18(1): p. 107–114.

42. Le Dily, F., et al., Distinct structural transitions of chromatin topological domains correlate with coordinated hormone-induced gene regulation. *Genes Dev*, 2014. 28(19): p. 2151–2162.

43. Trussart, M., et al., Defined chromosome structure in the genome-reduced bacterium Mycoplasma pneumoniae. *Nat Commun*, 2017. 8: p. 14665.

44. Buenrostro, J.D., et al., Transposition of native chromatin for fast and sensitive epigenomic profiling of open chromatin, DNA-binding proteins and nucleosome position. *Nat Methods*, 2013. 10(12): p. 1213–1218.

45. Serra, F., et al., Restraint-based three-dimensional modeling of genomes and genomic domains. *FEBS Lett*, 2015. 589(20 Pt A): p. 2987–2995.

46. Kirkpatrick, S., C.D. Gelatt, Jr., and M.P. Vecchi, Optimization by Simulated Annealing. *Science*, 1983. 220(4598): p. 671–680.

47. Metropolis, N. and S. Ulam, The Monte Carlo method. *J Am Stat Assoc*, 1949. 44(247): p. 335–341.

48. Trussart, M., et al., Assessing the limits of restraint-based 3D modeling of genomes and genomic domains. *Nucleic Acids Res*, 2015. 43(7): p. 3465–3477.

49. Fudenberg, G. and M. Imakaev, FISH-ing for captured contacts: towards reconciling FISH and 3C. *Nat Methods*, 2017. 14(7): p. 673–678.

50. Giorgetti, L. and E. Heard, Closing the loop: 3C versus DNA FISH. *Genome Biol*, 2016. 17(1): p. 215.

51. Fraser, J., et al., An overview of genome organization and how we got there: From FISH to Hi-C. *Microbiol Mol Biol Rev*, 2015 79(3): p. 347–372.

52. Solovei, I., et al., Spatial preservation of nuclear chromatin architecture during three-dimensional fluorescence in situ hybridization (3D-FISH). *Exp Cell Res*, 2002. 276(1): p. 10–23.

53. Sachs, R.K., et al., A random-walk/giant-loop model for interphase chromosomes. *Proc Natl Acad Sci USA*, 1995. 92(7): p. 2710–2714.

54. Heuser, J.E., The origins and evolution of freeze-etch electron microscopy. *J Electron Microsc (Tokyo)*. 2011. 60(Suppl 1): p. S3–S29.

55. Pettersen, E.F., et al., UCSF Chimera–a visualization system for exploratory research and analysis. *J Comput Chem*, 2004. 25(13): p. 1605–1612.

56. Zhang, Y. and J. Skolnick, Scoring function for automated assessment of protein structure template quality. *Proteins*, 2004. 57(4): p. 702–710.

57. Enright, A.J., S. Van Dongen, and C.A. Ouzounis, An efficient algorithm for large-scale detection of protein families. *Nucleic Acids Res*, 2002. 30(7): p. 1575–1584.

58. Cook, P.R., A model for all genomes: the role of transcription factories. *J Mol Biol*, 2010. 395(1): p. 1–10.

59. Baeza-Delgado, C., M.A. Marti-Renom, and I. Mingarro, Structure-based statistical analysis of transmembrane helices. *Eur Biophys J*, 2013. 42(2–3): p. 199–207.

60. Schmidl, C., et al., ChIPmentation: fast, robust, low-input ChIP-seq for histones and transcription factors. *Nat Methods*, 2015. 12(10): p. 963–965.

61. Junier, I., O. Martin, and F. Kepes, Spatial and topological organization of DNA chains induced by gene co-localization. *PLoS Comput Biol*, 2010. 6(2): p. e1000678.

62. Nagano, T., et al., Single-cell Hi-C reveals cell-to-cell variability in chromosome structure. *Nature*, 2013. 502(7469): p. 59–64.

63. Dekker, J. and L. Mirny, Biological techniques: Chromosomes captured one by one. *Nature*, 2013. 502(7469): p. 45–46.
64. Yardimci, G.G. and W.S. Noble, Software tools for visualizing Hi-C data. *Genome Biol*, 2017. 18(1): p. 26.
65. Dekker, J., M.A. Marti-Renom, and L.A. Mirny, Exploring the three-dimensional organization of genomes: interpreting chromatin interaction data. *Nat Rev Genet*, 2013. 14(6): p. 390–403.

11

Genome Structure Calculation through Comprehensive Data Integration

GUIDO POLLES, NAN HUA, ASLI YILDIRIM, AND
FRANK ALBER

11.1 INTRODUCTION

The three-dimensional organization of chromosomes and their locations in the nucleus greatly influences the functions of genes, including expression, silencing, and replication. The relative locations of tens of thousands of genes and millions of potential regulatory elements play an important role in orchestrating nuclear functions. For example, promoter–enhancer interactions and other regulatory interactions often act over considerable sequence distances in the kb to mb range or even between different chromosomes (Tolhuis et al., 2002; Dekker et al., 2017). Also the locations of genes and regulatory elements with respect to nuclear bodies, such as nucleoli, nuclear speckles, and lamina often influence their functions. Chromatin at the nuclear lamina is more likely to be silenced, while genes close to nuclear speckles tend to be highly expressed (Guelen et al., 2008; Dekker et al., 2017). For a better understanding of nuclear functions, it is therefore necessary to study the spatial folding patterns of chromosomes and the spatial organization of nuclear bodies in the nuclear context. Recent studies point to a hierarchical chromatin organization with compartments of chromatin in related functional states (Lieberman-aiden et al., 2009), topologically associating domains (TADs) (Dixon et al., 2012), sub-domains (Shen et al., 2012), and chromatin loops (Rao et al., 2014) (Figure 11.1). Chromosome conformation capture experiments revealed that chromatin is spatially compartmentalized into at least two major subcompartments, one associated with active, more open chromatin and one with inactive more compacted chromatin (Lieberman-aiden et al., 2009). The transcriptional repressor CTCF, cohesin, and other architectural proteins mediate chromatin interactions by stabilizing chromatin loops of about 100 kb sequence lengths, as well as the formation chromatin topological associated domains (Rao et al., 2014; de Wit et al., 2015; Sanborn et al., 2015; Fudenberg et al., 2016; Schwarzer et al., 2017).

Acquiring spatial information for a chromatin polymer containing billions of basepairs is a daunting task, especially when considering genome dynamics,

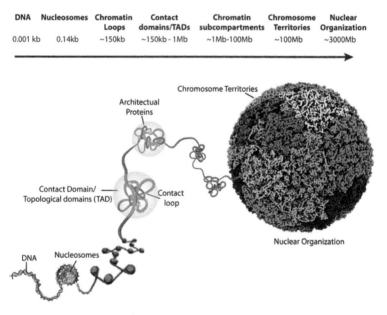

Figure 11.1 Hierarchical organization of genome structure.

which leads to structural variations from cell to cell in an isogenic sample. Recently, there has been a rapid development of new imaging and genomics technologies to shed light on genome organization in greater detail. New ensemble, as well as single-cell genomic, technologies probe chromatin interactions at an unprecedented resolution (Denker and De Laat, 2016). High-throughput imaging approaches localize the probabilistic distributions of chromatin probes and live-cell and super-resolution imaging visualize loci and sub-nuclear structures as well as trace the dynamics of chromatin loop formation (Dekker et al., 2017). To maximize synergy between different approaches and further enhance the field, a concerted effort is needed to coordinate the mapping of the dynamic nucleome organization.

Recently the National Institute of Health (NIH) formed the 4DN Nucleome consortium (https://www.4dnucleome.org/) to tackle this difficult mapping problem by coordinating efforts between many research groups (Dekker et al., 2017). The goal of the consortium is to develop a set of methods for mapping the structures and dynamics of the genome as well as produce integrated approaches to generate a first draft of a model of the 4D Nucleome. A central piece of these efforts is a joint analysis project in which many research groups pledged to produce a variety of different data on a selected set of tier 1 cell lines, comprising H1-ES human embryonic cells, hTert-HFF humane foreskin fibroblasts, as well as IMR90 lung fibroblasts, and GM12878 B-lymphocytes. Producing a wide array of data from many complementary experimental methods will facilitate benchmarking of the various experimental technologies and also will lay the foundation for integrating the various data to generate quantitative models of the nuclear organization for these cell types.

An important aspect of this project is therefore to establish computational tools that can integrate the wealth of information from various imaging and omics experiments to map three-dimensional (3D) structures compatible with the data. There are several benefits in calculating 3D structures from omics and imaging data. First, 3D structures are a natural way of integrating data types from complementary experimental methods. All data from omics and imaging experiments originate from 3D structures, often from a large population of cells. Therefore, when using an appropriate representation of experimental errors and uncertainties, one should be able to relate all data to an ensemble of representative 3D genome structures. Second, generating structures that are simultaneously consistent with different data sources allows a cross-validation of the data types. For example, one can directly assess 3D models generated from Hi-C data with spatial information from imaging experiments and analyze disagreement between the data. Third, 3D structures provide added value by revealing additional features not visible in the original input data set. For example, an ensemble of genome structures generated from Hi-C data may reveal specific higher order (i.e., multivariate) chromatin clusters that are observed in statistically significant subsets of cells, even though the ensemble-based Hi-C data provides only binary interactions (Dai et al., 2016).

To achieve these goals, it is essential to transform all experimental information into an accurate representation of chromatin structures. However, characterizing

3D genome structures at a meaningful resolution is a challenging task. Genome structures can substantially vary between single cells and are dynamic in nature. A probabilistic description is thus needed, surpassing traditional structural modeling that is based on the assumption of a single equilibrium structure, or a small number of metastable structures.

Here in this review, we first summarize various experimental methods that are used in the 4DN joint analysis project to shed light on aspects of the genome structural organization. Then we provide an overview of data-driven computational methods that use experimental data to generate structural models consistent with it. We specifically discuss methods for comprehensive data integration, which can reveal the folded states of complete genomes, and consider the structural heterogeneity across different cells. The ultimate goal for structural modeling is to provide a better understanding of how the chromosomal folding state relates to gene regulation and to provide insights on mechanisms and driving forces of chromatin folding.

11.2 EXPERIMENTAL SOURCES OF SPATIAL INFORMATION

To date, many experimental techniques are available to probe the spatial organization of the genome (Table 11.1). We categorize these techniques according to the type of spatial information they provide for genome modeling. For instance, data may provide information about a pairwise chromatin contact, multivariate chromatin contacts, probability distributions of locations or distances between loci, or the probability of specific contacts between chromatin regions or chromatin regions and nuclear bodies. When performing 3D structure modeling, the information class decides how the data is transformed into spatial restraints. In the following sections, we summarize some of the experimental methods according to this classification scheme.

In addition, a distinction is made between single-cell and ensemble methods. Single-cell experiments provide data specific to a single nucleus. The advantage of having a direct observation of a single physical structure comes with the downside that the variability of structures across different nuclei has to be analyzed by repeated experimental sampling. The throughput can be a limiting factor to capture relatively rare but functionally relevant events. In addition, low data coverage per cell may also be challenging. Ensemble methods, on the other side, provide aggregated data from a whole cell sample in a single experiment. Sample averages are statistically more accurate, but all the single cell information is lost, such as multivariate chromatin interactions in individual cells. Also, rare but functionally relevant features of individual cells may not be detectable in ensemble averaged data.

11.2.1 Frequencies of pairwise chromatin–chromatin interactions

Chromosome Conformation Capture (3C) methods (Dekker et al., 2002; Denker and De Laat, 2016), and their variants such as in situ Hi-C (Rao et al., 2014),

Table 11.1 Different Experimental Methods That Provide Information On 3D Genome Organization

Ensemble methods	Single cell methods		
	Omics methods	**Imaging methods**	
Hi-C (Lieberman-aiden et al., 2009)	GAM (Beagrie et al., 2017)	3D DNA FISH (Strongin, Groudine and Politz, 2014)	HIPMap (Shachar, Pegoraro and Misteli, 2015)
In situ Hi-C (Rao et al., 2014)	Chia-PET (Fullwood et al., 2009)	3D RNA FISH (Strongin, Groudine and Politz, 2014)	CASFISH (Chen et al., 2013; Deng et al., 2015; Ma et al., 2015, 2016)
TCC (Kalhor et al., 2012)	Micro-C (Hsieh et al., 2016)	Seq FISH (Lubeck et al., 2014)	MERFISH (Ma, Reyes-Gutierrez and Pederson, 2013; Miyanari, Ziegler-Birling and Torres-Padilla, 2013)
DNAse-Hi-C (Ramani et al., 2016)	Lamina DamID (Van Steensel and Henikoff, 2000)	Lamina DamID (Kind et al., 2013)	ChromEM/EMT (Ou et al., 2017)
Capture Hi-C (Mifsud et al., 2015)	TRIP (Akhtar et al., 2014)	cryoSXT (Do et al., 2015)	cryoET
COLA (Darrow et al., 2016)	3C (Dekker et al., 2002)	Super-resolution methods: 3DSIM, STORM, PALM, STED (Müller et al., 2010; Weiland, Lemmer and Cremer, 2011; Doksani et al., 2013; Beliveau et al., 2015; Boettiger et al., 2016; Nozaki et al., 2017)	LLS (Chen et al., 2014)
4C (Van De Werken et al., 2012)	5C (Dostie and Dekker, 2007)	LLS-PAINT (Legant et al., 2016)	

TCC (Kalhor et al., 2012), *Micro-C* (Hsieh et al., 2015), and *DNAse-Hi-C* (Ramani et al., 2016) as well as *Chromatin Interaction Analysis by Paired-End Tag Sequencing (ChiaPET)* (Fullwood et al., 2009; J. Zhang et al., 2012) probe the relative frequency of millions of binary chromatin interactions in a population of cells (Table 11.1) by averaging chromosome conformations from millions of nuclei. Most chromosome conformation capture protocols follow some general steps. First, an isogenic sample of cells is fixed by formaldehyde crosslinking. Second, the genetic material is digested either through a restriction enzyme (in Hi-C, in situ Hi-C, TCC) or through DNAse (DNAse-Hi-C), or micrococcal nuclease (Micro-C). Then re-ligation of proximal DNA fragments is induced. Due to the crosslinking step, many genomic regions close in 3D space will likely remain in spatial proximity even after the digestion step. This, in turn, will favor their ligation in the following step, regardless of their relative position in the sequence. Finally, newly formed ligation sites are extracted followed by pair-end sequencing of the extracted fragments. Alignment of the sequenced fragments to their position in the reference genome determines the contact frequency of chromatin regions that are in spatial proximity in the initial cell population. A variety of methods build on a similar schema, with different protocols or technologies applied at each step.

3C based methods map the proximity of chromatin without distinguishing the proteins that mediate these interactions. In contrast, *ChiaPET* (Fullwood et al., 2009; J. Zhang et al., 2012) relies on an immuno-precipitation step, yielding chromatin contact frequency maps for interactions mediated by only a specific protein of interest. For instance, *ChiaPET* has been applied to detect chromatin interactions mediated by architectural proteins CTCF, cohesion, or RNA polymerase II (Fullwood et al., 2009; J. Zhang et al., 2012; Tang et al., 2015).

Capture Hi-C is a variant of the Hi-C assay for probing interactions of specific genomic loci (Mifsud et al., 2015; Schoenfelder et al., 2015). Because sequencing is focused only on a subset of genomic loci, it typically achieves deeper sequencing and thus higher resolution than all-against-all genome-wide Hi-C assays. For example, *capture Hi-C* was used to examine the long-range regulatory interactions of tens of thousands of promoters in human cell types (Mifsud et al., 2015; Schoenfelder et al., 2015). By enhancing the signal provided by specific proteins or genomic loci, these methods can potentially enrich the signal of rare but functionally important interactions and therefore can convey additional functional information.

All ensemble-based chromosome conformation capture assays produce the relative frequency of specific interactions that are observed in a population of cells. Typically the pair-end sequenced reads are binned at a given resolution (> ~1 kb) and the resulting contact frequencies are normalized to correct systematic biases (Yaffe and Tanay, 2011; Imakaev et al., 2012). When used for 3D structural modeling, Hi-C contact frequencies have been used in several different ways. For example, Tjong et al. transformed contact frequencies into probabilities of observing a given chromatin contact in a cell population and used these probabilities to generate models which reproduce the frequency of direct contacts observed in the cell population (Tjong et al., 2016). Several additional approaches incorporate significant Hi-C contacts in form of direct contact constraints (Paulsen et al., 2017), whereas other approaches (Duan et al., 2010; Rousseau et al., 2011; Hu et al., 2013; Zhang et al., 2013; Varoquaux et al., 2014) relate contact frequencies to averaged distances between loci, either based on empirical curves obtained from imaging assays or on polymer physics considerations. These distances are then used to generate either a single consensus structure or an ensemble of possible structural solutions (see section below for more details).

Single-cell Hi-C methods (Nagano et al., 2013, 2017; Flyamer et al., 2017; Stevens et al., 2017) perform the Hi-C assay to individual nuclei and therefore enable the detection of pairwise DNA contacts present in the same cell. Because the experiment is performed on genetic material of only a single cell, only a fraction of all contacts can be detected. Despite the challenges, however, recent improvements in conformation capture protocols detect more than one million contacts per cell to be detected (Flyamer et al., 2017). The method has also been recently applied to thousands of individual cells and revealed insights into chromosome structural changes during the cell cycle (Nagano et al., 2017).

A second experimental strategy, *combinatorial single cell Hi-C*, is based on multiplex barcoding (Ramani et al., 2017). This method utilizes genetic "barcodes", i.e., short DNA sequences uniquely associated with each cell. Although applied to a large population of cells, barcodes enable the mapping of pairwise chromatin interactions to individual cells. After genome digestion, but before cell lysis, the sample is separated into multiple wells. A short barcode sequence, different for each well, is then added to the exposed DNA fragments. The whole sample is then mixed and separated again into multiple wells, and another random short sequence is added to the DNA in each well. This process is repeated multiple times, extending the barcode sequences, effectively creating a genetic identifier for each pair-end sequencing read that is unique to each cell. The probability for two different cells to acquire the exact same barcode by random chance drops exponentially with the number of iterations, and a small number of iterations is usually sufficient to ensure the uniqueness of the barcode per cell.

A large number of pairwise chromatin contacts can strongly constrain the configurational space, giving a "snapshot" of the genome configuration for the given cell. Because these data are typically only a small fraction of all chromatin interactions modeling efforts often are augmented with additional information, for example from ensemble Hi-C experiments. In addition, as in ensemble Hi-C it may not be possible in most cases to differentiate chromatin regions from the two homologous chromosome copies. In the paper of Nagano et al. (Nagano et al., 2013) modeling was performed for the single X chromosome of male cells. In a more recent paper, Carstens et al. adapted a Bayesian structure determination framework to include information about average chromosome extensions from *Fluorescent in situ hybridization* (FISH) data in the modeling process. In the same framework, they proposed a way to model diploid structures from data which does not separate between chromosome copies (Carstens, Nilges, and Habeck, 2016). In the study by Stevens et al., the modeling of whole cells was facilitated by the use of haploid embryonic mouse cells, effectively removing the degeneracy introduced by multiple chromosome copies (Stevens et al., 2017).

In their work, structures of 8 cells at 100 kb resolution were produced and analyzed. Interestingly, the resulting structures are robust even when only partial data is used. The model analysis confirmed that structural features like the compartmentalization of active and inactive chromatin are systematically found in each cell. At the same time, however, loops and TADs appear to be stochastic in nature and, although present in a population, they are subject to wide variations on a cell-to-cell basis.

11.2.2 Multivariate chromatin interactions from ligation free methods

In addition to ligation-based methods, such as Hi-C and its variants which provide a way to map pairwise contacts between genomic regions, ligation free methods allow the detection of "higher-order" interactions, i.e., the simultaneous co-localization of several chromatin regions in a given cell. One such method is *Genome Architecture Mapping (GAM)* (Beagrie et al., 2017), which sequences the genetic material of thin slices of individual fixed nuclei, obtaining a list of chromatin regions co-localized in the two-dimensional slab of a single cell. Using a large number of cell slices, mathematical modeling then enables the determination of the probability of pairwise and multivariate co-localization of chromatin regions in individual nuclei. GAM revealed more pronounced long-range and inter-chromosomal chromatin interaction patterns in comparison to Hi-C data (Beagrie et al., 2017). GAM data can be used in 3D structure modeling by incorporating spatial constraints for multivariate chromatin co-localizations. For instance, population-based modeling methods can be extended to reproduce the experimentally detected multivariate co-localization probabilities in a population of genome models.

A second method, *Split-Pool Recognition of Interactions by Tag Extension* (SPRITE), is based on split-pool barcoding of RNA and DNA (Quinodoz et al., 2017). After crosslinking, medium-sized particles of genetic material (i.e., clusters) are separated by sonication. The clusters are then uniquely barcoded by an iterative multiplex procedure before sequencing. The method can therefore provide data on hundreds of thousands of multivariate chromatin co-localizations in a population of cells. In structure modeling, SPRITE can provide single-cell multivariate contact constraints to ensure that specific chromatin clusters are present in individual cells of the population.

11.2.3 Probing chromatin proximity to nuclear bodies

Several experimental methods map the proximity of chromatin to nuclear bodies, such as the lamina located at the nuclear envelope (NE) or speckles, which are nuclear domains enriched in pre-mRNA splicing factors.

The proximity of chromatin to the nuclear envelope has been probed with Lamina *DNA adenine methyltransferase identification (DamID)* experiments (Vogel, Peric-Hupkes and van Steensel, 2007) by the van Steensel group, both as ensemble and single cell assays (Van Steensel and Henikoff, 2000;

Guelen et al., 2008; Kind et al., 2015; Yáñez-Cuna and van Steensel, 2017). Essentially, the method determines a molecular contact frequency of DNA to Dam-methylase fusion proteins. For instance, lamina-B fusion protein probes DNA proximity to the nuclear envelope (Yáñez-Cuna and van Steensel, 2017), whereas other proteins could measure DNA proximity to other nuclear compartments. Lamina-DamID data has been probed for a population of cells, in the form of lamina-binding "propensity" profiles, or, more recently, even for single cells (Kind et al., 2015).

Another method is *Tyramide Signal Amplification–Sequencing (TSA–Seq)*, which has recently been introduced by the laboratory of Andrew Belmont to probe the proximity of DNA loci to nuclear speckles (Dekker et al., 2017).

In 3D structure modeling, lamina-DamID data can be used to constraint specific loci to regions close to the NE. A recent study integrated lamina-DamID with Hi-C data to produce a population of genome structures that reproduce both, chromatin–chromatin contact probabilities as well as the probabilities of loci to be in spatial proximity to the NE (Li et al., 2017). Another study used lamina-DamID and Hi-C data in a resampling approach to ensure that lamina-associated domains (LADs) have a higher probability to be located at the nuclear periphery (Paulsen et al., 2017). Modeling highlighted differences in the positioning of specific domains in HeLa cells carrying mutations associated with known pathologies.

11.2.4 Imaging methods for mapping the spatio-temporal organization of the nucleus

Various imaging technologies visualize the structural organization and dynamics of the nucleus at spatio-temporal resolution. Microscopy tools are important in providing direct spatial relationships of genomic loci and nuclear bodies. Because imaging does not suffer from mappability or sequencing biases they provide important complementary information for cross-validating findings from genomics and proteomics mapping methods.

Fluorescent in situ hybridization (FISH) uses oligonucleotide probes or RNA-mediated recruitment of fluorescently labeled dCas9 (Deng et al., 2015) in fixed cells to probe spatial distances and interactions of genomic loci. FISH has provided key insights into the spatial organization of chromosomes and the spatial relationships between genomic loci (Boyle et al., 2011). For instance, FISH experiments have revealed the formation of chromosome territories (Bolzer et al., 2005; Cremer and Cremer, 2010), the sub-compartmentalization of chromosomes into functionally distinct regions (Boyle et al., 2011) and have been used to analyze spatial proximities between specific loci. Recently a *multiplex FISH* method traced the detailed structural folding patterns of entire chromosomes in a single cell at 1Mb resolution (Wang et al., 2016). These data were generated by imaging a multitude of probes simultaneously through multiplex imaging. Inherently, single-cell methods generally suffer from limited throughput to describe sufficient statistical variations of spatial features. *HIPMap* is a method for large-scale automatic analysis from high-throughput FISH imaging, which allows a detail description of the probability densities of pairwise and multivariate distances between genomic loci as well as radial positioning of genomic loci in a population of tens of thousands of cells (Shachar et al., 2015; Shachar, Pegoraro and Misteli, 2015).

Probability densities of multivariate distances from high-throughput imaging can be incorporated as spatial constraints in modeling efforts. For example, FISH imaging has been used to constrain average spatial distances in 3D structure modeling of chromosomes (Wang, Xu and Zeng, 2015). Ideally, one would generate populations of genome models that reproduce the observed probability densities.

Some imaging methods do not require crosslinking or cell fixation, and therefore allow observations in living cells. Live-cell imaging approaches, such as *CRISPR–dCas9 FISH* can provide detailed information about the dynamic behavior of chromatin regions. CRISPR–dCas9 FISH in live cells (Chen et al., 2013; Anton et al., 2014; Ma et al., 2015, 2016; Fu et al., 2016; Guan et al., 2017; Takei et al., 2017) can map the dynamic behavior of chromatin regions in real time. In addition, several other super-resolution live-cell imaging methods, such as *3DSIM, PALM, STORM, STED* can be used to probe the dynamics of loci (Müller et al., 2010; Weiland, Lemmer, and Cremer, 2011; Doksani et al., 2013; Beliveau et al., 2015; Boettiger et al., 2016; Nozaki et al., 2017).

Cryo soft X-ray tomography (SXT) reveals details of the nuclear ultrastructure, such as the size and shape of the nucleus, as well as its spatial compartments such as euchromatic and heterochromatic chromatin clusters, the location and sizes of nuclear bodies. SXT is an imaging technique for visualizing cells and their interior structures in 3D without the need for chemical fixation, staining, or dehydration, and consequently leads to images of the entire nucleus in its native state. An SXT tomogram is reconstructed from a series of 2D projection images. Contrast in SXT images is generated by variations of the linear absorption coefficient (LAC) for different nuclear regions and functional chromatin subcompartments (i.e., eu- and hetero-chromatin regions), membranes, and nuclear bodies such as the nucleolus, which can be easily detected in SXT reconstructions (Do et al., 2015). For instance, SXT allowed detailed segmentations of nuclear compartments, detection of pericentromeric chromosome clusters (Do et al., 2015; Le Gros et al., 2016; Tjong et al., 2016), and even the reconstruction of the detailed conformation of an X-chromosome in eight cells (Smith et al., 2014)

Chrom-EMT (CEMT) visualizes the chromatin ultrastructure across multiple scales by combining electron microscopy tomography with a labeling method (chromEM) to enhance the contrast of DNA. In this method, a fluorescent DNA-binding dye is able to photo-oxidize and polymerize diaminobenzidine (DAB), which in turn binds OsO_4 to allow detection of DNA and chromatin ultrastructures in electron tomograms of a nucleus of glutaraldehyde-fixed cells. This labeling makes it possible to detect chromatin folding patterns at nucleosome resolution inside a nucleus. Chrom-EMT images revealed that chromatin is a disordered 5 to 24-nm-diameter granular chain that is packed together at different concentration densities in the nucleus (Ou et al., 2017).

Cryo-electron tomography (CET) generates 3D reconstructions of cells in hydrated, close to native states at molecular resolutions (Mahamid et al., 2016; Oikonomou and Jensen, 2017). Recent advances in focused ion beam (FIB) technology make it possible to generate thin slices of mammalian cells that allow imaging of nuclear organization and folding patterns of chromatin in the nucleus. However, tracing of the chromatin fiber and reconstruction of individual complexes in the crowded nuclear environment are challenging due to the relatively low contrast and signal to noise ratio and distortions due to the missing wedge (Mahamid et al., 2016). However, new imaging technologies and advances in automation will make it possible to detect both structures and spatial positions of multiple large macromolecular complexes in individual cells at molecular resolution (Frazier, Xu and Alber, 2017).

After summarizing various experimental methods that provide data about the spatial genome organization, in the following sections we discuss data-driven computational methods that use experimental data to generate 3D genome structures consistent with it. In the last section of this review, we discuss methods that can integrate complementary sets of data and discuss recent examples.

11.3 COMPUTATIONAL METHODS FOR DATA-DRIVEN GENOME STRUCTURE MODELING

There are a number of approaches for data-driven genome structure modeling (Imakaev, Fudenberg, and Mirny, 2015; Serra et al., 2015) (Table 11.2). These methods use experimental data to generate genome or chromosome structures consistent with it. Approaches can be classified into three major groups based on the type of modeling, (Table 11.1): 1) Consensus models, which represent the data by a single "averaged" structure; 2) Resampling methods, which explore conformational variability of the structures; and 3) Population-based modeling, which perform a deconvolution of the ensemble data (Serra et al., 2015) (see a comparison in Table 11.2). In addition, models can differ in the structural granularity of chromatin representation, which can be described as i) DNA fiber and chromatin fiber models ranging from 10 basepair to 100 kb resolution; and ii) chromosome or genome models represented by positions of the chromatin domains at < ~1Mb resolution (i.e., Topological Associated Domains, henceforth TAD), or larger macro-domains and subcompartments at > ~5Mb resolution. Computational methods also differ in how the data is interpreted and translated into spatial constraints. So far, most modeling approaches use data from chromosome conformation capture experiments (e.g., Hi-C data) almost exclusively. For example, some methods interpret the contact frequency from ensemble Hi-C experiments into spatial distances between the corresponding loci, whereas other methods use contact constraints to enforce actual contacts between loci in a subset of structures in an ensemble so that the total number of contacts agrees with the relative contact frequency in the experiment (Table 11.2). We first discuss the different types of computational approaches.

11.3.1 Consensus structure methods

Consensus models generate a single structure, which represents a "best fit" of the data. Often this data is generated from a population of cells, for example ensemble

Table 11.2 Comparison Of Data-Driven 3D Genome Modeling Methods (from (Hua et al., 2017))

Method	Distance or Contact based	Genome coverage	Species	Model type
Duan et al. (Duan et al., 2010)	Distance	Whole Genome	Budding Yeast	Consensus
BACH (Hu et al., 2013)	Distance	All chromosomes	Mouse	Consensus
AutoChrom3D (Peng et al., 2013)	Distance	500–800 kb	Human	Consensus
ChromSDE (Zhang et al., 2013)	Distance	Chromosome 13	Mouse, Human	Consensus
PASTIS (Varoquaux et al., 2014)	Distance	All chromosomes	Mouse	Consensus
ShRec3D (Lesne et al., 2014)	Distance	30Mb, Chromosome 1	Human	Consensus
HAS (Zou, Zhang and Ouyang, 2016)	Distance	All chromosomes	Human	Consensus
3D-GENOME (Szalaj et al., 2016)	Distance	All chromosomes	Human	Consensus
MCMC5C (Rousseau et al., 2011)	Distance	142 kb(5C), 88.4Mb(Chr14, HiC)	Human	Resampling
TADbit (Baù and Marti-Renom, 2012)	Distance	500 kb(5C)	Human	Resampling
Junier et al. (Junier et al., 2012)	Contact	1.2Mb (chr11)	Human	Resampling
Gehlen et al. (Gehlen et al., 2012)	Contact	Whole Genome	Budding Yeast	Resampling
Meluzzi and Arya (Meluzzi and Arya, 2013)	Contact	135–270 kb	Pseudo	Resampling

(Continued)

Table 11.2 (Continued) Comparison Of Data-Driven 3D Genome Modeling Methods (from (Hua et al., 2017))

Method	Distance or Contact based	Genome coverage	Species	Model type
Trieu and Cheng (Trieu and Cheng, 2014)	Contact	All chromosomes	Human	Resampling
InfMod3DGen (Wang, Xu and Zeng, 2015)	Distance	All chromosomes	Yeast	Resampling
MOGEN (Trieu and Cheng, 2016)	Contact	Whole Genome	Human	Resampling
Di Stefano et al. (Di Stefano et al., 2016)	Contact	Whole genome	Human	Resampling
Chrom3D (Paulsen et al., 2017)	Contact	Whole genome	Human	Resampling
Kalhor et al. (Kalhor et al., 2012)	Contact	Whole genome	Human	Population
Giorgetti et al. (Giorgetti et al., 2014)	Contact	780 kb	Human	Population
MiChroM (Zhang and Wolynes, 2015; Di Pierro et al., 2016)	Contact	All chromosomes	Mouse	Population
PGS (Tjong et al., 2016)	Contact	Whole genome	Human	Population

Hi-C (Peng et al., 2013; Zhang et al., 2013). In this case, ensemble Hi-C contact frequencies are typically mapped to spatial distances, assuming an inverse relationship between contact frequencies and distances. A single 3D structure is then generated that minimizes the residual errors between modeled and expected distances by either optimizing a scoring function (Duan et al., 2010; Peng et al., 2013; Zhang et al., 2013; Lesne et al., 2014; Varoquaux et al., 2014), a likelihood function through Bayesian inference (Hu et al., 2013), or solving a generalized linear model (Zou, Zhang and Ouyang, 2016). Consensus models have been applied to whole genomes (Duan et al., 2010), individual chromosomes (Y. Zhang et al., 2012; Hu et al., 2013; Varoquaux et al., 2014; Szalaj et al., 2016;

Zou, Zhang and Ouyang, 2016), and chromatin globules (Peng et al., 2013; Lesne et al., 2014). A consensus model is typically fast to generate, and summarizes averaged structural features from ensemble data. However, chromosome conformations vary between cells and a single consensus structure cannot convey aspects of variability and dynamics and therefore may not represent instances of individual observable structures.

11.3.2 Resampling methods

Resampling methods (Table 11.2) calculate an ensemble of structures by independent optimizations of the same scoring function derived from the data. These methods introduce aspects of conformational variability because of the presence of multiple minima, unconstrained degrees of freedom, or thermodynamic fluctuations.

For instance, in a study by Junier et al., a thermodynamic ensemble of configurations of the human beta-globin locus were produced by enforcing CTCF loops; the strength of the interaction was tuned to reproduce the observed 3C contacts distributions (Junier et al., 2012). Meluzzi and Arya (Meluzzi and Arya, 2013) proposed a polymer model in which 3C contact probabilities can be reproduced by a thermodynamic ensemble of configurations in an opportunely shaped energy landscape. Other methods extended the frequency-to-distance mapping widely used in consensus modeling by allowing statistical deviations. For instance, MCMC5C uses the Markov chain Monte Carlo method to generate a representative sample from the posterior probability distribution over the space of structures from 5C (Chromosome Conformation Capture Carbon Copy) data (Rousseau et al., 2011). Another instance is InfMod3Dgen (Wang, Xu, and Zeng, 2015), which uses a polymer chain representation and a Bayesian framework in which the conformational energy from polymer physics is used as prior information to model an ensemble of chromatin structures based on Hi-C data.

A caveat in the resampling procedure comes from the fact that ensemble datasets can include conflicting data from mutually exclusive chromatin conformations. If the complete data is considered simultaneously in a single structure, there can be inconsistencies between data and models (i.e., restraints violations). As a consequence, some methods use only a subset of the data, for instance the most significant contact and non-contact information. TADbit is a package which includes a resampling modeling method (Baù and Marti-Renom, 2012; Serra et al., 2017). The procedure performs many independent optimizations from random starting configurations using the Integrative Modeling Platform (IMP) (Alber et al., 2007; Russel et al., 2012) by restraining the distance between two loci if their interaction frequency Z-score exceeds a cutoff (or keeping them apart if the frequency is below a lower cutoff). This protocol has been applied to model multiple systems, on different scales, including the alpha globulin locus in K562 and lymphoblastoid cells (Baú et al., 2011), the structure of the bacterial genome of *Caulobacter crescentus* (Umbarger et al., 2011), and genomic domains in *Drosophila* melanogaster (Serra et al., 2017).

Similarly, in another software tool called MOGEN, high and low valued frequency pairs in Hi-C maps are considered as contacts and non-contacts (Trieu and Cheng, 2014, 2016). In another study, Gehlen et al. randomly selected a subset from a pool of frequently observed interactions to explore the effect of specific genomic contacts in the yeast genome (Gehlen et al., 2012).

Di Stefano et al. (Di Stefano et al., 2016) selected significant intra-chromosome interactions in human lung fibroblast and stem cells to restrain the structural space of full diploid genomes at very high resolution (~3 kb). The structures are optimized using molecular dynamics, and reproduce the average nuclear positioning of active and lamina-associated genomic regions.

Finally, in a modeling framework developed by Paulsen et al. (Chrom3D), only statistically significant pair-wise interactions between TADs are considered in addition to significant contacts of TADs to the NE derived from lamina-DamID data. It simulates the positions of topologically associated domains (TADs) relative to each other and to the nuclear periphery by using a Monte Carlo optimization of a loss-score function (Paulsen et al., 2017).

11.3.3 Population-based deconvolution methods

Population-based deconvolution approaches generate a population of structures, in which the accumulated contacts overall structures reproduce the probability of contacts in the ensemble Hi-C data rather than each structure individually (Kalhor et al., 2012; Giorgetti et al., 2014; Zhang and Wolynes, 2015). As noted in the previous section, enforcing the full set of detected chromatin contacts may lead to violated restraints. Effectively, population-based methods distribute all expected Hi-C chromatin contacts across different structures in a population. As a result, they allow structures in different conformational states, where each state could contain only a coherent subset of chromatin contacts. For example, Giorgetti et al. developed a physical polymer model which uses an iterative Monte Carlo scheme to generate a thermodynamic ensemble of chromatin conformations (Giorgetti et al., 2014), and reproduced the 5C data well for a region spanning 780 kb. To achieve this goal, specific chromatin contact interaction potentials were optimized to mimic interactions statistically favoring or disfavoring the co-localization of chromatin regions according to the chromosome conformation capture data. The method was used to explore the repertoire of chromatin conformations within TADs.

In another approach, Zhang and Wolynes used the maximum entropy principle and molecular dynamics to model an ensemble of chromosome conformations consistent with a pseudo-Boltzmann distribution for an effective energy landscape that reproduces the experimentally measured pairwise contact frequencies from Hi-C data (Zhang and Wolynes, 2015). The approach was applied to model a population of mouse chromosomes with good agreement with experiment.

Additionally, an ensemble of structures for the *C. crescentus* bacterial genome was recently generated by combining restraint and population-based modeling approaches (Yildirim and Feig, 2018). In this study, a multiscale modeling protocol was performed guided by weak distance restraints derived from available Hi-C data (Le et al., 2013) and resulting models were then further reweighted in order to match the Hi-C data better considering the cell-to-cell variability. The approach resulted in an ensemble of highly structurally variable *C. crescentus* chromosome models that is in agreement with the Hi-C data.

Our group developed a population-based deconvolution method and software package (PGS – Population-based Genome Structure) for modeling structures of complete diploid genomes from Hi-C data (Kalhor et al., 2012; Tjong et al., 2016; Hua et al., 2017). The method performs a structure-based deconvolution of ensemble Hi-C data and generates a large population of distinct diploid 3D genome structures by maximizing the likelihood of observing the data. This is achieved by performing an iterative and step-wise optimization of both the assignment and satisfaction of chromatin contact constraints in the ensemble of genome configurations. Each iteration involves two steps: the constraint assignment (A-step) and the genome structure optimization (M-step). The latter uses a combination of the simulated annealing and conjugate gradient methods applied to each structure of the population (Hestenes and Stiefel, 1952; Kirkpatrick, Gelatt and Vecchi, 1983; Russel et al., 2012). The optimization hardness is increased in a step-wise manner by gradually adding more contact constraints during the iterative optimization process. The optimized genome structures can typically reproduce almost all the contacts from Hi-C experiments and avoid unphysical structures from simultaneous enforcement of conflicting data in the same structure.

We recently released the PGS software package (https://github.com/alberlab/pgs), which takes an experimental Hi-C contact frequency map and a segmentation of the genome sequence into chromatin domains (e.g., TADs) as input data. PGS then produces a population of 3D genome structures with the positions of each TAD in the nucleus, so that the probability of TAD domain contacts in the population reproduces those derived from the Hi-C experiment. The software automatically performs a basic analysis of the structures, including a report of the model quality using the contact probability agreement with experiments and structural features, such as the radial positions of individual chromatin domains in the nucleus.

The genome structures can provide rich sources of structural information, for example the presence of higher-order structural chromatin clusters (as described in (Dai et al., 2016)). For an assessment of the models, it is necessary to compare the predicted structures with independent experimental data which are not included as input information when modeling structures such as distances from 3D FISH experiments, chromatin-NE contact probabilities from lamina-DamID experiments or spatial features extracted from soft X-ray tomography experiments (Tjong et al., 2016). The method has been applied to model the diploid genomes of human lymphoblastoid cells (Kalhor et al., 2012; Tjong et al., 2016; Hua et al., 2017), mouse neutrophil cells (Kalhor et al., 2012; Tjong et al., 2016; Hua et al., 2017; Li et al., 2017), and *Drosophila melanogaster* embryonic cells at resolution ranging from 1Mb to 5Mb (Li et al., 2017) (Figure 11.2).

Since the 3D genome structures can vary dramatically from cell to cell, it is challenging to perform an analysis of structure–function relationships. In particular, it is necessary to distinguish functional chromatin interactions from

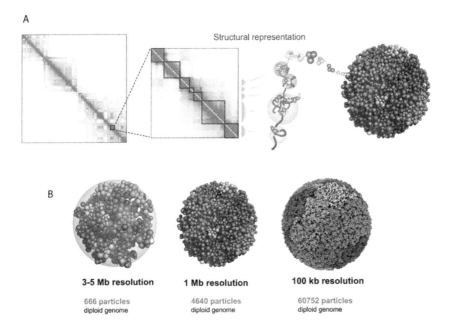

Figure 11.2 Structural representation of genome models. (A) Representation of the genome at the level of topological domains (right panel), which are defined by spheres of specific volumes. Domains boundaries are determined from the Hi-C contact map (left panel) using domain detection algorithms. (B) Structural representation of the genome at different levels of resolution. Also given are the number of particles used to model the genome.

the noise and also from random chromatin collisions in the models. As spatial patterns occurring in multiple structures are expected to be more likely of functional importance, we recently proposed an approach to comprehensively identify 3D chromatin clusters that occur frequently across a population of genome structures (Dai et al., 2016). Applying our method to genomes of human lymphoblastoid at Mb-scale resolution allowed the identification of an atlas of several thousand relatively stable inter-chromosomal chromatin clusters (Dai et al., 2016). Interestingly, a major portion of these clusters, given their enrichment in binding of specific regulatory factors, represents a spectrum of regulatory communities, with transcription communities being the most prevalent type. We revealed two major factors, centromere clustering and transcription factor binding, which significantly stabilize such communities. Finally, we show that the regulatory communities differ dramatically from cell to cell.

11.3.4 Comprehensive data integration through population-based genome structure modeling

Most models of genome structures have relied on just one data source, such as Hi-C, even though a single experimental method cannot typically capture all aspects of the spatial genome organization. However, integrating data from a wide range of technologies, each with complementary strengths and limitations will greatly increase the accuracy, resolution, and coverage of genome structure models. Moreover, integrative models will offer a way to cross-validate the consistency of data obtained from complementary technologies, for example from imaging and genomic technologies. However, it remains a major challenge to develop hybrid methods that can systematically integrate data from many different technologies to generate structural maps of the genome. Population-based deconvolution methods provide an ideal framework for comprehensive data integration. We have recently extended our expectation-maximization modeling framework to integrate variable data sources, for example combining contact frequency information from Hi-C and lamina-DamID data (Li et al., 2017). Here we describe a generalized framework that allows comprehensive integration of many data types (Figure 11.3).

To formalize the process, we can generally divide data into univariate, bivariate, and multivariate data types.

Univariate data describe features that depend on only a single chromosome region, such as the probability of a chromosome region to be in proximity to the lamina at the NE, which is derived from lamina-DamID data. Typically these data can be represented by a probability vector $U = \left(\left(u_I | I = 1,2,..,N \right) \right)$ with elements describing the probability that chromosome regions $I = 1,2,...,N$ share a given feature (e.g., the probability of a chromosome region I to be in proximity with the lamina at the NE). N is the number of all chromosomal regions in the genome (i.e., the total number of TAD domains or 100 kb bins).

Bivariate data describe features that depend on two chromatin regions, for example the probability of interaction between two chromosomal regions. Typically, these data are represented by a probability matrix $M = \left(m_{IJ} | I, J = 1,2,..,N \right)$; for example, a contact probability matrix derived from ensemble Hi-C data. Each

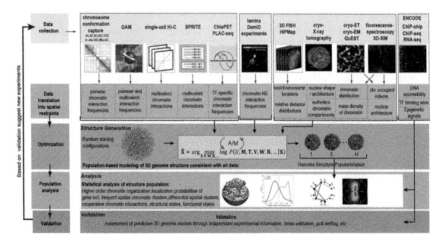

Figure 11.3 Genome structure calculation through data integration and population-based modeling.

element m_{IJ} is the probability that a given contact between chromosomal regions I and J exists in an ensemble of cells.

Multivariate data describe higher-order relationships between multiple chromosome regions, for example, the probability of observing several different chromosome regions to be in spatial proximity to each other. These data are represented by third order or higher-order tensors $\mathbf{T} = (t_{IJK} | I, J, K = 1, 2, .., N)$. The general goal is to generate a population of genome structures ($\mathbf{X} = \{\mathbf{X}_1, \mathbf{X}_2, ..., \mathbf{X}_S\}$ as a set of S diploid genome structures) that are statistically consistent with all available univariate (V), bivariate ($\mathbf{M}$), and multivariate ($\mathbf{T}$) data probabilities.

We formulate this genome structure modeling problem as a maximization of the likelihood $P(U, \mathbf{M}, \mathbf{T} | \mathbf{X})$. With known U, $\mathbf{M}$, and $\mathbf{T}$, we can calculate the structure population $\mathbf{X}$ such that the likelihood $P(V, \mathbf{M}, \mathbf{T} | \mathbf{X})$ is maximized. However, in a diploid genome, each domain has two homologous copies and in most cases the data expressed in the probability vectors U, matrices $\mathbf{M}$ and tensors $\mathbf{T}$ do not distinguish between homologous chromosome copies (i.e., the data is based on unphased data). Moreover, U, $\mathbf{M}$, and $\mathbf{T}$ are probabilities typically derived from ensemble experiments (for example, ensemble Hi-C experiments), therefore they cannot reveal which of the individual features co-exist in the same 3D structure.

To represent information derived from individual cells, we introduce latent variables ($\mathbf{V}, \mathbf{W}, \mathbf{R}$), which contain the information missing from ensemble data, namely which of the observed features belong to each of the S structures in the model population and also which homologous chromosome copies are involved. For example, for univariate lamina-DamID data, the latent variable is a binary matrix, $\mathbf{V} = (v_{is})_{2N \times S}$, which specifies which domain is located near the NE in each structure s of the population and also distinguishes between the two homologous TAD copies ($v_{is} = 1$ indicates that TAD i is located near the NE in structure s; $v_{is} = 0$ otherwise). For bivariate data we introduce the latent variables $\mathbf{W} = (w_{ijs})_{2N \times 2N \times S}$, which is a binary, 3rd-order tensor. For example, in the case of Hi-C data, we define $\mathbf{W}$ as the "contact indicator tensor", which contains the information about which specific domain contacts belong to each of the S structures in the model population and also which homologous chromosome copies are involved ($w_{ijs} = 1$ indicates a contact between domain spheres i and j in structure s; $w_{ijs} = 0$ otherwise). $\mathbf{W}$ is a detailed expansion of $\mathbf{M}$ into a diploid, single-structure representation of the data. The structure population $\mathbf{X}$ is consistent with $\mathbf{W}$. Therefore, the dependence relationship between these three variables is given as $\mathbf{X} \rightarrow \mathbf{W} \rightarrow \mathbf{M}$. Similarly, the dependence relationship between $\mathbf{X}$, $\mathbf{V}$, and U is given as $\mathbf{X} \rightarrow \mathbf{V} \rightarrow U$, because $\mathbf{X}$ is the structure population consistent with $\mathbf{V}$ and $\mathbf{V}$ is a detailed expansion of U at a diploid and single-structure representation of the data. Similarly, we define a latent variable $\mathbf{R}$ to define the missing information in the multivariate data type $\mathbf{T}$.

In addition, we can also consider additional information such as the nuclear volume or excluded volume constraints applicable to all domains that describe the non-overlapping volume of each domain.

Thus, the data integration optimization problem is expressed as:

$$\hat{\mathbf{X}} = \arg \max_{\mathbf{X}, \mathbf{V}, \mathbf{W}, \mathbf{R}} \log P(U, \mathbf{M}, \mathbf{T}, \mathbf{V}, \mathbf{W}, \mathbf{R} | \mathbf{X})$$

$$\text{subject to} \begin{cases} \text{nuclear volume constraint} \\ \text{excluded volume constraint} \end{cases}$$

The log likelihood can be expanded as

$$\log P(U,M,T,V,W,R \mid X) = \log P(U,M,T, \mid V,W,R) P(V,W,R \mid X)$$

$$= \log P(T \mid R) P(M \mid W) P(U \mid V) P(V,W,R \mid X)$$

Because there is no closed form of solution to the problem, we have developed a variant of the Expectation-maximization (EM) method to iteratively optimize this log likelihood (Tjong et al., 2016; Li et al., 2017). Each iteration consists of two steps (Figure 11.4a):

- Assignment step (*A-step*): Given the current model $X^{(t)}$ at the iteration step t, estimate the latent variables $R^{(t+1)}$, $W^{(t+1)}$, and $V^{(t+1)}$ by maximizing the log-likelihood over all possible values of R, W, and V.
- Modeling step (*M-step*): Given the current estimated latent variables $R^{(t+1)}$, $W^{(t+1)}$, and $V^{(t+1)}$, find the model $X^{(t+1)}$ that maximizes the log-likelihood function.

The detailed implementation of the A-step and M-step are described in detail in (Tjong et al., 2016; Hua et al., 2017; Li et al., 2017) (Figure 11.4). The step-wise optimization strategy also includes a gradual increase of the optimization hardness by adding contact constraints gradually with a decreasing contact probability threshold.

As a case study showcasing the benefits of data integration and 3D modeling, we review results from a study, which used the previously mentioned approach to determine the genome structure of *Drosophila melanogaster* embryonic cells by integrating ensemble Hi-C and lamina-DamID data (Li et al., 2017). The genome consists of 4 diploid chromosomes (chromosomes 2, 3, 4, and X) and was segmented at the level of TAD domains, which were represented by the positions of 1169 TAD spheres (Sexton et al., 2012; Li et al., 2017) (Figure 11.4). Chromatin contact probabilities M were derived from ensemble Hi-C experiments (Sexton et al., 2012) and the probabilities for domains to reside at the nuclear envelope V was derived from lamina-DamID experiments (Pickersgill et al., 2006). Following the procedure described above, the likelihood function $\hat{X} = \arg \max_{X,V,W} \log P(V,M,V,W \mid X)$ was optimized to generate a population of 10,000 genome structures that accurately reproduced the domain contact probabilities M and the NE-association probabilities V (Figure 11.4).

The contact probabilities (M) derived from the genome structure models were in excellent agreement with those from Hi-C experiments (average column-based Pearson's correlation coefficient (PCC) is 0.984) (Figure 11.5ab). The genome structures also satisfied almost all imposed contact constraints without constraint violations (e.g., 99.999% of all imposed contact restraints were

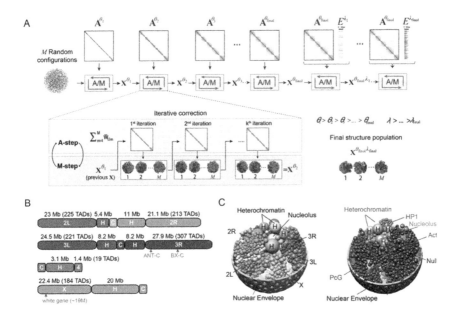

Figure 11.4 Overview of the population-based genome structure modeling approach and its application to the *Drosophila* genome. (A) The initial structures are random configurations. Maximum likelihood optimization is achieved through an iterative process with two steps, assignment (A) and modeling (M). We increase the optimization hardness over several stages by including contacts from the Hi-C matrix A with lower probability thresholds (θ). (B) Schematic of the *Drosophila* genome. Centromeres are labeled "C". Numbers indicate the length of the section in megabases (Mb). The heterochromatic region of each chromosome arm is labeled "H". (C) Snapshot of a single structure randomly picked from the final population (Left panel). (Right panel) The euchromatin domains are colored to reflect their epigenetic class: red–Active, blue–PcG, green–HP1, and dark–Null. Heterochromatin spheres are grey, and the nucleolus is pink. [This figure and caption are taken from (Li et al., 2017)].

satisfied at a tolerance of 0.05). The models also showed excellent agreement with the NE-association probabilities V derived from lamina-DamID experiments (Pearson's correlation of 0.95) (Figure 11.5cd).

11.3.5 Added value from 3D structure modeling

Embedding Hi-C and lamina-DamID data into 3D spatial models can provide substantial added value, namely predictions of structural features that are not visible from the individual input datasets (Figure 11.5). In the following section we highlight some examples showcasing the added values from 3D modeling the *D. melanogaster* genome structures. A quality measure of the structures can be obtained by analyzing how well contact probabilities can be predicted that were not included as input information in the optimization. When leaving out Hi-C data for any pair of TADs whose contact probability was lower than $m_{ij} = 0.06$,

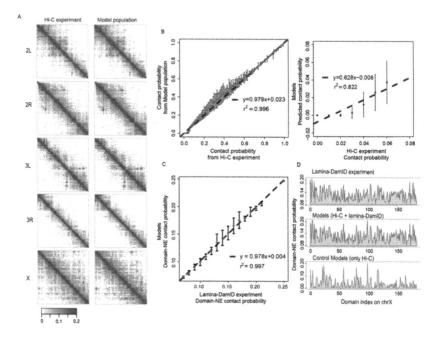

Figure 11.5 Reproduction of Hi-C and lamina-DamID data. (A) Heat maps of intra-arm contact probabilities from Hi-C experiments (left) and intra-arm contact frequencies from the structure population (right). (B) Agreement between the experimental data and model contact probabilities. (Left panel) The blue dot-line is the linear regression line between the average model contact probabilities of each bin and the midpoint Hi-C contact probabilities of the bins. Their Pearson's correlation is 0.998 with p-value < 2.2e−16. (Right panel) Close-up of the agreement between experiment and model for contacts with probabilities less than 6%, which are not used as constraints in our modeling procedure. In this range, Pearson's correlation is 0.907 with p-value = 4.87e-3. (C) The agreement between NE-association frequencies from lamina-DamID experiments and the model population. This figure is plotted in the same way as (B). The structure population well reproduces the input frequencies derived from lamina-DamID data, with a Pearson's correlation of 0.95 and p-value < 2.2e−16. (D) Comparison of experimental and model lamina-DamID frequencies on chrX. The top panel shows the input frequencies derived from the lamina-DamID signal, the middle panel shows the fraction of domains located at the NE in the structure population obtained by Hi-C and lamina-DamID data integration, and the bottom panel shows the fractions obtained in our control structure population generated using only Hi-C data. [Figure and caption are taken from (Li et al., 2017)].

the models were capable of predicting the missing data with good accuracy (Figure 11.5b, right panel). The models also allow predictions about the nuclear locations and interactions of pericentromeric heterochromatin. Due to technical limitations, no Hi-C measurements are available for interactions between repeat sequences, such as those of satellite repeats in pericentromeric heterochromatin.

However, through embedding the data into 3D models, it is possible to analyze the positions of genes with respect to the locations of pericentromeric heterochromatin in the model.

These models revealed distinct differences between heterochromatin localization probabilities for different chromosomes. For example, within the heterochromatin cluster, pericentromeric heterochromatin of chromosomes X and 4 were more often found closer to each other and were more peripheral in their nuclear locations than pericentromeric chromatin of chromosomes 2 and 3. These predictions were confirmed by FISH staining of heterochromatic repeat satellites.

Similarly, it was possible to predict the most probable locations of the nucleolus, which was also confirmed by FISH experiments. The models also provided significant insight into the distinct localization probability of euchromatic regions which were distinct for different chromosomes.

Surprisingly, the models also provided insights into the mechanisms of chromosome pairing. The genome of *D. melanogaster* is characterized by somatic homologous chromosome pairing in the interphase nucleus. The paired chromosomes touch only at a few specific interstitial sites. In the genome structure models, the pairing frequencies of homologous domains showed distinct and reproducible variations along the chromosomes. In the optimized structures, certain domain pairs of homolog TADs consistently have smaller average separations while others consistently have larger separations.

These pairing distance variations are TAD specific. Several proteins are known to affect somatic homolog pairing, one of which is Mrg15 (Smith et al., 2013). Mrg15 binds to chromatin and recruits the CAP-H2 proteins to cause homolog un-pairing. Interestingly, the genome models showed an anti-correlation between Mrg15 binding enrichment in a domain and its frequency of homologous pairing, even though this information is not imposed as input information. The higher the enrichment of Mrg15 in a domain, the lower is the fraction of paired homologs in the genome structures.

This observation also holds true if one focuses only on TADs in the same chromatin state, for example only TADs containing transcriptionally active open chromatin. Active domains observed with relatively high homolog pairing frequency are significantly more enriched with Mrg15 binding sites in comparison to active domains showing low levels of homolog pairing. It is remarkable that these models support the role of Mrg15 in disrupting homolog pairing, even though the structures were generated without any locus-specific constraints on the separation of homologous domains.

Also, the models reproduce the effect that polycomb associated chromatin (PcG) domains show significantly tighter homolog pairing than transcriptionally active domains, confirming previous experimental observations of tight homolog pairing to enhance gene silencing.

The question arises, why are these structural models able to reproduce observations about homolog chromosome pairing, even though the unphased Hi-C data and lamina-DamID data did not distinguish between chromosome copies and therefore could not provide any experimental clues about chromosome pairing? The chromosome pairing frequencies are naturally an indirect consequence of embedding all the data into 3D space of a limited nuclear volume and an excluded volume associated with all chromosome domains. The number of Hi-C contacts per chromatin domain and their location in the nucleus will affect the probability with which homolog domains approach each other in the model and therefore the observed pairing frequency reflects a complex interdependence of all the input constraints in the model. A 3D model allows a simple readout of such complex behaviors.

11.3.6 Data integration increases the accuracy of genome structure models

Finally, we highlight the benefits of integrating variable complementary data sources in this section. With an increasing amount of data, the quality of the model accuracy increases. For example, the accuracy of the NE-association probability in the models can be substantially improved when lamina-DamID data is combined with Hi-C data. A model that is only generated with Hi-C data reproduces the experimental lamina-DamID data with a Pearson's correlation of 0.64 with a p-value $< 2.2e\text{-}16$. When the lamina-DamID data is considered in the modeling, the data is reproduced with a Pearson's correlation of 0.95 and a p-value $< 2.2e\text{-}16$.

The models were also compared with the results from FISH experiments which measured the NE co-localization frequency of 11 loci. The comparison of the NE-association frequency in a model that is generated with only Hi-C data leads to a Spearman correlation coefficient of 0.376 (p-value 0.2542). In contrast, the model that integrates both Hi-C *and* lamina-DamID data reproduces the FISH data with a Spearman correlation of 0.642 (p-value 0.03312). This example highlights the improved accuracy of the models through the integration of complementary data. More importantly, apparently unrelated observables such as frequency of homolog domains are more accurate with the incorporation of lamina-DamID with Hi-C data. The Pearson's correlation between the Mrg15 binding signal and the average frequency of homolog pairing per domain is improved from -0.7 (p-value: 4.46e-4) to -0.81 (p-value: 7.59e-6) when lamina-DamID data was incorporated along with the Hi-C data during the generation of models.

11.4 CONCLUSIONS

Recent years have seen an enhanced interest in the spatial organization of genomes. Technological advances have revealed new insights and increased understanding of the genome folding processes and their functional relevance. A joint and coordinated effort of many research laboratories will facilitate the challenging task of producing quantitative models of the dynamic nucleome. Such models may help to study the role of genome structure in cell differentiation and disease; may allow a better understanding of the mechanistic principles of chromatin folding; and may facilitate the detection of the molecular machinery that shape variations of genome organization in different cells.

ACKNOWLEDGMENTS

This work was supported by the Arnold and Mabel Beckman Foundation (BYI program) (to F.A.), the National Institutes of Health (grant U54DK107981 and grant P41 GM109824 to F.A), and an NSF CAREER grant (1150287 to F.A.).

REFERENCES

Akhtar, W. et al. (2014) 'Using TRIP for genome-wide position effect analysis in cultured cells', *Nature Protocols*. Nature Publishing Group, 9(6), pp. 1255–1281. doi:10.1038/nprot.2014.072.

Alber, F. et al. (2007) 'Determining the architectures of macromolecular assemblies', *Nature*, 450(7170), pp. 683–694. doi:10.1038/nature06404.

Anton, T. et al. (2014) 'Visualization of specific DNA sequences in living mouse embryonic stem cells with a programmable fluorescent CRISPR/Cas system', *Nucleus*, 5(2), pp. 163–172. doi:10.4161/nucl.28488.

Baú, D. et al. (2011) 'The three-dimensional folding of the α-globin gene domain reveals formation of chromatin globules', *Nature Structural and Molecular Biology*, 18(1), pp. 107–115. doi:10.1038/nsmb.1936.

Baù, D. and Marti-Renom, M. A. (2012) 'Genome structure determination via 3C-based data integration by the Integrative Modeling Platform', *Methods*, 58(3), pp. 300–306. doi:10.1016/j.ymeth.2012.04.004.

Beagrie, R. A. et al. (2017) 'Complex multi-enhancer contacts captured by genome architecture mapping', *Nature*, 543(7646), pp. 519–524. doi:10.1038/nature21411.

Beliveau, B. J. et al. (2015) 'Single-molecule super-resolution imaging of chromosomes and in situ haplotype visualization using oligopaint FISH probes', *Nature Communications*, 6(1), p. 7147. doi:10.1038/ncomms8147.

Boettiger, A. N. et al. (2016) 'Super-resolution imaging reveals distinct chromatin folding for different epigenetic states.', *Nature*, 529(7586), pp. 418–422. doi:10.1038/nature16496.

Bolzer, A. et al. (2005) 'Three-dimensional maps of all chromosomes in human male fibroblast nuclei and prometaphase rosettes', *PLoS Biology*, 3(5), pp. 0826–0842. doi:10.1371/journal.pbio.0030157.

Boyle, S. et al. (2011) 'Fluorescence in situ hybridization with high-complexity repeat-free oligonucleotide probes generated by massively parallel synthesis', *Chromosome Research*, 19(7), pp. 901–909. doi:10.1007/s10577-011-9245-0.

Carstens, S., Nilges, M. and Habeck, M. (2016) 'Inferential Structure determination of chromosomes from single-cell Hi-C data', *PLOS Computational Biology*, 12(12), pp. 1–44. doi:10.1371/JOURNAL.PCBI.1005292.

Chen, B. et al. (2013) 'Dynamic imaging of genomic loci in living human cells by an optimized CRISPR/Cas system', *Cell*, 155(7), pp. 1479–1491. doi:10.1016/j.cell.2013.12.001.

Chen, B. C. et al. (2014) 'Lattice light-sheet microscopy: Imaging molecules to embryos at high spatiotemporal resolution', *Science*, 346(6208). doi:10.1126/science.1257998.

Cremer, T. and Cremer, M. (2010) 'Chromosome territories', *Cold Spring Harbor perspectives in biology*, p. a003889. doi:10.1101/cshperspect.a003889.

Cusanovich, D. A. et al. (2015) 'Multiplex single-cell profiling of chromatin accessibility by combinatorial cellular indexing', *Science*, 348(6237), pp. 910–914. doi:10.1126/science.aab1601.

Dai, C. et al. (2016) 'Mining 3D genome structure populations identifies major factors governing the stability of regulatory communities', *Nature Communications*, 7(May), p. 11549. doi:10.1038/ncomms11549.

Darrow, E. M. et al. (2016) 'Deletion of DXZ4 on the human inactive X chromosome alters higher-order genome architecture', *Proceedings of the National Academy of Sciences of the United States of America*, 113(31), pp. E4504-E4512. doi:10.1073/pnas.1609643113.

Dekker, J. et al. (2002) 'Capturing chromosome conformation', *Science*, 295(5558), pp. 1306–1311. doi:10.1126/science.1067799.

Dekker, J. et al. (2017) 'The 4D nucleome project', *Nature*, pp. 219–226. doi:10.1038/nature23884.

Deng, W. et al. (2015) 'CASFISH: CRISPR/Cas9-mediated in situ labeling of genomic loci in fixed cells', *Proceedings of the National Academy of Sciences*, 112(38), pp. 11870–11875. doi:10.1073/pnas.1515692112.

Denker, A. and De Laat, W. (2016) 'The second decade of 3C technologies: Detailed insights into nuclear organization', *Genes and Development*, 30, pp. 1357–1382. doi:10.1101/gad.281964.116.

de Wit, E. et al. (2015) 'CTCF binding polarity determines chromatin looping', *Molecular Cell*, 60(4), pp. 676–684. doi:10.1016/j.molcel.2015.09.023.

Di Pierro, M. et al. (2016) 'Transferable model for chromosome architecture', *Proceedings of the National Academy of Sciences of the United States of America*, 113(43), pp. 12168–12173. doi:10.1073/pnas.1613607113.

Di Stefano, M. et al. (2016) 'Hi-C-constrained physical models of human chromosomes recover functionally-related properties of genome organization', *Scientific Reports*, 6, p. 35985. doi:10.1038/srep35985.

Dixon, J. R. et al. (2012) 'Topological domains in mammalian genomes identified by analysis of chromatin interactions', *Nature*, 485(7398), pp. 376–380. doi:10.1038/nature11082.

Do, M. et al. (2015) 'Imaging and characterizing cells using tomography', *Archives of Biochemistry and Biophysics*, 581, pp. 111–121. doi:10.1016/j.abb.2015.01.011.

Doksani, Y. et al. (2013) 'Super-resolution fluorescence imaging of telomeres reveals TRF2-dependent T-loop formation', *Cell*, 155(2), pp. 345–356. doi:10.1016/j.cell.2013.09.048.

Dostie, J. and Dekker, J. (2007) 'Mapping networks of physical interactions between genomic elements using 5C technology', *Nature Protocols*, 2(4), pp. 988–1002. doi:10.1038/nprot.2007.116.

Duan, Z. et al. (2010) 'A three-dimensional model of the yeast genome', *Nature*, 465(7296), pp. 363–367. doi: nature08973 [pii]\n 10.1038/nature08973.

Flyamer, I. M. et al. (2017) 'Single-nucleus Hi-C reveals unique chromatin reorganization at oocyte-to-zygote transition', *Nature*, 544(7648), pp. 110–114. doi:10.1038/nature21711.

Frazier, Z., Xu, M. and Alber, F. (2017) 'TomoMiner and TomoMinerCloud: A software platform for large-scale subtomogram structural analysis', *Structure*, 25(6), p. 951–961.e2. doi:10.1016/j.str.2017.04.016.

Fu, Y. et al. (2016) 'CRISPR-dCas9 and sgRNA scaffolds enable dual-colour live imaging of satellite sequences and repeat-enriched individual loci', *Nature Communications*, 7, p. 11707. doi:10.1038/ncomms11707.

Fudenberg, G. et al. (2016) 'Formation of chromosomal domains by loop extrusion', *Cell Reports*, 15(9), pp. 2038–2049. doi:10.1016/j.celrep.2016.04.085.

Fullwood, M. J. et al. (2009) 'An oestrogen-receptor-α-bound human chromatin interactome', *Nature*, 462(7269), pp. 58–64. doi:10.1038/nature08497.

Gehlen, L. R. et al. (2012) 'Chromosome positioning and the clustering of functionally related loci in yeast is driven by chromosomal interactions.', *Nucleus*, 3(4), pp. 370–383. doi:10.4161/nucl.20971.

Giorgetti, L. et al. (2014) 'Predictive polymer modeling reveals coupled fluctuations in chromosome conformation and transcription', *Cell*, 157(4), pp. 950–963. doi:10.1016/j.cell.2014.03.025.

Guan, J. et al. (2017) 'Tracking multiple genomic elements using correlative CRISPR imaging and sequential DNA FISH', *Biophysical Journal*, 112(6), pp. 1077–1084. doi:10.1016/j.bpj.2017.01.032.

Guelen, L. et al. (2008) 'Domain organization of human chromosomes revealed by mapping of nuclear lamina interactions', *Nature*, 453(7197), pp. 948–951. doi:10.1038/nature06947.

Hestenes, M. R. and Stiefel, E. (1952) 'Methods of conjugate gradients for solving linear systems', *Journal of Research of the National Bureau of Standards*, 49(6), p. 409. doi:10.6028/jres.049.044.

Hsieh, T. H. S. et al. (2015) 'Mapping Nucleosome Resolution Chromosome Folding in Yeast by Micro-C', *Cell*, 162(1), pp. 108–119. doi:10.1016/j.cell.2015.05.048.

Hsieh, T. H. S. et al. (2016) 'Micro-C XL: Assaying chromosome conformation from the nucleosome to the entire genome', *Nature Methods*, 13(12), pp. 1009–1011. doi:10.1038/nmeth.4025.

Hu, M. et al. (2013) 'Bayesian Inference of Spatial Organizations of Chromosomes', *PLoS Computational Biology*, 9(1), p. e1002893. doi:10.1371/journal.pcbi.1002893.

Hua, N. et al. (2017) 'PGS: A dynamic and automated population-based genome structure software', *Cold Spring Harbor Labs Journals*. doi:10.1101/103358.

Imakaev, M. et al. (2012) 'Iterative correction of Hi-C data reveals hallmarks of chromosome organization', *Nature Methods*, 9(10), pp. 999–1003. doi:10.1038/nmeth.2148.

Imakaev, M. V., Fudenberg, G. and Mirny, L. A. (2015) 'Modeling chromosomes: Beyond pretty pictures', *FEBS Letters*, pp. 3031–3036. doi:10.1016/j.febslet.2015.09.004.

Junier, I. et al. (2012) 'CTCF-mediated transcriptional regulation through cell type-specific chromosome organization in the β-globin locus', *Nucleic Acids Research*, 40(16), pp. 7718–7727. doi:10.1093/nar/gks536.

Kalhor, R. et al. (2012) 'Genome architectures revealed by tethered chromosome conformation capture and population-based modeling', *Nature Biotechnology*, 30(1), pp. 90–98. doi:10.1038/nbt.2057.

Kind, J. et al. (2013) 'Single-cell dynamics of genome-nuclear lamina interactions', *Cell*. Cell Press, 153(1), pp. 178–192. doi:10.1016/j.cell.2013.02.028.

Kind, J. et al. (2015) 'Genome-wide maps of nuclear lamina interactions in single human cells', *Cell*, 163(1), pp. 134–147. doi:10.1016/j.cell.2015.08.040.

Kirkpatrick, S., Gelatt, C. D. and Vecchi, M. P. (1983) 'Optimization by simulated annealing', *Science*, 220, pp. 671–680.

Le Gros, M. A. et al. (2016) 'Soft x-ray tomography reveals gradual chromatin compaction and reorganization during neurogenesis in vivo', *Cell Reports*, 17(8), pp. 2125–2136. doi:10.1016/j.celrep.2016.10.060.

Le, T. B. et al. (2013) 'High-resolution mapping of the spatial organization of a bacterial chromosome', *Science (New York, N.Y.)*, 342(6159), pp. 731–734. doi:10.1126/science.1242059.

Legant, W. R. et al. (2016) 'High-density three-dimensional localization microscopy across large volumes', *Nature Methods*, 13(4), pp. 359–365. doi:10.1038/nmeth.3797.

Lesne, A. et al. (2014) '3D genome reconstruction from chromosomal contacts', *Nat Methods*, 11(11), pp. 1141–1143. doi:10.1038/nmeth.3104.

Li, Q. et al. (2017) 'The three-dimensional genome organization of *Drosophila melanogaster* through data integration', *Genome Biology*, 18(1), p. 145. doi:10.1186/s13059-017-1264-5.

Lieberman-aiden, E. et al. (2009) 'Comprehensive mapping of long-range interactions reveals folding principles of the human genome', *Science (New York, N.Y.)*, 326(October), pp. 289–294. doi:10.1126/science.1181369.

Lubeck, E. et al. (2014) 'Single-cell in situ RNA profiling by sequential hybridization', *Nature Methods*, pp. 360–361. doi:10.1038/nmeth.2892.

Ma, H. et al. (2015) 'Multicolor CRISPR labeling of chromosomal loci in human cells', *Proceedings of the National Academy of Sciences*, 112(10), pp. 3002–3007. doi:10.1073/pnas.1420024112.

Ma, H. et al. (2016) 'Multiplexed labeling of genomic loci with dCas9 and engineered sgRNAs using CRISPRainbow', *Nature Biotechnology*, 34(5), pp. 528–530. doi:10.1038/nbt.3526.

Ma, H., Reyes-Gutierrez, P. and Pederson, T. (2013) 'Visualization of repetitive DNA sequences in human chromosomes with transcription activator-like effectors', *Proceedings of the National Academy of Sciences*, 110(52), pp. 21048–21053. doi:10.1073/pnas.1319097110.

Mahamid, J. et al. (2016) 'Visualizing the molecular sociology at the HeLa cell nuclear periphery', *Science*, 351(6276), pp. 969–972. doi:10.1126/science.aad8857.

Meluzzi, D. and Arya, G. (2013) 'Recovering ensembles of chromatin conformations from contact probabilities', *Nucleic acids research*, 41(1), pp. 63–75. doi:10.1093/nar/gks1029.

Mifsud, B. et al. (2015) 'Mapping long-range promoter contacts in human cells with high-resolution capture Hi-C', *Nature Genetics*, 47(6), pp. 598–606. doi:10.1038/ng.3286.

Miyanari, Y., Ziegler-Birling, C. and Torres-Padilla, M. E. (2013) 'Live visualization of chromatin dynamics with fluorescent TALEs', *Nature Structural and Molecular Biology*, 20(11), pp. 1321–1324. doi:10.1038/nsmb.2680.

Müller, P. et al. (2010) 'COMBO-FISH enables high precision localization microscopy as a prerequisite for nanostructure analysis of genome loci.', *International journal of molecular sciences*, 11(10), pp. 4094–4105. doi:10.3390/ijms11104094.

Nagano, T. et al. (2013) 'Single-cell Hi-C reveals cell-to-cell variability in chromosome structure', *Nature*, 502(7469), pp. 59–64. doi:10.1038/nature12593.

Nagano, T. et al. (2015) 'Single-cell Hi-C for genome-wide detection of chromatin interactions that occur simultaneously in a single cell.', *Nature protocols*, 10(12), pp. 1986–2003. doi:10.1038/nprot.2015.127.

Nagano, T. et al. (2017) 'Cell-cycle dynamics of chromosomal organization at single-cell resolution', *Nature*, 547(7661), pp. 61–67. doi:10.1038/nature23001.

Nozaki, T. et al. (2017) 'Dynamic organization of chromatin domains revealed by super-resolution live-cell imaging', *Molecular Cell*, 67(2), pp. 282–293.e7. doi:10.1016/j.molcel.2017.06.018.

Oikonomou, C. M., Jensen, G. J. (2017). 'Cellular electron cryotomography: Toward structural biology in situ'. *Annual Review of Biochemistry*, 86, 873–896.

Ou, H. D. et al. (2017) 'ChromEMT: Visualizing 3D chromatin structure and compaction in interphase and mitotic cells', *Science*, 357(6349), p. eaag0025. doi:10.1126/science.aag0025.

Paulsen, J. et al. (2017) 'Chrom3D: Three-dimensional genome modeling from Hi-C and nuclear lamin-genome contacts', *Genome Biology*, 18(1), p. 21. doi:10.1186/s13059-016-1146-2.

Peng, C. et al. (2013) 'The sequencing bias relaxed characteristics of Hi-C derived data and implications for chromatin 3D modeling', *Nucleic Acids Research*, 41(19), p. e183. doi:10.1093/nar/gkt745.

Phan, S. et al. (2017) '3D reconstruction of biological structures: Automated procedures for alignment and reconstruction of multiple tilt series in electron tomography', *Advanced Structural and Chemical Imaging*, 2(1), p. 8. doi:10.1186/s40679-016-0021-2.

Pickersgill, H. et al. (2006) 'Characterization of the *Drosophila* melanogaster genome at the nuclear lamina', *Nature Genetics*, 38(9), pp. 1005–1014. doi:10.1038/ng1852.

Quinodoz, S. A. et al. (2017) 'Higher-order inter-chromosomal hubs shape 3-dimensional genome organization in the nucleus', *bioRxiv*, p. 219683. doi:10.1101/219683.

Ramani, V. et al. (2016) 'Mapping 3D genome architecture through in situ DNase Hi-C', *Nature Protocols*, 11(11), pp. 2104–2121. doi:10.1038/nprot.2016.126.

Ramani, V. et al. (2017) 'Massively multiplex single-cell Hi-C', *Nature Methods*, 14(3), pp. 263–266. doi:10.1038/nmeth.4155.

Rao, S. S. P. et al. (2014) 'A 3D map of the human genome at kilobase resolution reveals principles of chromatin looping', *Cell*, 159(7), pp. 1665–1680. doi:10.1016/j.cell.2014.11.021.

Rousseau, M. et al. (2011) 'Three-dimensional modeling of chromatin structure from interaction frequency data using Markov chain Monte Carlo sampling', *BMC Bioinformatics*, 12(1), p. 414. doi:10.1186/1471-2105-12-414\ r1471-2105-12-414 [pii].

Russel, D. et al. (2012) 'Putting the pieces together: Integrative modeling platform software for structure determination of macromolecular assemblies', *PLoS Biology*, 10(1). doi:10.1371/journal.pbio.1001244.

Sanborn, A. L. et al. (2015) 'Chromatin extrusion explains key features of loop and domain formation in wild-type and engineered genomes', *Proceedings of the National Academy of Sciences*, 112(47), p. 201518552. doi:10.1073/pnas.1518552112.

Schoenfelder, S. et al. (2015) 'Polycomb repressive complex PRC1 spatially constrains the mouse embryonic stem cell genome', *Nature Genetics*, 47(10), pp. 1179–1186. doi:10.1038/ng.3393.

Schwarzer, W. et al. (2017) 'Two independent modes of chromatin organization revealed by cohesin removal', *Nature*, 551(7678), pp. 51–56. doi:10.1038/nature24281.

Serra, F. et al. (2015) 'Restraint-based three-dimensional modeling of genomes and genomic domains', *FEBS Letters*, 589(20), pp. 2987–2995. doi:10.1016/j.febslet.2015.05.012.

Serra, F. et al. (2017) 'Automatic analysis and 3D-modelling of Hi-C data using TADbit reveals structural features of the fly chromatin colors', *PLoS Computational Biology*, 13(7), p. e1005665. doi:10.1371/journal.pcbi.1005665.

Sexton, T. et al. (2012) 'Three-dimensional folding and functional organization principles of the *Drosophila* genome', *Cell*, pp. 458–472. doi:10.1016/j.cell.2012.01.010.

Shachar, S. et al. (2015) 'Identification of gene positioning factors using high-throughput imaging mapping', *Cell*, 162(4), pp. 911–923. doi:10.1016/j.cell.2015.07.035.

Shachar, S., Pegoraro, G. and Misteli, T. (2015) 'HIPMap: A high-throughput imaging method for mapping spatial gene positions', *Cold Spring Harbor Symposia on Quantitative Biology*, 80, pp. 73–81. doi:10.1101/sqb.2015.80.027417.

Shen, Y. et al. (2012) 'A map of the cis-regulatory sequences in the mouse genome', *Nature*, 488(7409), pp. 116–120. doi:10.1038/nature11243.

Smith, E. A. et al. (2014) 'Quantitatively imaging chromosomes by correlated cryo-fluorescence and soft x-ray tomographies', *Biophysical Journal*, 107(8), pp. 1988–1996. doi:10.1016/j.bpj.2014.09.011.

Smith, H. F. et al. (2013) 'Maintenance of interphase chromosome compaction and homolog pairing in *Drosophila* is regulated by the condensin Cap-H2 and its partner Mrg15', *Genetics*, 195(1), pp. 127–146. doi:10.1534/genetics.113.153544.

Stevens, T. J. et al. (2017) '3D structures of individual mammalian genomes studied by single-cell Hi-C', *Nature*, 544(7648), pp. 59–64. doi:10.1038/nature21429.

Strongin, D. E., Groudine, M. and Politz, J. C. R. (2014) 'Nucleolar tethering mediates pairing between the IgH and Myc loci', *Nucleus*, 5(5), pp. 474–481. doi:10.4161/nucl.36233.

Szalaj, P. et al. (2016) '3D-GNOME: An integrated web service for structural modeling of the 3D genome', *Nucleic Acids Research*, 44(W1), p. gkw437. doi:10.1093/nar/gkw437.

Takei, Y. et al. (2017) 'Multiplexed dynamic imaging of genomic loci by combined CRISPR imaging and DNA sequential FISH', *Biophysical Journal*, 112(9), pp. 1773–1776. doi:10.1016/j.bpj.2017.03.024.

Tang, Z. et al. (2015) 'CTCF-mediated human 3D genome architecture reveals chromatin topology for transcription', *Cell*, 163(7), pp. 1611–1627. doi:10.1016/j.cell.2015.11.024.

Tjong, H. et al. (2016) 'Population-based 3D genome structure analysis reveals driving forces in spatial genome organization', *Proceedings of the National Academy of Sciences of the United States of America*, 113(12), pp. E1663–E1672. doi:10.1073/pnas.1512577113.

Tolhuis, B. et al. (2002) 'Looping and interaction between hypersensitive sites in the active β-globin locus', *Molecular Cell*, 10(6), pp. 1453–1465. doi:10.1016/S1097-2765(02)00781-5.

Trieu, T. and Cheng, J. (2014) 'Large-scale reconstruction of 3D structures of human chromosomes from chromosomal contact data', *Nucleic acids research*, 42(7), p. e52. doi:10.1093/nar/gkt1411.

Trieu, T. and Cheng, J. (2016) 'MOGEN: A tool for reconstructing 3D models of genomes from chromosomal conformation capturing data', *Bioinformatics*, 32(9), pp. 1286–1292. doi:10.1093/bioinformatics/btv754.

Umbarger, M. A. et al. (2011) 'The three-dimensional architecture of a bacterial genome and its alteration by genetic perturbation', *Molecular Cell*, 44(2), pp. 252–264. doi:10.1016/j.molcel.2011.09.010.

Van De Werken, H. J. G. et al. (2012) '4C technology: Protocols and data analysis', *Methods in Enzymology*, 513, pp. 89–112. doi:10.1016/B978-0-12-391938-0.00004-5.

Van Steensel, B. and Henikoff, S. (2000) 'Identification of in vivo DNA targets of chromatin proteins using tethered Dam methyltransferase', *Nature Biotechnology*, 18(4), pp. 424–428. doi:10.1038/74487.

Varoquaux, N. et al. (2014) 'A statistical approach for inferring the 3D structure of the genome', *Bioinformatics*, pp. i26–i33. doi:10.1093/bioinformatics/btu268.

Vogel, M. J., Peric-Hupkes, D. and van Steensel, B. (2007) 'Detection of in vivo protein – DNA interactions using DamID in mammalian cells', *Nature Protocols*, 2(6), pp. 1467–1478. doi:10.1038/nprot.2007.148.

Wang, S. et al. (2016) 'Spatial organization of chromatin domains and compartments in single chromosomes', *Science*, 353(6299), pp. 598–602. doi:10.1126/science.aaf8084.

Wang, S., Xu, J. and Zeng, J. (2015) 'Inferential modeling of 3D chromatin structure.', *Nucleic acids research*, 43(8), p. e54. doi:10.1093/nar/gkv100.

Weiland, Y., Lemmer, P. and Cremer, C. (2011) 'Combining FISH with localisation microscopy: Super-resolution imaging of nuclear genome nanostructures', *Chromosome Research*, 19(1), pp. 5–23. doi:10.1007/s10577-010-9171-6.

Yaffe, E. and Tanay, A. (2011) 'Probabilistic modeling of Hi-C contact maps eliminates systematic biases to characterize global chromosomal architecture', *Nature Genetics*, 43(11), pp. 1059–1065. doi:10.1038/ng.947.

Yáñez-Cuna, J. O. and van Steensel, B. (2017) 'Genome–nuclear lamina interactions: From cell populations to single cells', *Current Opinion in Genetics and Development*, pp. 67–72. doi:10.1016/j.gde.2016.12.005.

Yildirim, A. and Feig, M. (2018) 'High-resolution 3D models of Caulobacter crescentus chromosome reveal genome structural variability and organization', *Nucleic Acids Research*, 46(8), pp. 3937–3952. Available at: http://dx.doi.org/10.1093/nar/gky141.

Zhang, B. and Wolynes, P. G. (2015) 'Topology, structures, and energy landscapes of human chromosomes', *Proceedings of the National Academy of Sciences*, 112(19), pp. 6062–6067. doi:10.1073/pnas.1506257112.

Zhang, J. et al. (2012) 'ChIA-PET analysis of transcriptional chromatin interactions', *Methods*, 58(3), pp. 289–299. doi:10.1016/j.ymeth.2012.08.009.

Zhang, Y. et al. (2012) 'Spatial organization of the mouse genome and its role in recurrent chromosomal translocations', *Cell*, 148(5), pp. 908–921. doi:10.1016/j.cell.2012.02.002.

Zhang, Z. et al. (2013) '3D chromosome modeling with semi-definite programming and Hi-C data', *Journal of Computational Biology: A Journal of Computational Molecular Cell Biology*, 20(11), pp. 831–846. doi:10.1089/cmb.2013.0076.

Zou, C., Zhang, Y. and Ouyang, Z. (2016) 'HSA: Integrating multi-track Hi-C data for genome-scale reconstruction of 3D chromatin structure', *Genome Biology*, 17(1), p. 40. doi:10.1186/s13059-016-0896-1.

Modeling the Conformational Ensemble of Mammalian Chromosomes from 5C/Hi-C Data

GUIDO TIANA AND LUCA GIORGETTI

12.1 INTRODUCTION

Mammalian chromosomes are extraordinarily complex polymers, composed of $\sim 10^{11}$ atoms each. If the molecular structure of the chromatin fiber up to the kilobase (kb) scale has been widely studied *in vitro* and computationally (Bascom and Schlick, 2018), little is known about its fine-scale (sub-kb) structural properties *in vivo* (Belmont, 2014), although advances in electron microscopy have recently started to shed some light on this long-standing open question (Ou et al., 2017). On the other hand, the recent deployment and refinement of 3C-based techniques and Hi-C in particular have provided an extraordinary amount of information on how the chromatin fiber is folded inside a chromosome at larger genomic length scales ($\geq$10 kb). The introductory chapter by Job Dekker in this book (Chapter 1) gives a thorough overview of 3C methods, and the findings they have enabled. Here, we will only recall that 3C-based experiments have revealed that mammalian chromosomes are folded in a highly non-random manner, with organizational units spanning several orders of magnitude in genomic lengths. Dynamic loops of ~100 kb connecting sites bound by CTCF are embedded into a complex hierarchy of sub-megabase interaction domains including topologically associating domains (TADs), themselves merging into even larger multi-megabase compartments arising from the exclusive interactions of active and inactive chromosomal segments (Gibcus and Dekker, 2013).

An ever-growing body of experimental evidence suggests that chromosomal structures and TADs in particular play an important role in the control of gene expression, by partitioning the physical interactions between genes and their long-range regulatory sequences such as enhancers, and specifically favoring those that connect functionally related genetic elements. (Nora et al., 2012; Dixon et al., 2012; Galupa and Heard, 2018; Zhan et al., 2017). 3C-based data, however, represent average contacts over populations of millions of cells and do not give direct access to the actual conformations of the chromatin fiber in single cells, and how they evolve in time. Measurements based on DNA in situ hybridization (DNA FISH), which can measure the spatial position of selected genomic loci in single cells, revealed extensive variation in the three-dimensional conformation of chromosomes at all scales (Giorgetti and Heard, 2016). Measuring the cell-to-cell and temporal variability in chromosome conformation is therefore important to better understand how genes communicate with their enhancers, and how transcription is controlled in single cells.

Physical models of the chromatin fiber represent valuable tools in extracting the ensemble of conformations of the chromatin fiber that are "hidden" in 3C-based data, and in predicting its dynamics. Since loops and TADs, as well as enhancer–promoter interactions, occur over genomic length scales that are at least two orders of magnitude larger than those where the chromatin structure is partially understood (at least *in vitro* and computationally), physical models aiming at describing the long-range conformational properties of the genome include a significant degree of coarse-graining. Indeed, coarse-grained mechanistic models such as the loop-extrusion model (see Chapter 4) have provided important insights into the processes that might give rise to the structures observed in

3C-based experiments. However, they are not able to infer the single-cell conformations of the chromatin fiber that, once averaged over the cell population, give rise to the contacts measured at a specific genomic location in a specific 3C-based experiment. By contrast, the computational approach we describe in this chapter (which we have developed in (Giorgetti et al., 2014; Tiana et al., 2016; Zhan, Giorgetti, and Tiana, 2017)) rather focuses on the task of reconstructing the conformational ensemble of the chromatin fiber across a cell population, at the level of single TADs, starting from population-averaged 5C or Hi–C contact maps, without making any mechanistic assumption on the specific forces that drive chromosomal structures.

The model assumes that at the level of single TADs (within the 100 kb–1 Mb length scale), the set of conformations that the chromatin fiber adopts in a cell population can be described as the equilibrium ensemble of an interacting model polymer. Monomers in the polymer represent chromosomal segments that are long enough (well above the kb) that their internal molecular structure can be disregarded. Contrary to first-principle models, the model is completely agnostic on the molecular origins of the interactions between monomers. These latter, in fact, are only suggested by the experimental data (see Section 12.2) and eventually can be interpreted in terms of effective interactions accounting for the effects of specific proteins (e.g., CTCF) that mediate the spatial proximity of their DNA-binding sites, as well as more indirect effects due to nonspecific interactions in the nuclear milieu.

It is unlikely that an entire mammalian chromosome reaches equilibrium anytime during the ~20 hours of a cell cycle after expanding from the compact mitotic conformation. Indeed, the statistical properties of chromosomal contacts across several genomic length scales are compatible with those of an out-of-equilibrium (*crumpled*) globule (Lieberman-Aiden et al., 2009). On the other hand, however, time-lapse fluorescence microscopy experiments suggest that conformational changes within Mb-sized domains might take place on the time scale of minutes. Thus, the chromatin fiber within single TADs might have time to equilibrate in one single cell cycle (Tiana et al., 2016). Although the validity of the equilibrium ensemble cannot be formally proven, it is strongly supported by the fact that equilibrium models of TADs located at the X inactivation center in mouse embryonic stem cells provided realistic reconstructions of the 3D configuration of the chromatin fiber, including its cell-to-cell variability, and correctly predicted the structural effects deleting clusters of CTCF sites (see Section 12.4 and Giorgetti et al., 2014). This model was instrumental in demonstrating that single TADs arise from a multiplicity of conformations of the chromatin fiber, and that cell-to-cell differences in chromosome structure within TADs can correlate with gene transcription levels (Giorgetti et al., 2014).

12.2 SETTING UP THE MODEL

The general aim of our strategy is to build a coarse-grained polymer model of a sub-Mb chromosomal region, whose equilibrium ensemble reproduces experimental 3C-based data.

12.2.1 The experimental data

The input of the model is 3C-based contact matrices representing the three-dimensional interactions of the chromatin fiber across hundreds of kilobases of genomic sequences, such as 5C and Hi-C (Dekker et al., 2002) (see Chapter 1). 5C describes the chromosomal contacts across a target-enriched, selected genomic region spanning up to a few megabases. In the left panel of Figure 12.1, for example, we show 5C data from the central part of the X inactivation center in mouse embryonic stem cells, and highlight the two TADs that harbor the *Tsix* and *Xist* transcripts (Nora et al., 2012), which we simulated in (Giorgetti et al., 2014). 5C was performed in this case using the HindIII restriction enzyme and has an average genomic resolution of 3 kb. The 5C contact matrix elements C_{ij} are proportional to the number of times a ligation product between *single restriction fragments i* and *j* have been sequenced in the experiment. Since restriction fragments have different genomic sizes, the matrix representation in Figure 12.1 is not symmetric (each restriction fragment being represented with its actual genomic length). Hi-C data, in contrast, describe chromosomal interactions across the entire genome and their resolution strongly depends on high-throughput sequencing depth. In the Hi-C matrix subset shown in the right panel of Figure 12.1 for example, the data (Giorgetti et al., 2016) were binned at 20 kb resolution. It should be noted that Hi-C data are affected by experimental biases arising from the uneven mappability of genomic bins, as well as other factors that have been thoroughly discussed elsewhere (Imakaev et al., 2012; Yaffe and Tanay, 2011). Iterative data normalization procedures such as ICE (Imakaev et al., 2012) must be necessarily applied to the data before they can be used for modeling. The normalized Hi-C contact matrix elements C_{ij} are thus proportional to the number of times that any genomic sequences belonging to the *genomic bins i* and *j*

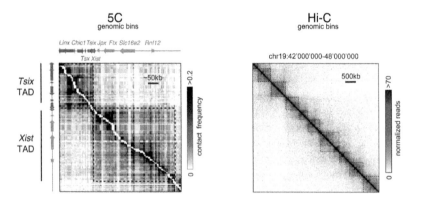

Figure 12.1 Left: 5C contact map of the *Tsix* and *Xist* TADs in the X inactivation center on the X chromosome in mouse embryonic stem cells (mESC). Dashed lines indicate the boundaries of the two TADs as described in (Nora et al., 2012) and (Giorgetti et al., 2016). Right: Iteratively corrected Hi-C interaction map of a region on chromosome 19 in mESCs. Dashed lines indicate TAD boundaries identified using the CaTCH algorithm (Zhan et al., 2017).

have been sequenced in the experiments. Note that the Hi-C matrix is symmetric since all genomic bins have the same length.

The central assumption that our model makes is that in both 5C and Hi-C, the matrix count C_{ij} (where i and j are single restriction fragments in 5C or larger genomic bins in Hi-C) is proportional to equilibrium contact probability between i and j, that is

$$p_{ij}^{exp} \equiv \langle \Delta(r_i, r_j) \rangle = \frac{C_{ij}}{z}, \tag{12.1}$$

where the contact probability p_{ij} between loci i and j is defined as the equilibrium average of a contact function $\Delta(r_i, r_j)$, which will be formally defined later and which is equal to 1 if the spatial positions r_i and r_j of the two loci are sufficiently close to each other, and zero otherwise. z is a proportionality constant, whose numerical value needs to be determined based on the experimental data. We generally assume $p_{ij} = 1$ if $C_{ij} > z$.

We usually set the numerical value of z to be equal to the average count of adjacent genomic segments (i.e., i and $i \pm 1$), as it is reasonable to assume that they are always in contact with each other (at least at their extremities). In the case of the 5C map displayed in Figure 12.1, for example, this procedure gives $z = 3097$ (Giorgetti et al., 2014). interestingly, this value coincides with the numerical value where the distribution of counts from non-adjacent segments (C_{ij} with $i \neq j \pm 1$) drops to zero at $C \approx 3000$ (cf. Figure 12.2). Thus, this value can be reasonably associated with the maximum contact probability based on the available contact data.

Since 5C and Hi-C experiments are performed in replicates, it is possible to associate a standard error of σ_{ij} to each count C_{ij}.

12.2.2 Geometry of the model

Due to the molecular complexity of the chromatin fiber (and the fact that its exact microscopic structure is unknown at sub-kilobase scale), we employ a

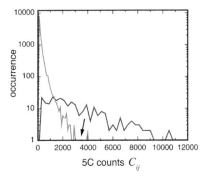

Figure 12.2 Histograms of 5C counts C_{ij} associated with adjacent ($i=j \pm 1$, green) and non-adjacent ($i>j+1$, blue) restriction fragments within the Tsix/Xist domains. The arrow marks the median of non-adjacent counts.

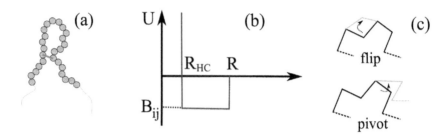

Figure 12.3 (a) Coarse-grained polymer model of the chromatin fiber. (b) The symmetrical square-well potential between beads. The interaction energy B_{ij} can be either negative (attractive) or positive (repulsive). (c) The elementary moves of the Monte Carlo algorithm.

coarse-grained approach and describe it as a chain of beads on a string (see Figure 12.3a). The position of each bead is identified by its Cartesian coordinates r_i. The distance $a \equiv |r_i - r_{i+1}|$ between consecutive beads is kept fixed and equal to the size of the experimental restriction fragment (in 5C) or bin (in Hi-C), expressed in kb. This is the most natural choice, as the model has the same resolution as the experimental data it aims at describing.

An important assumption that is implicitly made at this stage is that the linear density of the chromatin fiber is constant across the simulated region (i.e., all segments of a given number of kb display the same linear length). This is by no means obvious, as every bead represents a complex molecular system composed of DNA, nucleosomes, and DNA-binding proteins which can display differential internal rearrangements depending on the local abundance of histone modifications and the identity of the proteins bound to DNA. This assumption seems, however, to be reasonable in the case of the *Tsix* TAD (Giorgetti et al., 2016) where orthogonal experiments based on DNA FISH validated the conformations predicted by the model (see Section 12.4).

We will make no assumption initially on the numerical value of the base-pair density, as in the simulations all lengths will be expressed in terms of the bead distance a. This value will rather be obtained *a posteriori* by comparing the results of the simulations with independent experiments based on DNA FISH (see Section 3.2).

12.2.3 Choice of interactions: general principles

The core of the modeling strategy consists in finding interaction potentials between beads such that the equilibrium contact map of the model is as close as possible to the experimental data. To this aim, we rely on the principle of maximum entropy (Jaynes, 1957), which states that in the least-biased model, the associated probability distribution maximizes Shannon's entropy $S[P] = -\sum_{\{r_i\}} P \log P$ under the constraints

$$\sum_{\{r_k\}} P(\{r_k\}) \Delta(r_i, r_j) = p_{ij}^{exp},$$
(12.2)

i.e., that the average contact function matches the experimental data, and that the probability is normalized $\left(\sum_{\{r_k\}} P = 1\right)$. In other words, the model that maximizes the entropy is the one that guarantees that the minimum amount of subjective information (on top of the experimental data) is provided. The constrained maximization of the entropy gives

$$P\left(\{r_k\}\right) = \frac{1}{Z} \exp\left[-\sum_{ij} \lambda_{ij} \Delta\left(r_i, r_j\right)\right],\tag{12.3}$$

where λ_{ij} are the Lagrange multipliers which, in principle, set the averages by $\partial \log Z / \partial \lambda_{ij} = p_{ij}^{exp}$. However, this is a set of implicit equations involving the partition function, which is numerically intractable. A strategy to find the numerical values of the Lagrange multipliers is thus needed.

Together with Eq. (3), one can use the hypothesis that the polymer is at equilibrium to write the probability distribution according to Boltzmann, i.e.,

$$P\left(\{r_k\}\right) = \frac{1}{Z} \exp\left[-\frac{U\left(\{r_i\}\right)}{kT}\right],\tag{12.4}$$

where k is the Boltzmann constant and T is the temperature of the system. Comparing Eqs. (3) and (4) one finds

$$U\left(\{r_i\}\right) = \sum_{ij} B_{ij} \Delta\left(r_i, r_j\right),\tag{12.5}$$

with $B_{ij} = kT\lambda_{ij}$. Thus, the interaction potential of the maximum-entropy model is the sum of contact functions, modulated by energy parameters, proportional to the Lagrange multipliers of Eq. (3), which need to be determined.

12.2.4 Choice of interactions: implementation

The maximum entropy principle suggests the functional form of the interaction potential (Eq. 5), but does not provide any hint on how to find its numerical parameters. It asserts that the functional form of the potential should be the same as the quantity Δ_{ij} that is averaged in the experiment. Although we do not know exactly the functional form of the contact function in a 5C/Hi-C experiment, it is reasonable to assume a simple, spherical-well potential as depicted in Figure 12.3b, thus corresponding to a contact potential in the form

$$B_{ij}\Delta\left(\left|r_i - r_j\right|\right) = \begin{cases} +\infty & \text{if } \left|r_i - r_j\right| < R_{HC} \\ B_{ij} & \text{if } R_{HC} \leq \left|r_i - r_j\right| < R \\ 0 & \text{if } \left|r_i - r_j\right| \geq R \end{cases}\tag{12.6}$$

This choice introduces two further parameters, namely the hard-core radius R_{HC} and the interaction range R. We will consider them as meta parameters, build a maximum-entropy model for each of them, and eventually select the model which returns the best agreement with the experimental data.

In order to find the interaction matrix B_{ij}, we use an iterative Monte Carlo (MC) scheme to optimize the χ^2 between the experimental contact probabilities p_{ij}^{exp} and those (p_{ij}) that are back-calculated from the simulation as

$$p_{ij} \equiv \langle \Delta(r_i, r_j) \rangle. \tag{12.7}$$

The χ^2 is defined as

$$\chi^2 = \frac{1}{(N-3)(N-4)} \sum_{i<j-2} \frac{(p_{ij} - p_{ij}^{exp})^2}{\sigma_{ij}^2}, \tag{12.8}$$

where σ_{ij} is the standard deviation of the normalized counts obtained from replicate experiment, N is the length of the chain and only the pairs of beads separated by at least two other beads are considered in the sum.

We start from an initial guess for the interaction matrix, namely

$$B_{ij} = -\log \frac{p_{ij}^{exp}}{1 - p_{ij}^{exp}} - \frac{3}{2}\log|j-i|, \tag{12.9}$$

corresponding to the hypothesis of independent contact formation, with an entropy term borrowed from an ideal chain. We then perform a MC simulation (Section. 3.1) at temperature $T = 1$, corresponding to the temperature of the experiment in arbitrary units, recording the sampled conformations every n_{str} step. The optimal choice of n_{str} corresponds to the decorrelation time of the Markov chain generated by the Monte Carlo simulation; with the scheme described in 3.1 it corresponds to $n_{str} \approx 5000$. The initial χ^2 is calculated by evaluating the p_{ij} by Eq. (7) and inserting them into Eq. (8). In Eq. (9) we have also set the Boltzmann constant k equal to unity, thus, expressing temperature and energies in the same arbitrary units.

To optimize the interaction matrix B_{ij} so that the model reproduces the experimental contact probabilities, we perform a random minimization of χ^2 by varying a random matrix element by a random value extracted from a Gaussian distribution centered on zero and with standard deviation σ_R, then recalculate the new χ^2 and accept the change in B_{ij} only if χ^2 is decreased.

To make this optimization efficient, we use a reweighing scheme (Figure 12.4) similar to the one described in (Norgaard, Ferkinghoff-Borg, and Lindorff-Larsen, 2008). After each change $B_{ij} \rightarrow B'_{ij}$ in the interaction matrix, the new contact probabilities are calculated by reweighing the conformations $\{r_i\}$ visited in the Monte Carlo simulation performed with the original interaction matrix B_{ij}:

$$p'_{ij} = \frac{Z}{Z'} \sum_{\{r_i\}} \Delta(|r_i - r_j|) \exp[-B'_{ij} + B_{ij}], \tag{12.10}$$

where the new partition function is evaluated as

$$Z' = \sum_{\{r_i\}} \exp\left[(-B'_{ij} + B_{ij})\Delta(|r_i - r_j|)\right]. \tag{12.11}$$

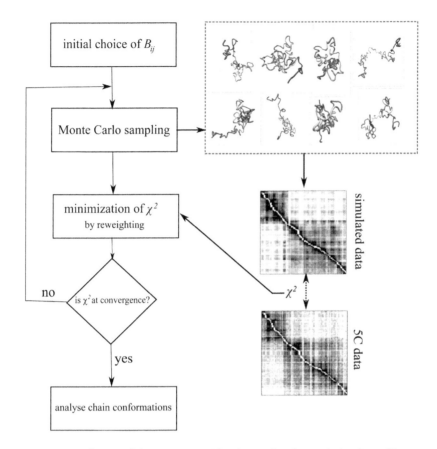

Figure 12.4 Scheme of the iterative MC scheme for the optimization of inter-action potentials. Adapted from (Giorgetti et al., 2014).

The random optimization procedure is repeated $n_{ro} \sim 10^3$ times, after which the conformations visited by the original Monte Carlo simulation are no longer representative of the equilibrium state of the new potential, and a new Monte Carlo sampling must is necessary. This procedure is repeated until the χ^2 has converged to a minimum.

Note that no angular potential is added to the model; however, stiffness can emerge from the simulation through the appearance of attractive potentials B_{ij} between near neighbors along the chain, if this is needed to satisfy the experimental contacts.

12.3 THE MODEL AT WORK

12.3.1 The sampling algorithm

The sampling of the conformational space of the polymer model is performed with a Metropolis Monte Carlo algorithm (Metropolis et al., 1953). Although in principle one can use molecular dynamics simulations, provided that the contact function Δ is approximated with a differentiable function, in our experience Monte Carlo simulations are more efficient for this specific problem.

We developed a Monte Carlo implementation which is suitable for studying coarse-grained models of most types of biopolymers (Tiana et al., 2014). The two types of elementary moves that are useful for chromatin models are random flips and random pivots (see Figure 12.3c), the frequency of the latter being 1/100 of that of the former.

For polymers of $\sim 10^2$ beads (corresponding to 300 kb, the size of the *Tsix* TAD), a reasonable degree of equilibration is obtained after 10^8 steps of the Monte Carlo algorithm, thus recording $\sim 10^3 - 10^4$ conformations.

12.3.2 Finding the interaction matrix

In Figure 12.5 we show, as an example, the results of the optimization process applied to the 5C data from the *Tsix* TAD in mouse embryonic stem cells. Along the optimization process, the χ^2 reaches a value of ≈ 10. Using lower-resolution Hi-C data it is easier to obtain $\chi^2 \approx 1$ (Zhan, Giorgetti, and Tiana, 2017); in this case the simulation is stopped when the χ^2 reaches 1, to avoid overfitting.

Multiple independent optimization runs converge to interaction matrices which are well correlated (cf. Figure 12.5c). This is not a trivial result, since the constrained entropy that is minimized by Eq. (3) is strictly convex (and thus

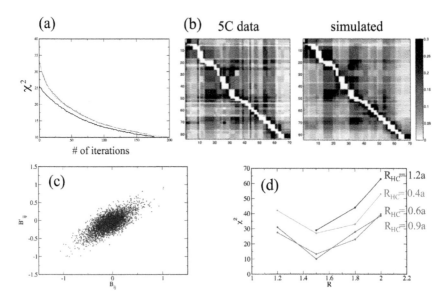

Figure 12.5 (a) The χ^2 between the 5C data of the *Tsix* domain and that back-calculated from the model during the optimization procedure. (b) Comparison between the experimental 5C map and that back-calculated from the model. (c) Comparison between energy matrices obtained from independent optimizations. (d) Dependence of χ^2 on R_{HC} and R.

guarantees a unique solution) only if the experimental data points C_{ij} are uncorrelated from each other (Pitera and Chodera, 2012). However, this is often not the case: for example, a polymeric effect, which assigns a large value to $p_{i,j+1}$ if beads i and j interact strongly, causes correlations in the contact matrix.

The optimal value of χ^2 obtained from the simulations depends on the choice of R_{HC} and R (see Figure 12.5d). For the model of the *Tsix/Xist* domains the lowest χ^2 is obtained using $R_{HC} = 0.6a$ and $R = 1.5a$ (where a is the distance between consecutive beads, which sets the length scale of the system). This result, obtained with a model in which beads correspond to 3 kb (corresponding to the experimental resolution), is not easy to extend to models with different binning size (e.g., to simulate Hi-C data), because a bin corresponds to a complex spatial arrangement of a flexible chain. The simplest approximation there is to assume the scaling law of an ideal chain; for Hi-C data with binning of 20 kb, for example, this leads to $R_{HC} = (3/20)^{1/2} \cdot 0.6a = 0.24a$ and $R = (3/20)^{1/2} \cdot 1.5a = 0.6a$ (Zhan, Giorgetti, and Tiana, 2017).

12.4 VALIDATION OF THE MODEL

In (Giorgetti et al., 2014), we validated the model by comparing its predictions to the results of independent experiments that were not used to train the model. The first validation was performed using 3D DNA with fluorescence in situ hybridization (3D DNA FISH) experiments in the *Tsix* TAD. In DNA FISH experiments, specific locations on chromosomes can be directly visualized in the cell nucleus using a fluorescence microscope, by hybridizing fluorescent DNA probes to their complementary target genomic sequence. When two or more probes labeled with different fluorophores are used, it is possible to measure distances between genomic locations as well as their cell-to-cell distribution, which can be compared with those predicted by the model.

Comparison between the predicted and the observed distance distributions from (Giorgetti et al., 2014) are displayed in Figure 12.6. It should be noted that several alternative (non-optimized) models can predict the average experimental distances between pairs of loci in the *Tsix* domain to a reasonable degree of approximation, including a set of models in which optimized interactions were randomly reshuffled. However, the full shape of the distribution is much more difficult to reproduce; for example, the p-value of the comparison between standard deviation in distances against randomly reshuffled models is $< 10^{-3}$ compared to $p = 0.05$ for the comparison of average distances (Giorgetti et al., 2014).

An interesting consequence of the comparison with FISH data is that it can define the numerical value of the length scale a. In fact, the comparison between calculated distance distributions (where lengths are expressed in units of a) and experimental distributions (where lengths are expressed in nm) requires the determination of the linear density of the fiber. This can in fact be extracted as the proportionality constant between the average predicted distances and the experimental ones, which results in $a = 53$ nm every 3 kb. The hard-core radius returning the best agreement with the contact probabilities was thus $R_{HC} = 0.6a = 32$

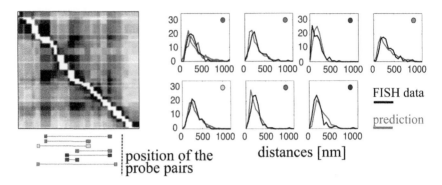

Figure 12.6 Comparison between the distribution of distances measured using 3D DNA FISH with probes in the *Tsix* TAD and those predicted by the model.

nm, in agreement with the width of 30 nm observed in chromatin fibers *in vitro* (Bian and Belmont, 2012), but larger than what was observed in recent electron microscopy experiments *in vivo* (Ou et al., 2017). It should be noted, however, that *a* represents an effective fiber diameter including the size of proteins bound to DNA and does not necessarily only describe DNA wrapped around nucleosomes.

A further, important, level of validation consists of testing the effects of genomic deletions in the model and in experiments. in (Giorgetti et al., 2014), we verified that the model correctly predicts the change in distances between pairs of loci within the *Tsix* TAD when genomic sequences corresponding to CTCF sites were deleted in real cells. In this case, the interaction potentials that had been initially optimized in the wild-type contact map were used without further optimization. Only the beads corresponding to the experimental mutation were deleted *in silico* (Giorgetti et al., 2014).

We further produced virtual mutations of every single bead in the *Tsix* TAD model, in order to predict the effect of systematically deleting sequences of ≈3 kb across the entire TAD. Importantly, deleting most beads leads to very little effects on the contact map, whereas deletion of a small number of them results in massive unfolding of the *Tsix* domain accompanied by increased interactions across the boundary with the neighboring *Xist* domain. These highly sensitive sites largely correlate with genomic locations bound by CTCF, which hinted at a fundamental role played by CTCF and its partner cohesin in establishing and maintaining TAD boundaries, which was later confirmed experimentally (Nora et al., 2017; Rao et al., 2017).

12.5 LESSONS WE LEARNED FROM THE MODEL

3C-based experiments provide information on the contact probabilities between chromosome loci, measured over a population of millions of cells. The model we described here essentially deconvolves single-cell conformations out of the population-averaged 3C-based data and can therefore give insights into otherwise inaccessible quantities, such as the extent of fluctuations and correlations in chromosome structure, the size and shape of single TADs, and their dynamical properties.

12.5.1 Chromosome domains display large conformational fluctuations

Quantitative analysis of the ensembles of conformations obtained by the simulation of TADs within the X inactivation center showed that the chromatin fiber fluctuates among very different (but not random) conformations (Tiana et al., 2016). The structural similarity between pairs of equilibrium conformations can be quantified by the distance root mean square difference (dRMSD), defined as

$$dRMSD\left(\left\{r_i^\alpha\right\},\left\{r_i^\beta\right\}\right) \equiv \left(\frac{1}{N(N-1)}\sum_{i<j}\left[\left|r_i^\alpha - r_j^\alpha\right| - \left|r_i^\beta - r_j^\beta\right|\right]^2\right)^{1/2}, \qquad (12.12)$$

or by the fraction q of common contacts, defined as

$$q\left(\left\{r_i^\alpha\right\},\left\{r_i^\beta\right\}\right) \equiv \frac{\sum_{i<j}\Delta\left(r_i^\alpha - r_j^\alpha\right)\Delta\left(r_i^\beta - r_j^\beta\right)}{\max\left[\sum_{i<j}\Delta\left(r_i^\alpha - r_j^\alpha\right), \sum_{i<j}\Delta\left(r_i^\beta - r_j^\beta\right)\right]}. \qquad (12.13)$$

The distribution of both quantities displays a broad, unimodal peak at large values of dRMSD, qualitatively similar to that of a random globule of a homopolymer and not compatible with a picture in which there is a dominant conformation of the fiber and small thermal fluctuations around it. On the other hand, however, equilibrium ensembles of chromosomal domains obtained from 5C and Hi-C maps are very different from those of a random globule, which would display a uniform decrease of contact probability moving away from the diagonal. Therefore, chromosome domains at the sub-Mb scale seem to display a hybrid behavior in between a random chain and a structured polymer. Indeed, dRMSD clustering of single conformations at the *Tsix* locus results in two large clusters of conformations, containing either elongated or compactly folded conformations (Figure 12.7a). Importantly, the coexistence of extremely different types of conformations was confirmed in 3D DNA FISH experiments where the entire *Tsix* TAD was imaged using structured illumination microscopy, a super-resolution technique allowing the lateral resolution of ~100 nm (Figure 12.7b). Strikingly, different degrees of TAD compaction correlate with the level of gene expression at the level of single cells (Giorgetti et al., 2014), suggesting that structural fluctuations could result in differential distances between promoters and their regulatory sequences within a single TAD, resulting in differential levels of transcription.

12.5.2 Size and shape of domains

Although the computational strategy described so far was initially based on 5C data, its technical implementation using Hi-C data is straightforward. Based on Hi-C data obtained in mouse embryonic stem cells (Giorgetti et al., 2016), we

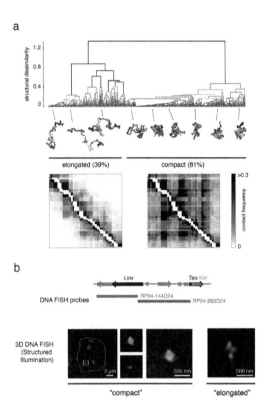

Figure 12.7 (a) Clustering of conformations in the equilibrium ensemble obtained after the optimization of interaction potentials shows that chromosome conformation is highly variable within the *Tsix* TAD. (b) Super-resolution 3D DNA FISH experiments confirm the presence of elongated and compact states at the same locus in single cells. Reproduced from (Giorgetti et al., 2014).

modeled each of the ≈2500 TADs in the mouse genome (Zhan, Giorgetti, and Tiana, 2017). This led to the discovery of interesting trends at the genome-wide level. We found that the population-averaged gyration radius of single TADs scales as $N^{1/2}$ where N is its genomic length were similar to what was expected from ideal chains. This can be expected from the Flory theorem for segments embedded in a large globule. However, other properties depart from this simplistic picture. First, the average number of intra-domain contacts scales linearly with N. Second, the degree of isotropy of the models is variable, not correlated with their mean size, and overall larger than what was expected for segments of a globule. Interestingly, many geometrical features of single TADs correlate with the expression level of the gees that they contain, with highly transcribed domains being overall more elongated than silent TADs (Zhan, Giorgetti, and Tiana, 2017).

12.5.3 Correlations in contact formation

The equilibrium ensembles generated by the model can also be used to predict the probability that multiplets of genomic loci are found in contact with each other simultaneously (Zhan, Giorgetti, and Tiana, 2017). This is another information which cannot be directly extracted from the 5C/Hi-C matrix. We focused on triplets of loci that are found in contact with each other significantly more often than what was expected from their pairwise contact probabilities. Operatively, defining with p_{ijk} the equilibrium contact probability of the triplet calculated from the model, the excess colocalization probability is $p^{*}_{ijk} \equiv p_{ijk} - (p_{ij}p_{jk} + p_{ik}p_{kj} + p_{ki}p_{ij})/3$, where p_{ij} is the equilibrium contact probability of the pair. A triplet is defined as co-localizing if the associated p^{*}_{ijk} displays a p-value lower than 5% in a negative control obtained reshuffling randomly the model interactions B_{ij}.

Based on our simulations, the ≈ 2500 domains of mouse embryonic stem cells display a diverse number of co-localizing triplets, ranging from zero to $\sim 10^{3}$. This is a small number if compared to the total number of possible triplets $\sim N^{3}/3! \sim 10^{5}$ and, interestingly, does not correlate with the activity of genes in the domains. These triplets are, however, enriched in CTCF, suggesting that the molecular mechanism employed by this protein is not just dimerization. Additionally, the statistical significance of triplets composed by three active promoters, as well as two active promoters and an enhancer is high, suggesting the sporadic occurrence of transient hubs containing multiple promoters and/or regulatory sequences. Interestingly, the triplet composed of the *Xite/Chicl/Linx* loci within the *Tsix* TAD belongs to those that are significantly colocalized, as previously observed experimentally [Giorgetti et al. (2014)].

12.5.4 Dynamical properties

Finally, the model can be used to predict the time-dependent dynamics of the chromatin fiber within single TADs. This task in principle presents two problems. First, the potentials in the model are step functions, and thus discontinuous. If this is not a problem for the Monte Carlo algorithm used to sample equilibrium conformations, it prevents the simulation of the dynamics of the polymer using, for example, Langevin equations. However, we have previously shown that a Monte Carlo algorithm can reproduce the correct dynamics of a system in the limit that its elementary moves are small (Tiana, Sutto, and Broglia, 2007). To comply with this requirement in the case of the polymer model of single TADs, only flips constrained to a width of 1 can be used as elementary moves of the polymer chain.

An additional problem is that the Monte Carlo algorithm does not provide an intrinsic time scale for the motion of the model, but measures time in terms of the number of steps of unknown time duration. We, however, got around this problem by equating the instantaneous diffusion coefficient of a monomer in the simulation with the experimental value that we measured in live-cell imaging of the *Chic1* locus within the *Tsix* TAD (Tiana et al., 2016). This resulted in 1 MC step = 0.015 s.

We then studied the time evolution of the *Tsix* TAD conformation by imposing an initial conformation of the model polymer and simulating multiple time trajectories starting from the same conformation. Despite the fact that the dynamics take place in a high-dimensional space of coordinates that is difficult to visualize, it is possible to map it to a quantity $q(t) \equiv q(r(t), r(0))$ (Eq. 13) that measures the fraction of contacts which were present in the initial conformation and are maintained at time t. We found that the dynamics $q(t)$ averaged over many replicated simulations starting from the same conformation is well described by the power law (Figure 12.8)

$$\overline{q(t)} \sim t^{-\alpha/2} \qquad (12.14)$$

with exponents α ranging from 0.5 to 1, depending on the initial conformation.

The power law behavior suggests that the system undergoes a sub-diffusive dynamic and does not have to cross large (i.e., $\ll kT$) barriers along its trajectory (this would result in an exponential decay). This can be easily understood considering the dynamics of the distance between pairs of loci. If the motion is purely diffusive, the distance Δr_i between two loci would undergo diffusion, and the associated probability would be Gaussian,

$$p(\{\Delta r_i\}, t) = \frac{1}{\prod_i (2\pi D_i t)^{1/2}} \exp\left[-\sum_i \frac{(\Delta r_i)^2}{2 D_i t} \right], \qquad (12.15)$$

where D_i is the diffusion coefficient associated with the i th pair. The average fraction of maintained initial contacts is then

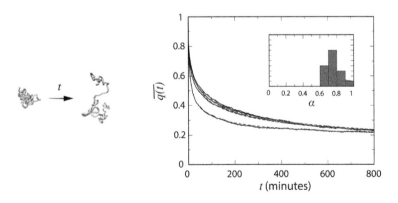

Figure 12.8 The dynamics of the model of a chromatin domain. Red dashed lines indicate power-law fits. In the inset, the distribution of exponents α of the fit.

$$\overline{q(t)} = \frac{1}{n_c}\sum_k \int_0^\infty d\Delta r_k \theta(\Delta r_k < R) \frac{1}{\prod_i (2\pi D_i t)^{1/2}} \exp\left[-\sum_i \frac{(\Delta r_i)^2}{2D_i t}\right]$$

$$= \frac{2}{\pi^{1/2} n_c}\sum_k \mathrm{erf}\left[\frac{R}{(2D_k t)^{1/2}}\right] \tag{12.16}$$

Where n_c is the number of contacts in the initial conformation, $\theta(c)$ is a function equal to 1 if condition c is true and 0 if it is false, and R is the interaction range of the potential. In the limit $t \ll R^2/\min_i D_i$, this can be approximated as

$$\overline{q(t)} = \frac{2}{\pi^{1/2} n_c}\sum_k \frac{R}{(2D_k)^{1/2}} \cdot \frac{1}{t^{1/2}} \tag{12.17}$$

implying an exponent 1/2 in Eq. (14).

If the loci are constrained by correlations imposed by the polymer, which however does not have to cross large energy barriers, a subdiffusive behavior is expected (Bouchaud and Georges, 1990), and Eq. (14) applies with $\alpha < 1$. If, on the other hand, the initial conformation were separated from the rest of conformational space by energy barriers comparable to kT, the formation of each contact could be regarded as a two-state dynamics, controlled by the master equation

$$\frac{dp_k}{dt} = u_k\left[1 - p_k(t)\right] - w_k p_k(t), \tag{12.18}$$

whose solution is

$$p_k(t) = A \exp\left[-(w_k + u_k)t\right] + B \tag{12.19}$$

and, thus, one would observe

$$\overline{q(t)} = \frac{1}{n_c}\sum_k \exp\left[-(w_k + u_k)t\right] + C, \tag{12.20}$$

where typically only the slowest exponential decay can be detected.

In principle, a non-exponential decay of $q(t)$ could be compatible with a glassy dynamic, namely the low-temperature behavior of a system with heterogeneous interactions. Typically, at low temperatures these systems display a small number (small with respect to the total number of allowed conformations) of populated thermodynamic states. In this case, depending on the details of the model, the dynamics can be described at low temperature by stretched exponentials (De Dominicis, Orland, and Lainee, 1985), logarithmic relaxation (Bryngelson and Wolynes, 1989; Shakhnovich and Gutin, 1989) or power laws (Koper and Hilhorst, 1987). However, this does not seem to be the case here, since the equilibrium properties of our model of the chromatin fiber are different from those of a glassy phase (cf. Sect. 5.1).

Figure 12.8 shows that the order of magnitude of the time $\overline{q(t)}$ needs to reach low values, i.e., the time it takes for the polymer to significantly move away from its initial conformation, is hundreds of minutes. This has important implications for transcriptional regulation, as it suggests that contacts between enhancers and their target genes within the same TAD could be lost and reformed multiple times within a single cell cycle, which lasts $\approx$ 16 h in mouse ESCs. This also implies that all contacts might appear and disappear dynamically, thus giving rise to the observed structural variability within a cell population.

12.6 CONCLUSIONS AND OUTLOOK

Our strategy for modeling chromosome structure using equilibrium polymer models has been instrumental in obtaining the first fundamental insights into the extent of cell-to-cell variability in chromosome conformation at the scale of TADs. Complemented with single-cell RNA and DNA FISH experiments, this led to the discovery that transcriptional fluctuations at the *Tsix* locus are coupled to structural fluctuations, possibly as a consequence of alternative spatial arrangements of genes and enhancers within the *Tsix* TAD. Despite the intense research efforts in the field, a few years later we still ignore the timescales over which such structural fluctuations are established. Many fundamental questions remain to be addressed, such as: How dynamic is chromosome conformation at the level of TADs and their sub-structures? How is the structural dynamics related to enhancer–promoter interaction frequencies? Do cell-to-cell and temporal variability in enhancer–promoter distances control transcription levels and heterogeneity, and if so, to what extent? Our model of the *Tsix* TAD, coupled with live-cell imaging of one genomic location within the same TAD, predicts that enhancer/promoter interactions dynamically assemble and disassemble several times during a cell cycle, with important implications for transcriptional regulation. Live-cell imaging experiments visualizing two or more loci within the TAD simultaneously would provide a much-needed test for this prediction. The recent revolution in genome editing technologies (notably spearheaded by the CRISPR/Cas9 approach) has brought the goal of imaging chromatin motion as well as RNA production in multiple colors simultaneously one step closer. Being able to evaluate if the structural dynamics in single cells reflects the predictions of our model will be an important test not only for our computational strategy, but more generally for the entire class of equilibrium polymer models based on population-averaged, crosslinking-based datasets.

ACKNOWLEDGMENTS

We are grateful to Edith Heard and Elphège Nora for the many insightful discussions on 5C-based polymer modeling strategy. Research in LG's laboratory is supported by the Novartis Foundation and an ERC Starting grant (759366, BioMeTRe).

REFERENCES

Bascom, G. D., and T. Schlick. 2018. "Mesoscale Modeling of Chromatin Fibers." In *Nuclear Architecture and Dynamics*, 123–147. Elsevier.

Belmont, A. S. 2014. "Large-scale chromatin organization: The good, the surprising, and the still perplexing." *Current Opinion in Cell Biology* 26:69–78.

Bian, Q., and A. S. Belmont. 2012. "Revisiting higher-order and large-scale chromatin organization." *Current Opinion in Cell Biology* 24:359–366.

Bouchaud, J.-P., and A. Georges. 1990. "Anomalous diffusion in disordered media: Statistical mechanisms, models and physical applications." *Physics Reports* 195:127–293.

Bryngelson, J., and P. Wolynes. 1989. "Intermediates and barrier crossing in a random energy model (with applications to protein folding)." *The Journal of Physical Chemistry* 93:6902–6915.

De Dominicis, C., H. Orland, and F. Lainee. 1985. "Stretched exponential relaxation in systems with random free energies." *Journal de Physique Lettres* 46:463–466.

Dekker, J., K. Rippe, M. Dekker, and N. Kleckner. 2002. "Capturing chromosome conformation." *Science* 295:1306–1311.

Dixon, J. R., S. Selvaraj, F. Yue, A. Kim, Y. Li, Y. Shen, M. Hu, J. S. Liu, and B. Ren. 2012. "Topological domains in mammalian genomes identified by analysis of chromatin interactions." *Nature* 485:376–380.

Galupa, R., and E. Heard. 2018. "Topologically Associating Domains in Chromosome Architecture and Gene Regulatory Landscapes during Development, Disease, and Evolution." *Cold Spring Harbor Symposia on Quantitative Biology*, 035030.

Gibcus, J. H., and J. Dekker. 2013. "The hierarchy of the 3D genome." *Molecular Cell* 49:773–782.

Giorgetti, L., R. Galupa, E. P. Nora, T. Piolot, F. Lam, J. Dekker, G. Tiana, and E. Heard. 2014. "Predictive polymer modeling reveals coupled fluctuations in chromosome conformation and transcription." *Cell* 157:950963.

Giorgetti, L., and Heard, E. 2016. "Closing the loop: 3C versus DNA FISH," *Biology* 17:215–224.

Giorgetti, L., B. R. Lajoie, A. C. Carter, M. Attia, Y. Zhan, J. Xu, C. J. Chen, et al. 2016. "Structural organization of the inactive X chromosome in the mouse." *Nature* 535:575–579.

Imakaev, M., G. Fudenberg, R. P. McCord, N. Naumova, A. Goloborodko, B. R. Lajoie, J. Dekker, and L. A. Mirny. 2012. "Iterative correction of Hi-C data reveals hallmarks of chromosome organization." *Nature Methods* 9:999–1003.

Jaynes, E. T. 1957. "Information theory and statistical mechanics." *Physical Review* 106:620–630.

Koper, G. J. M., and H. J. Hilhorst. 1987. "Power law relaxation in the random energy model." *EPL (Europhysics Letters)* 3:1213–1217.

Lieberman-Aiden, E., N. L. van Berkum, L. Williams, M. Imakaev, T. Ragoczy, A. Telling, I. Amit, et al. 2009. "Comprehensive mapping of long-range interactions reveals folding principles of the human genome." *Science* 326:289–293.

Metropolis, N., A. W. Rosenbluth, M. N. Rosenbluth, A. H. Teller, and E. Teller. 1953. "Equation of state calculations by fast computing machines." *Journal of Chemical Physics* 21:1087–1092.

Nora, E. P., A. Goloborodko, A. L. Valton, J. H. Gibcus, A. Uebersohn, N. Abdennur, J. Dekker, L. A. Mirny, and B. G. Bruneau. 2017. "Targeted degradation of CTCF decouples local insulation of chromosome domains from genomic compartmentalization." *Cell* 169:930–944.

Nora, E. P., B. R. Lajoie, E. G. Schulz, L. Giorgetti, I. Okamoto, N. Servant, T. Piolot, et al. 2012. "Spatial partitioning of the regulatory landscape of the X-inactivation centre." *Nature* 485:381–385.

Norgaard, A. B., J. Ferkinghoff-Borg, and K. Lindorff-Larsen. 2008. "Experimental parameterization of an energy function for the simulation of unfolded proteins." *Biophysical Journal* 94:182–192.

Ou, H. D., S. Phan, T. J. Deerinck, A. Thor, M. H. Ellisman, and C. C. O'Shea. 2017. "ChromEMT: Visualizing 3D chromatin structure and compaction in interphase and mitotic cells." *Science* 357:eaag0025.

Pitera, J. W., and J. D. Chodera. 2012. "On the use of experimental observations to bias simulated ensembles." *Journal of Chemical Theory and Computation* 8:3445–3451.

Rao, S. S. P., S.-C. Huang, B. Glenn St Hilaire, J. M. Engreitz, E. M. Perez, K.-R. Kieffer-Kwon, A. L. Sanborn, et al. 2017. "Cohesin loss eliminates all loop domains." *Cell* 171:305–320.e24.

Shakhnovich, E., and A. Gutin. 1989. "Relaxation to equilibrium in the random energy model." *Europhysics Letters* 9:569–574.

Tiana, G., L. Sutto, and R. A. Broglia. 2007. "Use of the Metropolis algorithm to simulate the dynamics of protein chains." *Physica A* 380:241–249.

Tiana, G., F. Villa, Y. Zhan, R. Capelli, C. Paissoni, P. Sormanni, E. Heard, L. Giorgetti, and R. Meloni. 2014. "MonteGrappa: An iterative Monte Carlo program to optimize biomolecular potentials in simplified models." *Computer Physics Communications* 186:93–104.

Tiana, G., A. Amitai, T. Pollex, T. Piolot, D. Holcman, E. Heard, and L. Giorgetti. 2016. "Structural fluctuations of the chromatin fiber within topologically associating domains." *Biophysical Journal* 110:1234–1245.

Yaffe, E., and A. Tanay. 2011. "Probabilistic modeling of Hi-C contact maps eliminates systematic biases to characterize global chromosomal architecture." *Nature Genetics* 43:1059–1065.

Zhan, Y., L. Giorgetti, and G. Tiana. 2017. "Modelling genome-wide topological associating domains in mouse embryonic stem cells." *Chromosome Research* 25:5–14.

Zhan, Y., L. Mariani, I. Barozzi, E. G. Schulz, N. Bliithgen, M. Stadler, G. Tiana, and L. Giorgetti. 2017. "Reciprocal insulation analysis of Hi-C data shows that TADs represent a functionally but not structurally privileged scale in the hierarchical folding of chromosomes." *Genome Research* 27:479–490.

13

Learning Genomic Energy Landscapes from Experiments

MICHELE DI PIERRO, RYAN R. CHENG, BIN ZHANG,
JOSÉ N. ONUCHIC, AND PETER G. WOLYNES

13.1 INTRODUCTION

Francis Crick famously advised biologists that if they wanted to understand biological function, they should first study biological structure. Doubtless, he came to this piece of wisdom from his early work with Watson that proposed the double helical structure of DNA whose beautiful symmetry by itself provided an explanation for heredity.

Crick's insight powered molecular biology for more than half a century. It has, nevertheless, been hard to strictly follow his advice on studying biology at the scale of chromosomes in living cells. The bottom-up viewpoint that gets at structure by starting from interatomic forces, which was partially employed by Crick and Watson for the DNA molecule, has an enormous range to cover to get all the way to the chromosome, from the nanometer scale of proteins to the meter. Whether the kind of symmetry considerations Watson and Crick employed so successfully in postulating the double-helix play any role at the cellular scale has also been unclear, aesthetic considerations aside. At the same time, the strict top-down approach, of just looking at chromosomes, is only just now starting to bear fruit, powered by recent advances in super-resolution microscopy techniques. Unlike the bottom-up approach based on forces, in any case, our yearning for an "explanation" of the structure would remain unsatisfied by mere observation, no matter how detailed.

In this chapter, we argue that the energy landscape theory, which has proven useful in understanding protein folding (Bryngelson, et al. 1995; Wolynes, 2015), provides a powerful bridge between the top-down and bottom-up perspectives when thinking about chromosome structure and function. In our view, there are many features of the energy landscape theory that recommend this method of attack. First, the energy landscape theory starts by acknowledging that all biomolecular constructs, not just large ones such as the chromosome, dynamically sample many structures. Sometimes, the range of structural variation is very small, so that a single snapshot does give a clear idea of what the molecule is doing, as it did for the DNA double-helix. That situation, arising from a funneled energy landscape, characterizes the smaller well-folded proteins, at least if you do not look too closely. But even proteins, as they function, often must depart far from the single snapshot view – they vibrate, crack, and locally disorder to carry out their responsibilities. When there are many structures, the statistical mechanical description employed by energy landscape comes to the fore by offering mathematical descriptions of the ensemble of structures through first quantifying the energetics and then simulating them by using coarse-grained force fields that can be employed at various length scales, ad libitum. The energy landscape theory can also quantify the extent of ordering and symmetry of the ensemble by employing statistical mechanical order parameters. Smaller biomolecular constructs are so battered by thermal motions that they remain mechanically near to equilibrium. For such systems near to thermal equilibrium, the energy landscape theory also provides a rigorous connection between dynamical pathways and thermodynamics. The rigor of this connection between kinetics and the landscape will break down, however, for the largest systems, where dynamical flows that violate detailed balance come into play (Wang, 2015). Certainly, some of the events in the cell's life history that set up chromosome structure during replication are such far-from-equilibrium processes. Nevertheless, we believe that much of the structure does not require very large-scale motions so that it is helpful to explore how far a quasi-equilibrium picture of chromosome structure formation and dynamics can be taken using an effective energy landscape. (We endorse in this way the viewpoint of Weiskopf about theory in the field of nuclear physics in the 1950's: Theories are like Austrian railway timetables before World War II. The timetables were not completely correct, but without them how would we have known how late the trains were?) In this spirit, the present chapter outlines the first steps we have undertaken to apply the energy landscape theory to the study of chromosomes. We have taken a top-down approach by starting from experimental structural data that are themselves highly coarse-grained.

Employing a maximum entropy Bayesian approach, we have used Hi-C data to construct coarse-grained energy landscapes for chromosomes. These experimentally informed landscapes, whatever their mechanistic limitations, essentially all reproduce the detailed information about the contacts between different genomic segments that has so far been provided by Hi-C. At the same time, simulation of these landscapes reproduces the results from independent experimental techniques as well, even though the data these provide were not used to build the landscape itself. The Hi-C crosslinking experiment gives information similar to that obtained for proteins using fluorescence resonance energy transfer and NMR spectroscopy. These experiments all measure the probability of forming specific physical contacts between locations on a chain molecule. By faithfully inverting these experimental data, the effective energy landscape characterizes, as well as we can, chromosome structures without introducing any other dynamical assumptions. Simulation of the landscape for the interphase chromosome shows the chromosome at interphase to be globally disordered but indicates that it does have a fluctuating order on moderate length scales (Zhang and Wolynes 2015). It is likely these locally ordered regions, by opening and closing, can act as structural switches for gene expression. In contrast, the landscape inferred for the mitotic chromosome when simulated gives rise to a globally ordered macroscopic structure with a beautiful hierarchical symmetry (Zhang and Wolynes 2016). The study of these effective energy landscapes that faithfully reflect the experimental data alone suggested to us a far simpler description of the landscape that is transferable from one chromosome to another, and from type to another. In other words, the principles learned using the energy landscape theory allow the construction of a force field that predicts detailed Hi-C data for many distinct human chromosomes, using epigenetic marks found through binding assays (Di Pierro, et al. 2017).

13.2 MAXIMUM ENTROPY PRINCIPLE AS A GENERAL FRAMEWORK FOR DATA-DRIVEN THEORIES

The use of Bayesian analysis and information theory to describe natural systems has a fraught philosophical history.

While some argue that information theory is logically prior to physics, many others (including, quite frankly, some of the present authors) would hold that physics is prior to information theory. For natural systems, we must first understand what constraints are so solidly settled in physics that they can be assumed to be given a priori. For bio-molecular problems, we would often seem safe in assuming that chain molecules remain connected and, therefore, have a fixed definite sequence. Even this assumption is not ironclad for the chromosome, because, during their life history, chromosomes can undergo recombination events. Likewise, we might regard it as trivially obvious that DNA molecules cannot lie on top of each other, because of excluded volume. Again, this constraint is generally valid, but topoisomerases do allow transient overlaps between chains to occur. Finally, do we believe that most of the degrees of freedom of the chromosome will be thoroughly "stirred" by thermal and biological motions? Whether the strong stirring motions assumption is true is quite unclear. Turbulence provides a clear caution to assuming that stirring always leads to a quasi-equilibrium state: vigorously stirring three-dimensional fluids, concentrates the kinetic energy in special modes so that the equipartition principle of indifference to where the energy is shared is strongly violated (Lee, 1952).

With all these caveats in mind, we believe it still makes sense to apply Bayesian arguments to chromosome structures by assuming that, besides the weak constraints just mentioned, the ensemble of structures resembles a Boltzmann ensemble for a chain molecule moving on some coarse-grained free energy surface.

In order to use the Maximum Entropy Principle of information theory to build an energy landscape for chromatin, we start from the landscape associated with a generic polymer with weak excluded volume and then introduce information specific to chromosomes as it emerges from experimental evidence present in the literature.

The starting point describes a homo-polymer with a potential $U_{HP}(\vec{r})$, which encodes the reliable a priori knowledge. This potential includes soft-core interactions and energies for bonds and angles energies. Additional features must be added to this potential so that the sequence-specific experimental data for a chromosome can be fit. The total potential characterizing the ensemble is then $U(\vec{r})$. Using this model we can write the canonical average for a generic function $\varphi(\vec{r})$ representing an experimental observable with the relation:

$$\varphi = \langle \varphi(\vec{r}) \rangle_U = \frac{\int \varphi(\vec{r}) e^{-\beta U(\vec{r})} d\vec{r}}{\int e^{-\beta U(\vec{r})} d\vec{r}} = \int \varphi(\vec{r}) \pi(\vec{r}) d\vec{r}$$

In this equation, $\vec{r}$ is the many-dimensional vector characterizing the positions in Cartesian space of all the loci in the chromosome, and $\pi(\vec{r})$ is the probability density for the model and $\beta = 1/k_B T$.

Following the Maximum Entropy approach introduced by Jaynes (Jaynes, 1957), one then seeks a probability density $\pi^{ME}(\vec{r}) = \frac{1}{Z} e^{-\beta U_{ME}(\vec{r})}$ that reproduces the experimental values of a certain set of n observables $\varphi_i(\vec{r})$ but that minimizes the relative information entropy. This input information leads to the following constraint equations:

$$c_0 = \int \pi^{ME}(\vec{r}) d\vec{r} - 1$$

$$c_1 = \int U^{HP}(\vec{r}) \pi^{ME}(\vec{r}) d\vec{r} - E_0$$

$$c^{DATA}_i = \int \varphi_i(\vec{r}) \pi^{ME}(\vec{r}) d\vec{r} - \varphi_i^{exp} \qquad i = 1, ..., n$$

Setting each one of these quantities equal to zero provides the maximum entropy solution; the first equation ensures that the probability will be normalized, the second fixes the average potential energy E_0 and the last n equations ensure that the expectation values from the maximum entropy ensemble and the experimentally measured values coincide for each one of the considered observables.

To find the probability density $\pi^{ME}(\vec{r})$ we need to maximize the information entropy

$$S = -\int \pi^{ME}(\vec{r}) \ln\left(\pi^{ME}(\vec{r})\right) d\vec{r}$$

subject to the input data constraints. This approach is based on the fact that a constrained maximization of the entropy is tantamount to minimizing the information built into the distribution, other than what is contained in the constraints themselves and our a priori notions about chain connectivity, etc.

Using Lagrange multipliers to perform the constrained maximization yields the following extremum condition

$$\frac{\partial S}{\partial \pi^{ME}} - \lambda_0 \frac{\partial c_0}{\partial \pi^{ME}} - \lambda_1 \frac{\partial c_1}{\partial \pi^{ME}} - \sum_{i=1}^{n} \lambda_i^D \frac{\partial c_i^{DATA}}{\partial \pi^{ME}} = 0$$

which in turn leads to the probability distribution

$$\pi^{ME}(\vec{r}) = \frac{e^{-\lambda_1 U_{HP}(\vec{r}) - \sum_{i=1}^{n} \lambda_i^D \varphi_i(\vec{r})}}{\int d\vec{r}\, e^{-\lambda_1 U_{HP}(\vec{r}) - \sum_{i=1}^{n} \lambda_i^D \varphi_i(\vec{r})}}$$

Renaming the parameter λ_1 to the familiar name of β, thus, setting the energy scale, we see that the potential energy function associated with the maximum entropy probability then takes the form:

$$U_{ME}(\vec{r}) = U_{HP}(\vec{r}) + \frac{1}{\beta} \sum_{i=1}^{n} \lambda_i^D \varphi_i(\vec{r})$$

The last step needed before obtaining the least biased data-driven energy function is to adjust the parameters λ_i so that the energy scale is set and expectation values of the observables are made to coincide with their experimental values in the canonical ensemble:

$$\langle \varphi_i \rangle_{\vec{\lambda}} = \frac{1}{Z(\vec{\lambda})} \int e^{-\beta U_{ME}(\vec{r}, \vec{\lambda})} \varphi_i(\vec{r}) d\vec{r} = \varphi_i^{exp}.$$

The normalization here is the partition function

$$Z(\vec{\lambda}) = \int e^{-\beta U_{ME}(\vec{r}, \vec{\lambda})} d\vec{r}.$$

To carry out the fitting we can define a convex objective function θ as

$$\theta\left(\vec{\lambda}\right) = \ln Z\left(\vec{\lambda}\right) + \beta \sum_{i=1}^{n} \lambda_i^D \varphi_i^{\text{exp}}.$$

The partial derivatives of this objective function are

$$g_i\left(\vec{\lambda}\right) = \frac{\partial \theta}{\partial \lambda_i} = -\beta \langle \varphi_i \rangle_{\vec{\lambda}} + \beta \varphi_i^{\text{exp}}$$

It is easy to see that once all the constraint equations $\langle \varphi_i \rangle_{\vec{\lambda}^*} = \varphi_i^{\text{exp}}$ are satisfied for a certain parameter vector $\vec{\lambda}^*$ then the target function has achieved a stationary value. There is, however, no guarantee that a unique stationary solution exists. To find the solution iteratively, we can calculate the Hessian matrix of second derivatives of the target function:

$$B_{ij}\left(\vec{\lambda}\right) = \frac{\partial^2 \theta}{\partial \lambda_i \partial \lambda_j} = \beta^2 \left(\langle \varphi_i \varphi_j \rangle_{\vec{\lambda}} - \langle \varphi_i \rangle_{\vec{\lambda}} \langle \varphi_j \rangle_{\vec{\lambda}} \right)$$

This matrix is positive semi-definite; as a result, if a stationary point exists then it is a minimum.

It is easy to find the optimal set of parameters $\vec{\lambda}^*$ numerically, then, by using Newton's method. This procedure involves iterating the following scheme over l

$$\lambda^{l+1} = \lambda^l - B^{-1}\left(\lambda^l\right) g\left(\lambda^l\right)$$

until the parameter set $\{\lambda^l\}$ converges adequately.

13.3 INTERPRETING DNA–DNA LIGATION ASSAYS AND DECODING THE STRUCTURE OF CHROMOSOMES

We see that by assuming that the set of chromosome structures from a population of cells can be described by a quasi-Boltzmann distribution, we can use experimental structural data, even if highly averaged and incomplete, to derive an effective energy landscape for chromosomes. DNA–DNA ligation assays, for example, Hi-C (Lieberman-Aiden, et al. 2009), offer the largest data sets on how the structure correlates with genomic location. Therefore, Hi-C data represent very promising experimental constraints for the determination of the structural ensembles of chromosomes. Genome-wide ligation assays measure the frequency at which any given pair of genomic segments in the chromosomes comes into close spatial proximity inside the nucleus. Chromosomes are first crosslinked using formaldehyde to freeze interactions between loci. Subsequently, the DNA is cut into fragments using restriction enzymes. A ligase is then used to randomly ligate DNA at a low concentration so that ligation between crosslinked fragments is favored in comparison to freely floating fragments. The ligated interacting loci are then amplified by polymerase chain reaction, sequenced, and their number quantified.

For the interpretation of Hi-C experiments and for defining the mathematical constraints that go along with these chemically complex experiments it seems that the probability of crosslinking should be the key phase space observable. Little is known about the precise functional relationship between crosslinking probability and geometric distances in the original DNA. Rigorously inferring this relationship between geometry and chemistry would depend on multiple factors related to the experimental set-up and the reagents. Everything is further complicated by the coarse-grained nature of both the data and the corresponding polymer model, which we have defined in the previous section as the tabula rasa. It is, however, safe to assume that the probability for a crosslinking event to take place between two loci i and j is a decreasing function of their geometric distance r_{ij}.

We postulate that the probability of crosslinking can be approximated by the function

$$f\left(r_{ij}\right) = \frac{1}{2}\left(1 + \tanh\left[\mu\left(r_c - r_{ij}\right)\right]\right)$$

This functional form is a switch function; at a short distance, the probability of crosslinking is close to unity (sure event) while at a long distance the probability of a crosslinking event is zero. Such a shape for the probability is consistent with the fact that crosslinking is only possible in case of physical proximity. The normalization of f, as well as the parameter μ, may depend not only on the experimental protocol but also through chemistry on the sequence. We assume the sequence effects have been normalized properly.

Typically, DNA–DNA ligation assays collect crosslinking events over a set of many cells. The sampling of the crosslinking probability is therefore averaged over a set of different chromosome conformations to produce the measured probabilities:

$$p_{ij} = \left\langle f\left(r_{ij}\right)\right\rangle = \frac{\int f\left(r_{ij}\right)e^{-\beta U(\vec{r})}d\vec{r}}{\int e^{-\beta U(\vec{r})}d\vec{r}}$$

$$= \int f\left(r_{ij}\right)\pi^{ME}\left(\vec{r}\right)d\vec{r}$$

where we have used the canonical ensemble and where $\beta = 1/k_B T$, $\vec{r}$ is the vector characterizing the positions in Cartesian space of all the loci in the chromosome, $U(\vec{r})$ is the potential energy of the system that is found using the maximum entropy principle.

The most general energy function consistent with the experimental knowledge turns out to be then:

$$U_{ME}\left(\vec{r}\right) = U_{HP}\left(\vec{r}\right) + \frac{1}{\beta}\sum_{i=1}^{n}\lambda_i\, f\left(r_{kl}\right)$$

As we have discussed, the remaining step in determining the energy function above is to learn the optimal set of parameters by iterating the scheme described before until the simulated and experimental contact probabilities agree to a desired accuracy. This procedure for learning an effective potential from structural data has strong analogies to how optimal statistical potentials are determined from databases of protein structures using Z-score optimization (Goldstein, et al. 1992), an approach already motivated by the energy landscape theory.

Once the energy function has been properly inferred from the experimental data, molecular dynamics simulations enable the study of the structure and dynamics of the chromosome in full three-dimensional detail. These simulations have led to results that are not at all obvious just by inspecting the two-dimensional data sets obtained from the ligation assays themselves.

In carrying out simulations, it is not obvious how to find the equilibrium properties of our data-driven energy function since long molecules move very slowly and if there is strong excluded volume, they also can entangle with themselves. We note that chromatin in vivo can overcome topological constraints, and the extent to which this entangling can occur will influence the conformational ensemble of the chromatin polymer. Inside the cell, equilibrium is encouraged by the activity of topoisomerases, which are enzymes that cut DNA so to change its topology. These enzymes, which are absolutely needed for replication, are generally present in the cell nucleus. To mimic their effect, equilibrium is achieved in simulations by allowing the chromatin polymer to occasionally cross its own chains; that is, in the tabula rasa model one uses relatively "soft" repulsive interactions that allow an overlap with only a small energy penalty, enough to keep particles from remaining for a long time on top of each other but nevertheless weak enough to allow the chains to pass through each other.

The analysis of the three-dimensional configurations sampled by simulating the interphase energy landscape has revealed many interesting features that are by no means immediately evident in the two-dimensional input data. Interphase chromosomes simulated with the landscape model turn out to be mostly unknotted, having at most a trefoil knot, while an equilibrated chain with a trivial energy landscape should contain many quite complex knots. In the data-driven energy landscape, the chain becomes locally more rigid than in the tabula rasa model, thus, preventing the formation of knots (Figure 13.1). Such rigidity appears to arise from the chain coiling upon itself, forming structures that resemble a new level of supercoiling beyond that which is familiar at the DNA level. While interphase chromosomes are disordered at a global scale, sampling many structures, it is possible to study the dominant local structure in chromosomes. These local minimum energy structures can be found by reducing the effective temperature of the Boltzmann distribution. At length scales shorter than a megabase, these quenched structures resemble those sampled at a higher temperature; but on larger length scales, differences arise because at high temperature order is lost through fluctuations between ordered and disordered local states.

Some topologically associated domains (TADs) display distinctly two-state-like free energy profiles, reminiscent of the folding-unfolding transitions in small

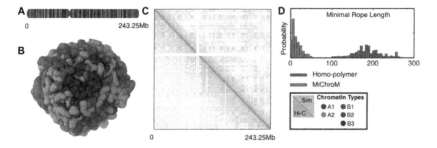

Figure 13.1 (A) Chromosome 2 of human lymphoblastoid cells (GM12878) is modeled as a sequence of discrete chromatin types, similar to a sequence of amino acids for a protein. (B) Given such a sequence, the Minimal Chromatin Model (Di Pierro, et al. 2016) can be used to generate an ensemble of 3D conformations. (C) The simulated ensemble (top right triangle) reproduces the compartmentalization patterns observed in Hi-C experiments (bottom left triangle from (Rao, et al. 2014)). (D) The complexity of the knots in the chromosome is studied by calculating the minimal rope length associated with each polymer conformation by the algorithm Shrink-On-No-Overlap (Pieranski, 1998) The chromosomes simulated with MiChroM have significantly fewer knots (red histogram) than a control model constituted by a typical random homo-polymer (blue histogram). The most common knot in the MiChroM chromosome is the trefoil knot (3_1 corresponding to a minimal rope length of 16.33 (Pieranski, 1998)) and the vast majority of the conformations contain knots simpler than a 7_1, corresponding to a minimal rope length of 30.70 (Pieranski, 1998).

proteins. The local free energy landscape of each of the TADs varies along the genome sequence. The free energy profiles for some TADs are much more clearly bistable than the profiles for other regions. These activated transitions between folded (closed) and unfolded (open) TADs suggest a possible mechanism for temporally modulating or controlling distal interactions in gene transcription.

The energy landscape for chromatin exhibits a structural complexity that reflects its biological heterogeneity as well as possibly experimental error. A simpler description of the typical local order in chromosomes can be found by averaging over different regions, so as to create an "ideal" chromosome landscape. The resulting model then describes a uniformly self-interacting chromosome that has an energy function that is invariant to translation along the sequence. The quenched structures of the ideal chromosome energy landscape resemble those formed from the complete data-driven landscape locally but are more simply patterned. Strikingly, on the global scale, the quenched ideal chromosome displayed a regular hierarchically layered fiber of fibers resembling the familiar metaphase chromosome as visualized in the light microscope (Kireeva, et al. 2004). In some sense, this is expected for any sequence translation invariant Hamiltonian as a mathematical consequence of sequence translation symmetry. It is interesting that Crick postulated such a structure for the chromosome based on symmetry and some nuclear digestion experiments decades ago (Bak, et al. 1977). Apparently, the metaphase and replication are governed by symmetry principles.

When we used the maximum entropy strategy to invert Hi-C data that are directly obtained for the metaphase chromosome itself we do indeed find structures that resemble the quenched ideal chromosome. The mitotic chromosome as determined by the experimentally informed landscape is not a crystal, but is clearly an anisotropic liquid crystal that breaks rotational symmetry to form a cylinder (Figure 13.2). It is also chiral. This breaking of symmetry predicted analyzing the energy landscape of mitotic chromosomes has been also seen in microscopic studies of sister chromatids, which are mirror image pairs directly after the genome is duplicated (Delatour and Laemmli, 1988).

The success obtained by the data-driven energy landscapes and the resulting symmetry ideas open the way to the possibility of specifying a transferable force field describing the physical principles behind chromatin folding. The ideal chromosome landscape encodes the symmetry principles by capturing the generic tendency of chromatin to order locally like liquid crystals.

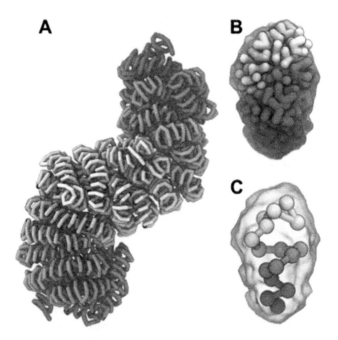

Figure 13.2 (A) The chromosomal structure formed at a low information theoretic temperature by the Ideal Chromosome. The Ideal Chromosome potential was obtained in this case from the direct inversion of metaphase DNA–DNA ligation assays. (B) A structure obtained by the direct inversion of contact maps from a mitotic chromosome. Like that of the quenched ideal chromosome, the super-helical structure is evident. (C) Coarse-grained representation of the structure in panel B. We see that chiral symmetry is spontaneously broken: the mitotic chromosome possesses handedness.

13.4 LEARNING THE PRINCIPLES OF CHROMATIN FOLDING: THE MINIMAL CHROMATIN MODEL

In this section, we describe a physical theory motivated by symmetry principles that explains how local interactions between genomic loci can lead to the conformations of human chromosomes in interphase. This theoretical energy landscape model for chromatin folding has been called the Minimal Chromatin Model (MiChroM) (Di Pierro, et al. 2016) to highlight the intent to include in the model as little detailed sequence information as possible. The MiChroM energy function again relies on the maximum entropy principle to parameterize the structural consequences of biochemical interactions between DNA and the nuclear proteome, as discussed below in detail. The MiChroM approach assumes that chromosomes fold under the action of a cloud of proteins each of which binds to DNA with a different selectivity and can mediate DNA–DNA interactions; any physical process involved in chromatin folding is assumed to operate through such protein-mediated contact interactions. Importantly, the free energy landscape approach is able to model the structural effect of biochemical interactions even though the precise identity of the interacting biomolecules remains unknown.

The first assumption made about the nature of the biochemical interactions between DNA and proteins is that the genome can be partitioned into intervals where only one of a few "types" of chromatin is found. Each type of chromatin we assume is marked by some characteristic histone modifications and interacts with a characteristic combination of nuclear proteins. As a result, when two segments of chromatin come into contact, the effective free energy change due to this contact depends, at first order, on the chromatin type assigned to each segment. This assumption is supported by both biochemical and structural data. For instance, the binding patterns of nuclear proteins have revealed five distinct types of chromatin in *Drosophila* cells (Filion, et al. 2010). Similarly, high-resolution contact mapping experiments (Hi-C) have revealed that, globally, human chromatin is indeed segregated into compartments. Analysis of early Hi-C experiments revealed that loci typically exhibited one of two long-range contact patterns, a finding that was initially interpreted as there being two spatial neighborhoods, dubbed the A and B compartments (Lieberman-Aiden, et al. 2009). More recently, higher resolution experiments indicate the presence of six distinct interaction patterns, corresponding then to six sub-compartments (A1, A2, B1, B2, B3, and B4) in human lymphoblastoid cells (GM12878) (Rao, et al. 2014). A similar compartmentalization of the genome has been observed in many organisms [including mouse (Dixon, et al. 2012; Rao, et al. 2014) and *Drosophila* (Sexton, et al. 2012; Eagen, et al. 2015; Li, et al. 2017)], and has been confirmed by microscopy (Wang, et al. 2016). Crucially, the contact patterns seen at any given locus are cell type–specific and are strongly associated with particular chromatin marks. MiChroM postulates that the contact patterns seen in DNA–DNA ligation assays are the result of a process of some sort of phase separation in which chromatin with similar biochemical properties coalesce together in the nucleus.

In the full MiChroM model, it is also assumed that certain specific pairs of loci bind together to form particularly frequent contacts. In high-resolution Hi-C maps of the human genome, these strong contacts are visible as fairly distinct local peaks in the contact probability (Rao, et al. 2014). The majority of these contacts are intra-chromosomal and are associated with the presence of CCCTC-binding factor (CTCF) and the cohesin protein complex; these contacts have been usually called "loops" in the literature. It must be noted, however, that some strong intra-chromosomal contacts also exist in absence of CTCF or cohesin. Recently, it has been suggested that some strong inter-chromosomal contacts are also associated with the presence of super-enhancers and bound transcription factors (Rao, et al. 2017); contacts of this sort have been referred as "links". In a far-from-equilibrium process, it is thought that cohesin forms loops by extruding DNA through its ring (Nasmyth, 2001; Alipour and Marko 2012; Sanborn, et al. 2015; Fudenberg, et al. 2016), while the protein CTCF acts to control the extrusion by marking the loci at which it should stop (Rao, et al. 2014). Loops associated with the presence of CTCFs typically enclose a few hundred kilobases of DNA. There is evidence that such structures are involved in diverse regulatory functions, including activation, repression, and insulation (Phillips and Corces, 2009). MiChroM makes no assumption about the particular mechanism of loop formation and any non-equilibrium aspect is thought to still be describable as a very strong constraining potential. The tendency of chromatin to form particularly long-lived contacts at special loci along the genome is encoded in the model as a change in the effective free energy of a chromatin configuration when the two loci anchoring a loop are in contact.

MiChroM embraces the idea of there being an ideal chromosome potential by assuming that every time any pair of loci comes into contact, whatever their type, there is a gain/loss of effective free energy and that this contact energy depends on the genomic distance between the two loci. As already stressed, the ideal chromosome potential describes the local structure of chromatin in the absence of compartmentalization. This last physical assumption is supported by the widespread evidence that chromatin often behaves like a liquid crystal (Boy de la Tour and Laemmli, 1988; Naumova, et al. 2013), and is consistent with the existence of some sort of higher-order fiber in chromatin (Maeshima, et al. 2010; Grigoryev, et al. 2016), although the idea behind it is more general.

The energy terms reflecting the three physical assumptions explained above are themselves directly related to three collective phase-space observables, which can be defined using the input probability of crosslinking. One of these collective variables is the average number of crosslinks between two chromatin structural types k and l; this quantity for a chromosome configuration $\vec{r} \equiv \{\vec{r}_i\}$ is:

$$T_{kl}(\vec{r}) = \sum_{\substack{i \in \{\text{Loci of Type } k\} \\ j \in \{\text{Loci of Type } l\}}} f(r_{ij})$$

The expectation value of T_{kl} is a proxy for the energy associated with kl-type contacts. Therefore, constraining T_{kl} to its experimental value is similar to constraining the effective contact energy between types to an appropriate value.

Another relevant collective variable is the total number of crosslinking events between loci that are known to form a loop. This quantity is

$$L(\vec{r}) = \sum_{(i,j)\in\,\{\text{Loops Sites}\}} f(r_{ij})$$

This quantity determines the looping energy. The number of contacts that occur for each fixed genomic distance* d:

$$G_d(\vec{r}) = \sum_i f(r_{i,\,i+d})$$

constrains the ideal chromosome energy terms. We can now constrain the expectation value of the phase-space observables defined above to their experimentally determined values extracted from the DNA–DNA ligation assays. This results in three classes of constraints that capture the assumptions made about the process of chromatin folding.

$$c_0 = \int \pi^{\text{MiChroM}}(\vec{r})d\vec{r} - 1$$

$$c_1 = \int U_{HP}(\vec{r})\pi^{\text{MiChroM}}(\vec{r})d\vec{r} - E_0$$

$$c^{kl}_T = \int T_{kl}(\vec{r})\pi^{\text{MiChroM}}(\vec{r})d\vec{r} - T^{\text{exp}}_{kl} \qquad \{\forall k,l \in \text{Set of Types}: l \geq k\}$$

$$c_L = \int L(\vec{r})\pi^{\text{MiChroM}}(\vec{r})d\vec{r} - L^{\text{exp}}$$

$$c^d_G = \int G_d(\vec{r})\pi^{\text{MiChroM}}(\vec{r})d\vec{r} - G^{\text{exp}}_d \qquad 3 \leq d \leq d_{\text{cutoff}}$$

The constraints above define the maximum entropy probability distribution and the MiChroM energy function:

$$U_{\text{MiChroM}}(\vec{r}) = U_{HP}(\vec{r}) + \sum_{\substack{k \geq l \\ k,l \in \text{Types}}} \alpha_{kl} \sum_{\substack{i\in\{\text{Loci of Type } k\} \\ j\in\{\text{Loci of Type } l\}}} f(r_{ij})$$

$$+\chi \sum_{(i,j)\in\,\{\text{Loops Sites}\}} f(r_{ij}) + \sum_{d=3}^{d_{\text{cutoff}}} \gamma_d \sum_i f(r_{i,\,i+d})$$

The potential energy is composed of 4 terms: the initial homo-polymer potential, type-to-type interactions, loop interactions, and an ideal chromosome term. The first three terms in the energy function depend only on Cartesian distance while

---

* If we label loci of size Δd in a chromosome by using an increasing integer index i then the distance between a locus i and a locus j is represented by $d = j - i$ in units Δd.

the last term depends both on Cartesian and genomic distance. All terms only act through direct physical contact, a very appealing feature that is consistent both with the idea that chromosome organization arises from the differential binding of proteins on chromatin and with the fact that crosslinking can only act at short range in space.

In developing the MiChroM model the Lagrange multipliers α's, χ, and γ's, were determined using Newton's iteration method to adjust the parameter set to reproduce as well as possible the collective observables for the Hi-C map of chromosome 10 of GM12878 cells (Rao, et al. 2014). Chromosome 10, which is 136 Mbp long, was modeled as a polymer containing 2,712 monomers, each representing 50 kb of DNA. The compartment annotations from Hi-C maps were used to assign to each monomer a corresponding chromatin type. Similarly, the positions of loops between pairs of monomers were extracted from the same Hi-C maps in an automated way (Rao, et al. 2014).

The final MiChroM energy function contains just 27 parameters to specify the entire chromatin map, which contains much more information. Once this function has been learned using data from one chromosome, it is possible to perform molecular dynamics simulations of chromatin using the classification of loci into chromatin types and the location of loops as input; a procedure comparable to simulating protein folding using an amino acid sequence and disulfide bond positions as input.

The fitted structural ensemble of chromosome 10 found with the MiChroM energy function closely reproduces the experimental data sets. Apparently, the minimal amount of information introduced in the model (much less than in the full inversion that we discussed in the previous section) is sufficient to characterize adequately the current data sets. Additionally, the simple physical processes that were postulated seem to be sufficient to reproduce the complexity of genome organization of the single chromosome 10 with today's experimental accuracy. The contact map obtained from simulations is extremely well correlated to the experimental contact map (Pearson's r = 0.95). The correspondence of the "checkerboard" patterns that were previously attributed to compartmentalization is visually obvious. The scaling relationship between the probability of forming contacts and genomic distance, which was often used to justify the non-equilibrium fractal globule model, is also reproduced with great accuracy by this effective equilibrium model. In general, all features reasonably larger than the 50 kb resolution of the model are correctly recapitulated by MiChroM.

Without any further modification, once having been trained only on chromosome 10, the MiChroM energy landscape was able to predict the conformational ensembles of all the remaining autosomes of the same cell line (GM12878). Using the experimental input of the monomer type sequences and the loop annotations, the model produced chromosomal structural ensembles that were again consistent with the experimental contact maps of every chromosome studied. Notably, for all these chromosomes, none of which was ever used in the training phase, the correlation between simulation and experiment was as good as for the training chromosome (typically Pearson's r $\geq$ 0.95). The transferability of the effective energy function clearly highlights the effectiveness of the assumptions made and, more broadly, suggests that the physical mechanisms encoded in MiChroM provide an appropriate physical basis for chromatin folding.

The conformational ensembles of the chromosomes display compact chromosomal territories. The phase separation of the chromatin types is evident in the sampled ensembles; similarly, highly expressed genes tend to co-localize. In simulations, segments of chromatin belonging to the same structural type segregate forming liquid droplets, which rearrange dynamically by splitting and fusing. This simple process of phase separation is then sufficient to explain the emergence of the patterns of compartmentalization observed in DNA–DNA ligation assays and microscopy experiments. Interestingly, in the simulations, A-type chromatin tends to be less densely packed and lies preferably at the periphery of the chromosomal territory. This observation is consistent with previous findings using both microscopy and Hi-C (Hubner and Spector, 2010; Rao, et al. 2014; Boettiger, et al. 2016).

Like the structures formed from the complete Hi-C data inversion, MiChroM chromosomal conformations turn out to be largely devoid of knots. This topological feature follows directly from the quasi-equilibrium energy landscape based on the experimental ligation assay data. If unusual far-from-equilibrium physics is required to prevent knot formation it is somehow mimicked quite well by an energy landscape. Remarkably, the simple equilibrium mechanism underlying the energy landscape approach produces ensembles of structures that are almost knot-free, even though topoisomerases in the nucleus (and the weak excluded volume in simulations) allow topology changes.

When simulating multiple chromosomes, despite the existence of extensive contacts, the polymers occupy non-overlapping regions of space. The phase separation of chromatin types now extends across these chromosomal territories. The intra- and inter-chromosomal patterns of interaction are still correctly sampled, suggesting that MiChroM may encode correlations beyond the single chromosome scale and might be applicable to study features of the nucleus as a whole.

The Minimal Chromatin Model assumes that chromosomes fold through protein-mediated contact interactions and offers a simple strategy for recapitulating the energy landscape created by such interactions. This model forms transient contacts rather than permanent ones, which is consistent with the fact that most of the experimentally observed contacts between two genetic loci only occur in a small fraction of cells at a given time (Lieberman-Aiden, et al. 2009; Bantignies, et al. 2011). In humans, a handful of chromatin structural types define a sequence that encodes the energy landscape of chromosomes; knowledge of this sequence is sufficient to predict the arrangement of interphase DNA in vivo. The MiChroM Hamiltonian can reliably be transferred from one chromosome to other chromosomes; suggesting the plausibility of the proposed energetic mechanism, even if the underlying biochemical details are presently unclear.

The classification of loci into chromatin types and the positions of specific chromatin loops, which are inputs to MiChroM, was initially extracted from the Hi-C data; this input is however strongly associated with epigenetic features (histone modifications and bound CTCF motifs in convergent orientation) that can be directly and inexpensively assayed by ChIP sequencing. In the next section,

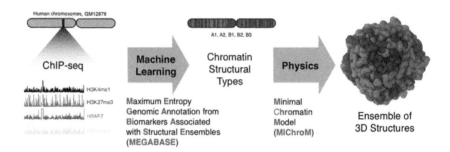

Figure 13.3 Schematic representation of the MEGABASE + MiChroM computational pipeline (Di Pierro, et al. 2017).

we explain how it is possible to exploit these associations along with MiChroM to predict in silico the three-dimensional structure of whole genomes starting from one-dimensional genomics data, which are often already publicly available (Figure 13.3).

13.5 BREAKING THE CODE OF CHROMATIN FOLDING: USING MACHINE LEARNING TO UNRAVEL THE RELATIONSHIP BETWEEN EPIGENETICS AND GENOME ARCHITECTURE

Hi-C studies (Rao, et al. 2014) of human chromosomes show that the presence of certain biomarkers located at a chromatin locus, such as histone tail methylation or acetylation, correlates well with which compartment that locus will occupy. For example, chromatin that belongs to the B1 sub-compartment typically exhibits an enhanced presence of the H3K27me3 histone tail methylation. Does detailed information about such epigenetic modifications and DNA-binding proteins that decorate chromatin, suffice to predict the fold of chromosomes? If so, readily available one-dimensional data about how the genome is modified could give us information about chromosome structure in many different cells and cell types, without going to the expense of Hi-C studies.

The chromatin immunoprecipitation-sequencing (ChIP-Seq) data sets for many human chromosomes have already been collected by the Encyclopedia of DNA Elements (ENCODE) Consortium (Dunham, et al. 2012). Each ChIP-Seq experiment probes the presence of a particular epigenetic modification or nuclear binding protein along the genome. The ChIP-Seq biochemical assay first crosslinks the proteins that are bound to DNA and then shears the DNA into fragments. Fragments of interest are then immuno-precipitated using antibodies designed to bind specific proteins. The fragments are then sequenced to find the genomic position where the proteins were bound. Generally, each ChIP-Seq experiment reports the number of counts for observing a particular biomarker at each locus of chromatin normalized by the number of counts from a control experiment. This ratio is referred to as the signal enrichment.

A ChIP-Seq data track for a transcription factor typically displays narrow peaks on the order of a few base pairs, which reveal the binding sites of that particular protein on the DNA. Other biomarkers, such as epigenetic modifications of the histones, tend to exhibit diffuse enrichment patterns on the genome. Rather than modeling high-resolution ChIP-Seq data, which can be strongly influenced by the presence or absence of a single signal peak, we integrate (sum) the signal enrichment for each experiment over windows of 50 kilobases for each locus. These re-binned smoothed data are less sensitive to the presence of any single signal peak. Each 50 kb window contains roughly ten nucleosomes. Furthermore, the re-binned signal enrichment for each experiment is discretized into 20 states ranging from 1 (low) to 20 (high) for simplicity.

The chromatin sequence is thus coarse-grained and re-encoded as a sequence indicating which biomarkers are strongly or weakly enriched at each locus. (Rao, et al. 2014) explored the relationship between compartmentalization and ChIP-Seq data by calculating the average signal enrichment of select biomarkers for loci belonging to the A1, A2, B1, B2, B3, or B4 sub-compartments. This work showed that the sub-compartments A1 and A2 tended to exhibit higher levels of epigenetic modifications, such as histone tail methylation and acetylation, compared to the B1, B2, and B3 sub-compartments. The one exception is that B1 loci on average exhibit higher levels of H3K27me3 modification.

While the existence of a correspondence between structural types and epigenetic marks is clear, the precise way epigenetic marks encode subtype is not immediately evident. A closer examination of the full distribution of signal enrichment for each sub-compartment reveals a rather complicated picture. For example, while the average acetylation level for chromatin of the A1 sub-compartment is generally higher than the acetylation level for B compartment chromatin, the distribution of acetylation levels for any sub-compartments overlap strongly with those of other sub-compartments. The difference between the means of the distributions is roughly as large as the square root of the variance of either distribution. The long tail of the distribution for A1 that extends towards high signals is responsible for the higher average levels of acetylation. A similar picture exists for other biomarkers; additional examples can be found in (Di Pierro, et al. 2017). It seems clear that simply classifying chromatin structural types based on any single ChIP-Seq enrichment signal being high or low would be insufficient to predict the compartmentalization observed in the experiment. Chromatin compartmentalization cannot be attributed to the presence or absence of any single biomarker. Yet it also seems there may be a more subtle but still predictable set of correlations that might be used to assign sub-compartment types.

To learn such an association between structure and biochemistry, we first align the structural annotation from DNA–DNA ligation experiments with the ChIP-Seq data tracks. The structural and biochemical characterization of each chromatin locus l can then be combined into a state vector:

$$\vec{\sigma}(l) = (C(l),\ \mathrm{Exp}_1(l), \mathrm{Exp}_2(l), \ldots, \mathrm{Exp}_L(l))$$

where C is the sub-compartment annotation for the GM12878 cell line (Rao, et al. 2014) (i.e., A1, A2, B1, B2, or B3) and the components labeled by Exp with subscripts ranging from 1 to L denote the signals for each of the ChIP-Seq experiments at the same locus.

The ChIP-Seq signals of adjacent chromatin loci are highly correlated. We would expect measurement noise to be uncorrelated. Hence, the correlated biochemical state of adjacent chromatin loci may provide additional information about their local structural properties. To allow for this information to be incorporated, we append to the measured state vector of any locus l the biochemical state of the adjacent loci (i.e., l-2, l-1, l+1, l+2). A database for learning the rules of association can then be set up. This database consists of a set of state vectors:

$$\vec{\sigma}(l) = (CST(l), \mathrm{Exp}_1(l-2), \ldots, \mathrm{Exp}_L(l-2), \mathrm{Exp}_1(l-1), \ldots, \mathrm{Exp}_L(l-1), \mathrm{Exp}_1(l), \ldots,$$

$$\mathrm{Exp}_L(l), \mathrm{Exp}_1(l+1), \ldots, \mathrm{Exp}_L(l+1), \mathrm{Exp}_1(l+2), \ldots, \mathrm{Exp}_L(l+2))$$

Classifying all chromatin loci in this manner allows us to collect M vectors, $\{\vec{\sigma}^{(s)}\}_{s=1\ldots M}$.

To quantify the information encoded in the collection of state vectors, we once again employ the Principle of Maximum Entropy to formulate a probabilistic model for the association of biochemical input data and the local chromatin sub-compartment type. That is we seek a $P(\vec{\sigma})$ that is consistent with the experimentally sampled data, i.e., $\{\vec{\sigma}^{(s)}\}_{s=1\ldots M}$. We can think of this construction as reflecting a local energy function for certain biochemical marks to show up in a compartment.

This model is constrained to reproduce the single-site and pairwise frequencies of the data set,

$$\sum_{\substack{\sigma_k \\ k \neq i}} P(\vec{\sigma}) = f_i(\sigma_i)$$

and

$$\sum_{\substack{\sigma_k \\ k \neq i,j}} P(\vec{\sigma}) = f_{ij}(\sigma_i, \sigma_j)$$

Here, $f_i(\sigma_i)$ and $f_{ij}(\sigma_i, \sigma_j)$ denote the single-site and pairwise frequencies, respectively, and i and j are indices of the state vector $\vec{\sigma}$. As discussed earlier, the solution to maximizing the entropy given these constraints corresponds to a Boltzmann distribution for some energy function:

$$P(\vec{\sigma}) = \exp(-H(\vec{\sigma})) / Z$$

where

$$H(\vec{\sigma}) = -\sum_{i=1}^{N-1} \sum_{j=i+1}^{N} J_{ij}(\sigma_i, \sigma_j) - \sum_{i=1}^{N} h_i(\sigma_i).$$

and J_{ij} and h_i parameters are the Lagrange multipliers. The J_{ij} interactions capture local pairwise correlations between epigenetic markers and between markers and chromatin types, while the h_i parameters are related to the individual frequencies of chromatin types and markers.

To determine the parameters of the probability distribution (i.e., the J_{ij} and h_i parameters) one must then determine a set of parameters that are most consistent with the data $\{\vec{\sigma}^{(s)}\}_{s=1...M}$. This training can be done with the pseudo-likelihood maximization Direct Coupling Analysis (plmDCA) approach (Ekeberg, et al. 2013), which uses the pseudo-likelihood approximation of Besag (Besag, 1975) to construct a probabilistic model for sequences composed of discrete labels. In the present case, these sequences contain the combined information of the epigenetic markers and the sub-compartment types.

Instead of maximizing the likelihood of observing an entire sequence of data, $\{\vec{\sigma}^{(s)}\}_{s=1...M}$, which would be computationally difficult, in practice the algorithm maximizes an approximate form of the likelihood of observing $\vec{\sigma}^{(s)}$ called a pseudo-likelihood based only on each local neighborhood:

$$P\left(\vec{\sigma} = \vec{\sigma}^{(s)}\right) \approx \prod_{i=1}^{N} P\left(\sigma_i = \sigma_i^{(s)} \,|\, \sigma_j = \sigma_j^{(s)} \text{ for all } j \neq i\right).$$

For a pairwise Markov random field,

$$P\left(\sigma_i = \sigma_i^{(s)} \,|\, \sigma_j = \sigma_j^{(s)} \text{ for all } j \neq i\right) = \frac{\exp\left[h_i\left(\sigma_i^{(s)}\right) + \sum_{\substack{j=1 \\ j \neq i}}^{N} J_{ij}\left(\sigma_i^{(s)}, \sigma_j^{(s)}\right)\right]}{\sum_{\{\sigma_i\}} \exp\left[h_i\left(\sigma_i\right) + \sum_{\substack{j=1 \\ j \neq i}}^{N} J_{ij}\left(\sigma_i, \sigma_j^{(s)}\right)\right]}$$

where the normalization in the denominator is summed over the collection of 5 labels A1, A2, B1, B2, B3 as well as the 20 signal states assigned to ChIP-Seq data.

Maximizing the pseudo-likelihood of observing the collection of M training state vectors, $\{\vec{\sigma}^{(s)}\}_{s=1...M}$, is equivalent to minimizing the negative of the log of the pseudo-likelihoods with respect to the parameters **J** and **h**:

$$\ell_{PL} = -\frac{1}{M} \sum_{s=1}^{M} \sum_{i=1}^{N} \log P\left(\sigma_i = \sigma_i^{(s)} \,|\, \sigma_j = \sigma_j^{(s)} \text{ for all } j \neq i\right)$$

To avoid overlearning from noisy data, the minimization is actually performed supplemented by a regularization term added to l_{PL}:

$$R = \lambda_J \sum_{i=1}^{N-1} \sum_{j=i+1}^{N} \sum_{a}^{25} \sum_{b}^{25} J_{ij}(a,b)^2 + \lambda_h \sum_{i=1}^{N} \sum_{a}^{25} h_i(a)^2$$

where the parameters $\lambda_J = \lambda_h = 0.01M$ were used. The model remains robust for a wide range of parameter values ranging from 0.0001-0.1. Hence, in practice we minimize the object function, i.e., $\ell_{PL} + R$, with respect to the parameters $\mathbf{J}$ and $\mathbf{h}$ to find a model that optimally represents the training set of data:

$$\{\mathbf{J},\mathbf{h}\} = \arg\min_{\{\mathbf{J},\mathbf{h}\}}\left(\left(\ell_{PL}(\mathbf{J},\mathbf{h}) + R_{L2}(\mathbf{J},\mathbf{h})\right)\right)$$

The conjugate gradient method is used to iteratively minimize the objective function until convergence is reached.

This inferred probabilistic model dubbed Maximum Entropy Genomic Annotation from Biomarkers Associated with Structural Ensembles (MEGABASE) was introduced in (Di Pierro, et al. 2017). Once the model is obtained, it can be used to predict the chromatin structural types for a locus l given only the experimental ChIP-Seq measurements of neighboring loci $(l-2, l-1, l, l+1, l+2)$:

$$CST(l) = \arg\max P(CST \mid \mathrm{Exp}_{1,\ldots,L}(l-2, l-1, l, l+1, l+2))$$

This procedure is equivalent to finding the chromatin type that minimizes the inferred local energy function, $H(\vec{\sigma})$, given a set of experimental ChIP-Seq measurements.

The most probable sequence of chromatin types predicted in this way can then be used as direct input for further molecular dynamics simulations using the MiChroM potential, which generates an ensemble of 3D structures from the predicted sub-compartment assignments. The de novo prediction of chromosome architecture for human lymphoblastoid cells was extensively compared against DNA–DNA ligation and fluorescence in situ hybridization data (Di Pierro, et al. 2017). The comparison demonstrated that there is sufficient information encoded in the biochemical data to accurately predict chromosomal structures. The broad agreement between theory and experiment points to there being a fairly definite, albeit complex, sequence-to-structure relationship between epigenetic modifications made to chromosomes and the forces that lead to their three-dimensional structure.

Since the MEGABASE annotation can be found by using as input biochemical data about epigenetic marks alone, it supports the idea that phase separation of distinct chromatin types is either caused by the marked proteins or partially causes epigenetic marks to be made at some point in the cell's life. The cartoon in Figure 13.4 illustrates a possible physical process responsible for compartmentalization.

While the chromosomal contact maps predicted by using MEGABASE along with MiChroM agree remarkably well with the experiment, a number of mismatches exist between the type annotation provided by MEGABASE and the sub-compartment annotations from the experiment; this is entirely expected. While the compartment annotation was made from observing chromatin loci

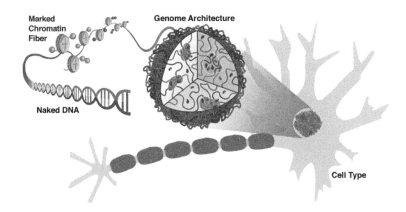

Figure 13.4 Proteins and epigenetic markings (shown in red and blue) decorate the naked DNA giving it distinct biochemical properties; these markings differ between cell types. Biochemically similar segments of the chromatin phase separate together, forming micro-phases that shape the cell-type-specific 3D architecture of the genome. The resulting patterns of interactions between loci predict accurately DNA–DNA ligation experiments. Our results indicate that epigenetic markings carry enough information to determine the global organization of the genome.

in spatial proximity to one another, MEGABASE relies entirely on biochemical information only from each particular locus. A locus of chromatin may end up in a compartment composed of biochemically dissimilar chromatin because it is physically connected to those segments, a sort of local frustration (Bryngelson, et al. 1995; Wolynes, 2015).

Since correlation is not necessarily causation, the success of MEGABASE does not rigorously imply that biomarkers cause compartmentalization. It is possible some structural environments encourage particular epigenetic markings. Nevertheless, some theoretical (Potoyan and Papoian, 2012) and experimental studies (Shogren-Knaak, et al. 2006; Tessarz and Kouzarides, 2014; Wilkins, et al. 2014) show that epigenetic modifications can change chromosomal structure locally. In our view, it is likely that the causality link is indeed oriented from epigenetics to structure in a similar manner as the information flow of the computational pipeline of MEGABASE and MiChroM.

13.6 CONCLUSION

In the few pages of this chapter, we have retraced the steps that have already led to a usefully predictive physical theory of chromatin folding. The path taken resembles the one traveled not so long ago in setting the foundations of the energy landscape theory for protein folding. Three main factors contribute to the success of the endeavor and possibly characterize, at large, the modern approach to developing physical theories in biology. The first factor is the rigorous mathematical language of physics; embedding a new theory in the construct of statistical mechanics, thermodynamics, and symmetry principles allows one to immediately access and employ a vast array of ideas and methods, such as, for example, the maximum entropy method. This method allows one to see patterns in the data that are otherwise inaccessible, e.g., the absence of knots and the presence of liquid crystalline order. Specific biological insights are also necessary for a new theory of genome organization. These insights in the present case are the existence of chromatin types as well as the existence of special looping interactions. Of course, the success of such a mathematical understanding of chromosomes relies on the existence of the large data sets that are now available to molecular biologists.

Combining these three elements, we see the energy landscape theory of the chromosome allows us to suss out unexpected three-dimensional structural principles from two-dimensional experimental data. In many ways, the theory for chromatin folding outlined in this chapter is similar to molecular mechanics models in computational chemistry; just as with proteins, a one-dimensional sequence (amino acids there, epigenetic markers here) leads predictably to a well-defined ensemble of three-dimensional conformations. Much as for coarsegrained force fields in chemistry, the number of parameters in MiChroM is small, but, nevertheless, the number of parameters is large enough to require clever optimization methods to find them; these methods can be formulated using energy landscapes. The level of detail in the model has already proved enough to produce specific, quantitative predictions. The quality of these predictions not only validates the physical approach, but also provides new computational tools to use in further investigations of the structure–function relation at the chromosomal level.

ACKNOWLEDGEMENTS

Work at the Center for Theoretical Biological Physics was sponsored by the National Science Foundation (Grants PHY-1427654). J.N.O was also supported by the Welch Foundation (Grant C-1792) and the NSF INSPIRE award (MCB-1241332). Additional support to P.G.W. was also provided by the D.R. Bullard-Welch Chair at Rice University (Grant C-0016). B.Z. was supported by National Science Foundation Grants MCB-1715859.

REFERENCES

Alipour E, Marko JF. 2012. Self-organization of domain structures by DNA-loop-extruding enzymes. *Nucleic Acids Research* 40:11202–11212.

Bak AL, Zeuthen J, Crick FHC. 1977. Higher-order structure of human mitotic chromosomes. *Proceedings of the National Academy of Sciences of the United States of America* 74:1595–1599.

Bantignies F, Roure V, Comet I, Leblanc B, Schuettengruber B, Bonnet J, Tixier V, Mas A, Cavalli G. 2011. Polycomb-dependent regulatory contacts between distant Hox loci in *Drosophila*. *Cell* 144:214–226.

Besag J. 1975. Statistical-analysis of non-lattice data. *Statistician* 24:179–195.

Boettiger AN, Bintu B, Moffitt JR, Wang SY, Beliveau BJ, Fudenberg G, Imakaev M, Mirny LA, Wu CT, Zhuang XW. 2016. Super-resolution imaging reveals distinct chromatin folding for different epigenetic states. *Nature* 529:418–422.

Boy de la Tour E, Laemmli UK. 1988. The metaphase scaffold is helically folded: sister chromatids have predominantly opposite helical handedness. *Cell* 55:937–944.

Bryngelson JD, Onuchic JN, Socci ND, Wolynes PG. 1995. Funnels, Pathways, and the Energy Landscape of Protein-Folding - a Synthesis. *Proteins-Structure Function and Genetics* 21:167–195.

Delatour EB, Laemmli UK. 1988. The metaphase scaffold is helically folded: Sister chromatids have predominantly opposite helical handedness. *Cell* 55:937–944.

Di Pierro M, Cheng RR, Aiden EL, Wolynes PG, Onuchic JN. 2017. De novo prediction of human chromosome structures: Epigenetic marking patterns encode genome architecture. *Proceedings of the National Academy of Sciences of the United States of America* 114:12126–12131.

Di Pierro M, Zhang B, Aiden EL, Wolynes PG, Onuchic JN. 2016. Transferable model for chromosome architecture. *Proceedings of the National Academy of Sciences of the United States of America* 113:12168–12173.

Dixon JR, Selvaraj S, Yue F, Kim A, Li Y, Shen Y, Hu M, Liu JS, Ren B. 2012. Topological domains in mammalian genomes identified by analysis of chromatin interactions. *Nature* 485:376–380.

Dunham I, Kundaje A, Aldred SF, Collins PJ, Davis C, Doyle F, Epstein CB, Frietze S, Harrow J, Kaul R, et al. 2012. An integrated encyclopedia of DNA elements in the human genome. *Nature* 489:57–74.

Eagen KP, Hartl TA, Kornberg RD. 2015. Stable chromosome condensation revealed by chromosome conformation capture. *Cell* 163:934–946.

Ekeberg M, Lovkvist C, Lan YH, Weigt M, Aurell E. 2013. Improved contact prediction in proteins: Using pseudolikelihoods to infer Potts models. *Physical Review E* 87.

Filion GJ, van Bemmel JG, Braunschweig U, Talhout W, Kind J, Ward LD, Brugman W, de Castro IJ, Kerkhoven RM, Bussemaker HJ, et al. 2010. Systematic protein location mapping reveals five principal chromatin types in *Drosophila* cells. *Cell* 143:212–224.

Fudenberg G, Imakaev M, Lu C, Goloborodko A, Abdennur N, Mirny LA. 2016. Formation of chromosomal domains by loop extrusion. *Cell Reports* 15:2038–2049.

Goldstein RA, Lutheyschulten ZA, Wolynes PG. 1992. Optimal protein-folding codes from spin-glass theory. *Proceedings of the National Academy of Sciences of the United States of America* 89:4918–4922.

Grigoryev SA, Bascom G, Buckwalter JM, Schubert MB, Woodcock CL, Schlick T. 2016. Hierarchical looping of zigzag nucleosome chains in metaphase chromosomes. *Proceedings of the National Academy of Sciences of the United States of America* 113:1238–1243.

Hubner MR, Spector DL. 2010. Chromatin dynamics. *Annual Review of Biophysics*, 39:471–489.

Jaynes ET. 1957. Information theory and statistical mechanics. *Physical Review* 106:620–630.

Kireeva N, Lakonishok M, Kireev I, Hirano T, Belmont AS. 2004. Visualization of early chromosome condensation: A hierarchical folding, axial glue model of chromosome structure. *Journal of Cell Biology* 166:775–785.

Lee TD. 1952. On some statistical properties of hydrodynamical and magneto-hydrodynamical fields. *Quarterly of Applied Mathematics* 10:69–74.

Li QJ, Tjong H, Li X, Gong K, Zhou XJ, Chiolo I, Alber F. 2017. The three-dimensional genome organization of *Drosophila* melanogaster through data integration. *Genome Biology* 18:145.

Lieberman-Aiden E, van Berkum NL, Williams L, Imakaev M, Ragoczy T, Telling A, Amit I, Lajoie BR, Sabo PJ, Dorschner MO, et al. 2009. Comprehensive mapping of long-range interactions reveals folding principles of the human genome. *Science* 326:289–293.

Maeshima K, Hihara S, Eltsov M. 2010. Chromatin structure: Does the 30-nm fibre exist in vivo? *Current Opinion in Cell Biology* 22:291–297.

Nasmyth K. 2001. Disseminating the genome: Joining, resolving, and sepa-rating sister chromatids during mitosis and meiosis. *Annual Review of Genetics* 35:673–745.

Naumova N, Imakaev M, Fudenberg G, Zhan Y, Lajoie BR, Mirny LA, Dekker J. 2013. Organization of the mitotic chromosome. *Science* 342:948–953.

Phillips JE, Corces VG. 2009. CTCF: Master weaver of the genome. *Cell* 137:1194–1211.

Pieranski P. 1998. In search of ideal knots. In: Stasiak A, Katritch V, Kauffman LH, editors. *Ideal knots*. Singapore; River Edge, NJ: World Scientific. p. 20–41.

Potoyan DA, Papoian GA. 2012. Regulation of the H4 tail binding and folding landscapes via Lys-16 acetylation. *Proceedings of the National Academy of Sciences of the United States of America* 109:17857–17862.

Rao S, Huang S-C, Glenn St. Hilaire B, Engreitz JM, Perez EM, Kieffer-Kwon K-R, Sanborn AL, Johnstone SE, Bochkov ID, Huang X, et al. 2017. Cohesin loss eliminates all loop domains, leading to links among superenhancers and downregulation of nearby genes. *Cell* 171:305–320.

Rao SSP, Huntley MH, Durand NC, Stamenova EK, Bochkov ID, Robinson JT, Sanborn AL, Machol I, Omer AD, Lander ES, et al. 2014. A 3D map of the human genome at kilobase resolution reveals principles of chromatin loop-ing. *Cell* 159:1665–1680.

Sanborn AL, Rao SSP, Huang SC, Durand NC, Huntley MH, Jewett AI, Bochkov ID, Chinnappan D, Cutkosky A, Li J, et al. 2015. Chromatin extrusion explains key features of loop and domain formation in wild-type and engi-neered genomes. *Proceedings of the National Academy of Sciences of the United States of America* 112:E6456–E6465.

Sexton T, Yaffe E, Kenigsberg E, Bantignies F, Leblanc B, Hoichman M, Parrinello H, Tanay A, Cavalli G. 2012. Three-dimensional folding and func-tional organization principles of the *Drosophila* genome. *Cell* 148:458–472.

Shogren-Knaak M, Ishii H, Sun JM, Pazin MJ, Davie JR, Peterson CL. 2006. Histone H4-K16 acetylation controls chromatin structure and protein interactions. *Science* 311:844–847.

Tessarz P, Kouzarides T. 2014. Histone core modifications regulating nucleosome structure and dynamics. *Nature Reviews Molecular Cell Biology* 15:703–708.

Wang J. 2015. Landscape and flux theory of non-equilibrium dynamical systems with application to biology. *Advances in Physics* 64:1–137.

Wang X, Tucker NR, Rizki G, Mills R, Krijger PH, de Wit E, Subramanian V, Bartell E, Nguyen XX, Ye J, et al. 2016. Discovery and validation of subthreshold genome-wide association study loci using epigenomic signatures. *Elife* 5: e10557.

Wilkins BJ, Rall NA, Ostwal Y, Kruitwagen T, Hiragami-Hamada K, Winkler M, Barral Y, Fischle W, Neumann H. 2014. A cascade of histone modifications induces chromatin condensation in mitosis. *Science* 343:77–80.

Wolynes PG. 2015. Evolution, energy landscapes and the paradoxes of protein folding. *Biochimie* 119:218–230.

Zhang B, Wolynes PG. 2016. Shape transitions and chiral symmetry breaking in the energy landscape of the mitotic chromosome. *Phys Rev Lett* 116:248101.

Zhang B, Wolynes PG. 2015. Topology, structures, and energy landscapes of human chromosomes. *Proceedings of the National Academy of Sciences of the United States of America* 112:6062–6067.

Physical 3D Modeling of Whole Genomes: Exploring Chromosomal Organization Properties and Principles

MARCO DI STEFANO, JONAS PAULSEN, EIVIND HOVIG, AND CRISTIAN MICHELETTI

14.1 INTRODUCTION

Modeling chromosome conformation is arguably one of the most active, and rapidly growing, areas in theoretical and computational biophysics. This upsurge of genome modeling studies was prompted by the rapid development of experimental procedures, *in primis* chromosome conformation capture techniques [1–8]. These techniques crosslink proximal genomic regions, and ligate the interacting DNA, thereby mapping the propensity of various genomic *loci* to interact in 3D space. A related technique can be used to detect interactions at the single-cell level [9–11]. Despite their recent introduction, these quantitative techniques have already reshaped our view about how chromosomes are organized in the nucleus, how they are reconfigured in different cell lines or at different stages of the cell cycle, and about their structure–function interplay [7, 10, 12–14]. For these reasons, already from their very first applications, chromosome conformation capture methods [1, 15] have been used as phenomenological constraints in structural models. The modeling step served the dual purpose of interpreting the experimental data and formulating verifiable quantitative predictions.

The challenge of using experimental data, and particularly distance (or proximity) restraints, to establish three-dimensional models is common in molecular biophysics, especially for protein structure determination [16]. Such approaches are typically articulated over three steps, possibly repeated in an iterative fashion.

First, one defines a general polymer model that, while purposely lacking any chemical specificity, can still capture the salient physical properties of the target biomolecule, e.g., contour length, thickness, bending, and torsional rigidities. Next, a stochastic sampling procedure, such as Monte Carlo or constant-temperature molecular dynamics, is used to explore the conformational ensemble of the aspecific polymer model. Finally, out of the explored structural repertoire, one singles out the conformers whose properties best match those measured experimentally (overall gyration radius, domain organization, locality of contacts, etc.). If the sought properties are unlikely to emerge spontaneously in the unbiased ensemble, as is the case for secondary structure elements or specific sets of contacts, the phenomenological modeling can be alternatively accomplished by gradually introducing *ad hoc* biased interactions in the model to promote better and better agreement with the target experimental data. In both cases, a judicious combination of the aspecific physics-based properties of the polymer and the *ad hoc*, specific phenomenological constraints, can yield a viable structural ensemble for the biomolecule of interest. Transferring these approaches to entire genomes is a formidable task for the stratification of several challenges: the sheer size of genomic systems, the high packing density of chromosomes in the nuclear environment, and the paucity of experimental data compared to the large number of degrees of freedom needed to pin down detailed chromosome models.

Here, we shall discuss and revisit recent studies where we tackled such challenges [17–21]. Our approach was largely based on the use of steered molecular dynamics and on coarse-grained models of human chromosomes inside the nucleus to enforce the spatial proximity of *loci* that are known [7, 20] or assumed [17] to be in contact. The viability of the models was established *a posteriori* by verifying that they were stable against introducing additional constraints [8], and that the models could reproduce known structural and functional features that had not been used as input for the phenomenological constraints. The advantages of such strategy are twofold. On the one hand, it provides a three-dimensional structure of the chromosomes, thus, giving an evocative and realistic representation of the sought genome arrangement. On the other hand, the physical model exposes in a direct manner a number of functionally oriented features that are already embodied in the phenomenological constraints, albeit in an implicit form. As a notable illustration of the latter point, we will examine the intriguing connection between gene coregulation and gene colocalization [17]. Specifically, we will show that models of the human chromosome 19 obtained by imposing the spatial proximity of *loci* known to be coregulated [22] acquire large-scale features that are compatible with those of Hi-C maps [7]. The result proves the viability of the hypothesis of hubs where genes are efficiently coexpressed because they are brought in spatial proximity, and shows that structural models of the genome can be obtained by using data other than Hi-C as sources for the imposed constraints. This is an important point for the prospective use of combining phenomenological data from diverse sources to improve the resolution of the models.

In this chapter, we will give a detailed account of the aforementioned modeling studies, giving ample space to the methodological challenges.

14.2 GENOME-WIDE HUMAN CHROMOSOME MODELS

In this section, we describe the modeling approach by presenting the three main steps of its implementation. The first is the definition of a general polymer model representing the chromosomes. The second is the identification of those pairs of intra-chromosomal *loci* that correspond to statistically significant entries in Hi-C maps. The third is the use of steered molecular dynamics to colocalize these pairs on chromosome models packed inside a nucleus. In addition to these major elements, we will cover other more specific technical issues. For example, we discuss the strategy we used to organize the chromosome models inside the nucleus, taking into account the experimental radial positions [23], or the strategy we applied to match the resolution of the experimental Hi-C maps (100kb) with the intrinsic granularity of the general polymer model.

14.2.1 General polymer model

To represent chromosomes, we used an aspecific bead-spring model [24] that is entirely general, apart from capturing the salient physical properties of the 30 nm chromatin fiber [25]. The model was introduced by Rosa et al. [26, 27] to address the out-of-equilibrium physical mechanisms participating in the formation of chromosome territories [28, 29]. Since then it has been used in various contexts, from studying the organization of chromatin in the form of crumpled (or fractal) globules [30, 31] to testing the coregulation–colocalization hypothesis [17] and the presently discussed human genome modeling [20].

The potential energy of the polymer model consists of three terms:

$$H = U_{LJ} + U_{FENE} + U_{KP}. \tag{14.1}$$

The first term is a truncated and shifted, purely repulsive Lennard–Jones potential, which accounts for the excluded volume effects:

$$U_{LJ} = \sum_{(i,j)} \begin{cases} 4\varepsilon\varepsilon_{ij} \left[\left(\dfrac{\sigma}{d_{i,j}}\right)^{12} - \left(\dfrac{\sigma}{d_{i,j}}\right)^{6} + 1/4 \right] & \text{if } d_{i,j} \leq 2^{1/6}\sigma, \\ 0 & \text{if } d_{i,j} > 2^{1/6}\sigma, \end{cases} \tag{14.2}$$

where $\varepsilon = k_B T$ is the Lennard–Jones amplitude, k_B is the Boltzmann constant, T is the temperature, ε_{ij} is equal to 10 if $|i-j|=1$, and 1 otherwise, $\sigma = 30$ nm is the thickness of the chain and corresponds to ~3 kb, and $d_{i,j}$ is the modulus of $\vec{d}_{i,j} = \vec{r}_i - \vec{r}_j$ which is the distance vector between monomers i and j at positions $\vec{r}_i$ and $\vec{r}_j$, respectively.

The second term, which enforces the chain connectivity, is a FENE potential:

$$U_{FENE} = -\sum_{i} 150\varepsilon \left(\frac{R_0}{\sigma}\right)^2 \left[1 - \left(\frac{d_{i,i+1}}{R_0}\right)^2 \right], \tag{14.3}$$

where $R_0 = 1.5\sigma$ is the maximum bond length.

The last term is a Kratky–Porod potential, and is needed to reproduce the bending rigidity of the chromatin fiber, whose persistence length, l_p, is about 150 nm [32]:

$$U_{KP} = \sum_{i} \frac{\varepsilon l_p}{\sigma} \left(1 - \frac{\vec{d}_{i,i+1} \cdot \vec{d}_{i+1,i+2}}{d_{i,i+1} d_{i+1,i+2}} \right), \tag{14.4}$$

where we set $l_p = 5\sigma = 150$nm.

The three terms in Eq. 14.1 account for the intra-molecular interactions of each individual model chromosome. The inter-molecular potential is, instead, simply given by the excluded-volume Lennard–Jones interactions (Eq. 14.2) between pairs of overlapping monomers in two chromosomes.

There are two practical advantages of using this polymer model. First, the potentials in Eq. 14.1 are short range, which makes it possible to simulate with an affordable computational time even large systems, such as the entire human genome. Second, it allows for a precise control of the topological properties of the chains. In fact, the combined action of the FENE and LJ potentials restrains the fluctuations in bond length and this prevents chain crossing events that could yield to *cis* or *trans* entanglement, such as knots or links.

14.2.2 Initial conditioning of the model nucleus

Generating the initial state for the simulations of the model nucleus is a challenge in itself because of the tight confinement of the 23 pairs of chromosomes, totaling ~2m in contour length, inside the nucleus, approximately a ~10 μm wide sphere. Packing the model chromosomes at this high density, corresponding to 10% volume fraction, while avoiding steric clashes is a non-trivial task. Even more so considering that to model interphase conditions, the chromosomes should not mix randomly, but rather form separate territories [33] with only limited intermingling [34]. Additionally, one wishes to account for phenomenological propensities of different chromosomes to be at definite radial distances from the center of the nucleus [23]. Finally, one should also account for the fact that regions such as centromeres, which contain highly repetitive elements, should form compact domains [35–37].

To satisfy these multiple requirements, we devised a specific protocol for the initial nuclear positioning of the model chromosomes. These consisted of two copies of the human autosomes (chromosomes 1 to 22), and of a single copy of the sex chromosome X, i.e., we neglected the inactive copy of chromosome X (in females), or the small chromosome Y (in males).

First, following ref. [26], we organized each chromosome in a rod-like solenoid made of stacked rosettes. This arrangement is meant to mimic the linear, compact, and ordered organization of mitotic chromosomes, where intra- and inter-chain entanglement is expectedly minimal.

The rod-like chromosomes are then phenomenologically positioned inside a spherical nucleus with a diameter equal to 10 μm. Specifically, the rods are placed with their midpoints at the correct phenomenological distance from the nuclear sphere center, while the orientation of their axis is picked randomly. The positioning proceeds from the longest to the shortest rod: for each newly added chromosome, 10,000 random positioning attempts are carried out and, out of these, we retain the one that: (i) is free of steric clashes with previously placed rods, and (ii) whose midpoint radial positioning best matches the experimental average distance from the nuclear center reported in ref. [23].

We note that this positioning scheme does not guarantee that all rods fit entirely inside the nuclear sphere. In fact, the longest chromosome (chromosome 1), whose solenoidal molecule has a length of ~14.5 μm, inevitably has parts that protrude outside the ~10 μm-wide nucleus. These protrusions were eliminated during the initial stages of the system dynamical evolution, as described in the following section.

14.2.3 Molecular dynamics simulations

The free dynamics of the chains was described with an underdamped Langevin equation:

$$m\ddot{r}_{i\alpha} = -\partial_{i\alpha}H - \gamma\dot{r}_{i\alpha} + \eta_{i\alpha}(t). \tag{14.5}$$

Here, H is the system potential energy, index i runs overall monomers in the systems, α runs over the x,y,z Cartesian components, m is the monomer mass, γ is the friction coefficient, and η is a stochastic noise satisfying the standard fluctuation dissipation relations: $\langle \eta_{i\alpha} \rangle = 0$ and $\langle \eta_{i\alpha}(t) \eta_{j\beta}(t') \rangle = 2\kappa_B T \gamma \, \delta_{ij} \delta_{\alpha\beta} \delta(t-t')$.

The dynamical evolution was integrated with the LAMMPS simulation package [38], using values for the bead mass, $m = 1.0$, and for the friction coefficient $\gamma = m/(2\tau_{LJ})$ from [24]. The integration time step was set equal to $0.006\tau_{LJ}$, where $\tau_{LJ} = \sigma(m/\varepsilon)^{1/2}$ is the characteristic Lennard–Jones time.

During the very first stages of the Langevin dynamics, the system of rod-like chromosomes was subjected to the action of a radial potential that exerted a compressive force exclusively on the monomers protruding beyond the nuclear sphere. Applying this confining potential for $1{,}200\tau_{LJ}$ was typically sufficient to bring all chromosomes entirely inside the spherical nucleus. During this compression step, each centromere was concomitantly compactified by attractive Lennard–Jones potentials acting on its constitutive beads.

After this stage, the system was allowed to evolve spontaneously for a total of 2×10^7 integration steps, corresponding to $1.2\ 10^5\tau_{LJ}$.

14.2.3.1 MAPPING OF THE SIMULATION TIME TO ACTUAL TIME UNITS

It is informative to establish an approximate mapping of the characteristic Lennard–Jones time, τ_{LJ} to actual time units, so to develop a feeling for the order of magnitude of the actual time scales addressable in simulations.

The simplest possible mapping can be made by matching the nominal diffusion coefficient of individual monomers (beads) in our simulations, $D = k_B T$, with the one expected from Stoke's law, $\dfrac{k_B T}{6\pi\eta\sigma/2}$. Equating the two expressions for the diffusion coefficient yields $\sqrt{m} = \dfrac{6\pi\eta\sigma^2}{\sqrt{\varepsilon}}$, which in turn gives $\tau_{LJ} = \dfrac{6\pi\eta\sigma^3}{\varepsilon}$. For our model system $\sigma = 30\text{nm}$ and $T = 300\text{K}$, so that $\varepsilon = k_B T = 4.2\ pN\ nm$. The remaining parameter is the viscosity, η.

If we were dealing with polymeric chains with much smaller size than chromosomes, e.g., globular proteins of a few hundred amino acids, then η could be set to the nominal viscosity of the nucleosol (about 1 cP). However, it has been shown that the motion of entire chromosomes and of their subparts is controlled by a much higher effective viscosity [39], $\eta \sim 200$. This enhancement arguably arises from the fact that the obstacles in the crowded and structured nucleosol environment may offer a higher hindrance to the motion of chromosomes than to protein-sized molecules.

Using the larger effective viscosity, one estimates that the duration of the simulated relaxation dynamics, $1.2\ 10^5\tau_{LJ}$ corresponds to timespans in the hour range, in realtime.

Over this timespan, the initial rod-like chromosomes rearrange significantly, filling most of the nuclear sphere. The final arrangement, shown in Figure 14.1A is qualitatively similar to that observed in imaging experiments, such as FISH [23, 33], on interphase genomes, which established that individual chromosomes

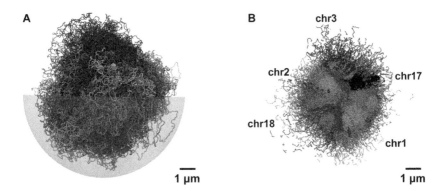

Figure 14.1 Genome-wide models of human chromosomes in lung fibroblast cells (IMR90). (A) Typical genome-wide spatial arrangement of the human chromosomes after the phenomenological chromosome positioning based on the FISH data in ref. [23] and the steering process based on the Hi-C data from ref. [7]. (B) Tomographic slice of 150 nm of the chromosomes conformation in panel A. Different colors correspond to different chromosomes. Reproduced with permission from ref. [20].

fill local regions of space with only modest intermingling at their boundaries [34]. This is indeed analogous to what is seen in Figure 14.1B, which presents a cut-through view of the system.

14.2.4 Statistical models to identify constraints from Hi-C datasets

The utilization of Hi-C data as constraints in 3D genome modeling may require a conversion of chromosome-interaction frequencies to Euclidean distances. Several strategies for defining distance constraints from Hi-C data have been used. Common to these strategies is the assumption that a high number of detected Hi-C interactions between pairs of regions is indicative of close proximity in 3D space. In one strategy, an inverse relationship between the Hi-C interactions and spatial proximity is assumed. This relationship can be encoded through a simple mapping function [40], or by more sophisticated curve-fitting methods [41]. Both of these strategies, however, assume that the (population-based) Hi-C data stem from an unbiased probabilistic sampling from a single, stable conformation. It is increasingly realized that such assumption is not strictly valid, as structural variability between single cells is commonly observed [42]. To accommodate this variability, two strategies have been used. In population-based modeling, the goal is to define a scoring function on an entire population of structural models that is simultaneously optimized [43]. In statistics-based modeling, the constraints are selected based on statistical tests that identify interactions in the Hi-C data that are significantly frequent in the population. The structural models are then constrained based on the stable contacts.

We have used the zero-inflated negative binomial method (ZiNB) to define such contacts:

$$P_{\delta_{ij}}(X \geq n_{ij} \mid \hat{\theta},\ \hat{p},\ \hat{\pi}) = (1 - \hat{\pi}\ )NB(X \geq n_{ij} \mid \hat{\theta},\ \hat{p}). \qquad (14.6)$$

In this expression, p, θ, and π are determined for groups of genomic distances (δ_{ij}) between bins in the Hi-C data (see Figure 14.2). This statistical model captures the probability of observing a given number of contacts (n_{ij}) between two bins (i and j) conditional on the genomic distance between them (δ_{ij}), while taking into account the sparsity of the data at high resolutions. The p parameter gives an estimate of the probability of observing a zero in the negative binomial model. Expectedly, this will gradually increase for increasingly higher genomic distances, since the data are typically more sparse away from the diagonal. The θ parameter captures the dispersion, or the extra variance observed with respect to a Poisson distribution. This parameter will typically decrease with

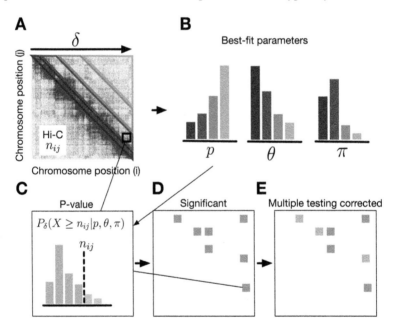

Figure 14.2 Overview of the statistical method used to detect significant Hi-C interactions. (A) Input Hi-C interaction matrix. The matrix is divided into bins according to increasing genomic distances (δ). (B) For each genomic distance bin, best-fit parameters (p, θ, π) of the Zero-inflated Negative Binomial (ZiNB) distribution are fitted. (C) For each pixel in the Hi-C interaction matrix, a P-value is calculated utilizing the parameter set given by the genomic distance bin the pixel belongs to. D–E: From the P-values, statistically significant entries of the matrix are identified (D) and multiple testing correction is performed (E). The resulting significant interactions are used as pairwise constraints in the 3D modeling.

increasing genomic distance, since the contact frequencies in the Hi-C data are higher near the diagonal. The π parameter accounts for to the additional zeros in the data that cannot be explained by the negative binomial distribution alone. This parameter will typically be highest at short genomic distances, since the distribution will approximate a normal distribution for large expected contact frequencies. The zero entries will, in that case, have to be modeled through the parameter π. For larger genomic distances (and therefore lower expected inter-action frequencies), zero entries can be incorporated in the negative binomial model directly.

The ZiNB model is applied to each enrty of Hi-C matrices, obtaining indi-vidual P-values for all possible pairwise interaction. Due to the large number of P-values, these have to be corrected for multiple testing before selecting a subset of the significant pairwise interactions as constraints in the modeling.

For more coarse-grained systems such as domain-level modeling, taking data sparsity into account is less important, meanwhile, it becomes necessary to also incorporate the general propensity of observing contacts for the domains overall. This can be done in the noncentral hypergeometric (NCHG) model [19, 21]:

$$P\left(X \geq n_{ij}\right) = \sum_{x \geq n_{ij}} P(x \mid n, n_i, n_j, \omega_{ij})$$

$$= \frac{\displaystyle\sum_{x \geq n_{ij}} \binom{n_i}{x}\binom{2n - n_i}{n_j - x}\omega_{ij}^x}{\displaystyle\sum_{x} \binom{n_i}{x}\binom{2n - n_i}{n_j - x}\omega_{ij}^x}. \tag{14.7}$$

In this expression, the model incorporates both the total number of contacts observed for each domain (n_i and n_j), as well as the total number of observed contacts for a chromosome (n). The dependency of contact frequencies on genomic distance is captured by the ω_{ij} parameter, which is estimated from the expected number of contacts for given genomic distances:

$$\omega_{ij} = \frac{\lambda_{ij}\left(2\lambda - \lambda_i - \lambda_j + \lambda_{ij}\right)}{\left(\lambda_i - \lambda_{ij}\right)\left(\lambda_j - \lambda_{ij}\right)}, \tag{14.8}$$

where λ_{ij} is the expected number of contacts between domain i and j for a given genomic distance (δ_{ij}).

14.2.5 Steered molecular dynamics simulations

To enforce the colocalization of the regions corresponding to the statistically sig-nificant (target) pairings, we used a protocol based on steered molecular dynamics.

First, based on the mapping of 100 kbp (the Hi-C map resolution) with 33 beads in our model, we identified the specific stretches of the model chromosomes that correspond to the target pairs. For each pair of stretches, A and B, we introduced a time-dependent harmonic potential

$$U_H = \frac{1}{2}k(L_{AB},t)(d_{A,B} - d_{eq})^2 \qquad (14.9)$$

where $d_{A,B}$ is the distance of the centers of mass of the segments A and B, and d_{eq} and $k(L_{AB},t)$ are respectively the reference distance, and the spring constant of the harmonic constraint. The former was set equal to $d_{eq} = 60$ nm, comparable to the size of proteins that bridge chromatin strands in Hi-C experiments.

The spring constant, k, was instead varied during the steering dynamics. Specifically, it was ramped up gradually with time according to:

$$k(L_{AB},t) = k^0(L_{AB}) \cdot t / (6000\ \tau_{LJ}) \qquad (14.10)$$

until an overstretching of the chain was detected, in which case the ramping was stopped. The overstretching condition used to stop the ramping of the potential was that the harmonic force between Hi-C constrained regions remained equal or below 300 σ/ε.

Note that in Eq. 14.10, the amplitude $k^0(L_{AB})$ depends on the sequence distance of the A and B regions, L_{AB}. This dependence was introduced to account for the fact that, in the absence of constraints, regions at small genomic distances obviously have a higher chance of being in contact than regions at higher sequence separations, and hence a stronger constraining potential has to be used in the second case.

We accordingly first divided the target contacts into sets covering different ranges of genomic distances in steps of 10 Mbp ([0:10]Mbp, [10:20] Mbp etc.). Next, for each set, we computed the probability distribution of the spatial distance of the various target pairs in the relaxed configuration, $P_{relaxed}(d_{AB})$. To a first approximation, the addition of a harmonic biasing potential of amplitude k as in Eq. 14.9 would modify the contacting probability to: $P_{biased}(d_{AB}) \approx P_{relaxed}(d_{AB}) \cdot \exp(-k(d_{AB} - d_{eq})^2 / (2\ k_B T))$. We, therefore, established $k^0(L_{AB})$ as the potential amplitude, k, needed to have a probability larger than 90% to bring the pair into contact, that is at a distance smaller than $d_c = 2d_{eq} = 120$ nm.

14.3 STEERED CONFORMATIONS

The first relevant question that can be addressed within our framework is to what extent the target contacts derived from Hi-C matrices can be actually established within a steered-MD trajectory. This is clearly an important question *per se* from the modeling perspective, but is also relevant biologically, because it can reveal the degree of polydispersity, or structural diversity, captured within Hi-C maps. In fact, if the latter was too high, it would become unfeasible to satisfy a significant proportion of the target constraints within any given simulation.

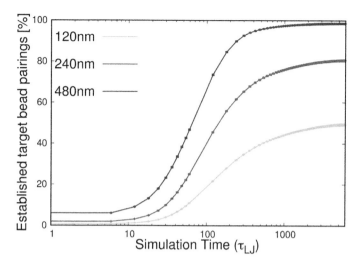

Figure 14.3 Time-dependent percentage of established Hi-C constraints. The curves show the increase of the amount of satisfied Hi-C constraints, based on [7]. The proximity criterion between segments of 100 kb (33 model beads) is assessed on the distance between the closest pair of beads for various cutoff distances: 120 nm, 240 nm, and 480 nm. This contact definition is mimicking the mechanism of interaction capture in a typical Hi-C experiment. Reproduced with permission from ref. [20].

The compliance of the genomic model to the steering protocol is shown in Figure 14.3. The figure shows the time evolution of the percentage of target pairs that have been brought closer than 120, 240, and 480 nm. The steady increase of the curves, which have been averaged over 10 trajectories, reflects the progressive action of the steering dynamics. On average, at the end of the simulation, 80% of the target contacts are established within 240 nm, and about all of them within 480 nm.

14.3.1 Functional insight from structural models

Profiling the number of imposed constraints that are actually satisfiable is a necessary first assessment of the viability of the modeling scheme. Much more interesting and informative, however, is probing the obtained configurations for features that have not been used as input for the constraints. Of particular interest are functionally oriented features encoded in the genome spatial organization. This includes, for instance, fluorescence in situ hybridization (FISH) measurements of the preferential radial positioning of *loci* that have high or low gene content, that are up or down regulated, that are associated with the nuclear lamina, and so on. Note that such features are not manifestly encoded in Hi-C matrices. It is precisely for this reason that verifying whether the correct nuclear positioning of these regions emerges after, and only after, imposing the phenomenological constraints would be highly significant from the biophysical point of view.

14.3.2 Nuclear positioning of functional regions

As a prerequisite for the structural-functional analysis, we first examined to what extent the imposed proximity constraints, which are much fewer than the number of degrees of freedom of the model genome, suffice to pin down specific structural features. We addressed this point by considering various portions of each chromosome and measuring how much their radial positioning, that is their distance from the center of the spherical nucleus, varied across ten independent steering simulations.

The observed variability of the radial placement is summarized in the color-coded map of Figure 14.4. One notes interesting patterns: the degree of variability does not show obvious correlations with the length of chromosomes. Rather it depends on the density of Hi-C constraints. In particular, the chromosomes with the highest number of constraints per unit length (16, 19, 7, and 9) are also those with the least variable radial positioning. This pleasing feature, which is not necessarily expected *a priori*, is an encouraging basis for investigating more detailed aspects of the model genome organization.

We then moved on to examine the radial positioning of several functional regions before and after the steering process. Specifically, we considered *loci* corresponding to gene-rich and gene-poor regions, lamina-associated domains (LADs), regions enriched with activating or repressing histone modifications and of *loci* associated with positive (gpos) or negative (gneg) bands of the Giemsa

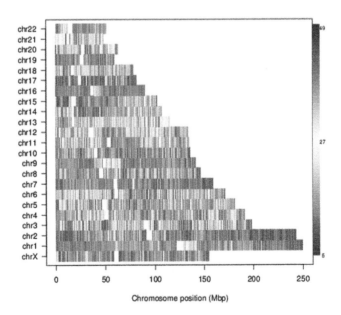

Figure 14.4 Chromosome-wise variability of radial bead position. The color scale indicates the standard deviation of the radial positions of beads in the 10 genome-wide models in IMR90 human cells with phenomenological initial chromosome positioning. Reproduced with permission from ref. [20].

staining technique. The latter are known to correspond to regions that are respectively poor or rich on AT content which, in turn, is reflective of heterochromatic or euchromatic regions [44].

The circular plot of Figure 14.5A clarifies that prior to the steering process, these *loci* are about evenly distributed (compared to other regions) across spherical shells at various radial distances. This clarifies that the initial placement of chromosomes, despite being based on the known preferred radial positioning, does not bias significantly the radial location of specific functional regions compared to any other region.

This uniformity is lost during the steering process when the aforementioned functional *loci* systematically acquire specific positional preferences. In particular, as it is shown in Figure 14.5, regions that are rich in genes, in epigenetic markers controlling gene activation and repression (H3K4me3, H3K9me3, and H3K27me3), or generally in euchromatin (negative Giemsa bands) show a marked preference to occupy the nuclear center only after steering. The opposite preference is, instead, clearly visible in regions that are gene poor or associated with heterochromatin (positive Giemsa bands) and LADs.

All these preferential positionings introduced by steering are correct, in that they match known positional propensities observed *in vivo*. Gene-rich chromosomes (e.g., chromosome 19) are consistently visualized in the center of the nucleus by FISH experiments, while gene-poor ones (such as chromosome 18) retain a peripheral positioning [33, 45, 46]. In ref. [44], chromosomal regions associated with positive (or negative) Giemsa bands of chromosome 7 are found at the nuclear periphery (or center) in human cells. Finally, it is known that

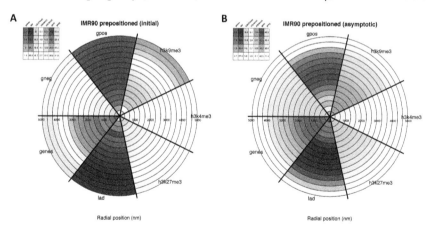

Figure 14.5 Radial positioning of functionally related features in the model nucleus. The circular histograms give the percentage of beads associated with functional features relatively to the total number of beads in a given radial shell cumulated over the 10 simulation replicates. The results are shown for regions associated to H3K9me3 (orange), H3K4me3 (yellow), H3K27me3 (green), LADs (blue) and genes (red), and negative (cyan) and positive (purple) Giemsa staining bands before (A) and after steering (B). Adapted with permission from ref. [20] and its related supplementary material.

chromosome regions which interact preferentially with the nuclear lamina are located at the periphery of the nucleus. These domains result from anchoring of chromosomal regions to a protein meshwork (the nuclear lamina) coating the inner nuclear membrane. Genomic regions anchor to the nuclear lamina in large domains (0.1–100 kilobases) that are observed in single cells, but also display cell-to-cell variation [47].

This is an important point. Not so much for the found positioning *per se*, because it largely supports previously known biological facts [47]. But rather, the key point is that the results demonstrate that this genuinely functional information is actually implicitly encoded in the significant Hi-C constraints. In fact, though it is not immediately evident from the Hi-C map, the preferential positioning can be extracted from it and made explicit thanks to a physics-based modeling.

14.3.3 Model refinement and interphase-mitotic reconfiguration

As we discussed, the steering process based on the limited number of phenomenological constraints sufficed to establish a number of specific, and biologically correct, structural features in the model interphase genome.

This observation poses, in turn, two questions, both related to the plasticity of the steered conformations: (i) can they be reorganised by adding additional phenomenological constraints? (ii) can they be reconfigured into locally compact configurations, thus mimicking the rearrangements that interphase chromosomes sustain when transitioning to their mitotic forms [13, 48]?

For the former question, we applied to the steered conformations an additional set of distance constraints based on the *in situ* Hi-C study of IMR90 cells of ref. [8]. The set consisted of 8,040 interactions, that are largely complementary to those used so far [7]. In fact, they were obtained using different *in situ* Hi-C experiments at a higher nominal resolution (~1 *kb*) and have a higher degree of locality. Specifically, the median sequence separation of the contacts in [8] is 220 kilobases, which is about 200 times shorter than the contacts in [7] (46.8 Megabases).

As shown in Figure 14.6, the steered system responded well to the addition of the local target constraints, which were established with about the same compliance as the original set. Pleasingly, the formation of these new contacts did not disrupt previously established ones and, especially, all aforementioned functionally oriented radial placements were preserved [20].

Additionally, we compared the large-scale structural organization of the chromosomes obtained by combining the two sets of target contacts with the genomic model of Kalhor et al. [6]. We chose this term of comparison because it established the macro-domain organization of human chromosomes using a variety of phenomenological constraints other than Hi-C. It is, therefore, complementary to ours. We restricted considerations to chromosome 19, which is the one with the highest gene density, and found that the domains found or ref. [6] had a significant overlap with the spatial clusters in our steered model, see Figure 14.7. This

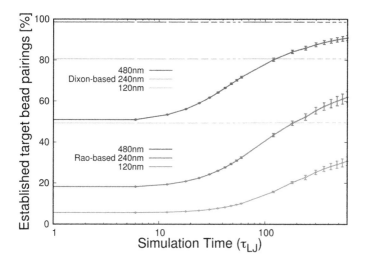

Figure 14.6 Time-dependent percentage of established Hi-C constraints during the additional steering dynamics. The curves show the percentage of the maintained constrained based on ref. [7], and the newly established from ref. [8]. The same contact criterion, and cutoff distances of Figure 14.3 apply. Reproduced with permission from ref. [20].

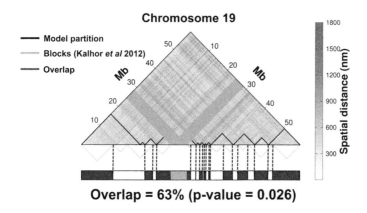

Figure 14.7 Partition of chromosome 19 in macrodomains. The boundaries of the optimal 13 macrodomains are overlaid on the upper triangle of the map of spatial distances between 100 kb regions on chromosome 19, and juxtaposed to the borders of the 13 blocks identified in ref. [6]. The overlapping regions (shown in blue) of the two partitions span 63% of the chromosome excluding the centromere (shown in gray). Reproduced with permission from ref. [20].

result has two relevant implications: (i) it provides an independent confirmation of the correctness of the large-scale organization in the physics-based genomic modeling and (ii) it shows the feasibility of adding various layers of constraints to progressively refine the model itself.

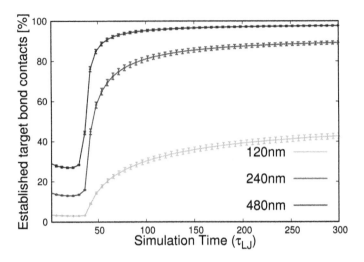

Figure 14.8 Time-dependent percentage of established mitotic constraints. The panel shows the increasing percentage of satisfied constraints during the interphase⟶mitotic recondensation dynamics. The same cutoff distances of Figure 14.3 are used. Reproduced with permission from ref. [20].

Next, we studied whether the model chromosomes could be straightforwardly reconfigured to a state rich in local contacts, as are mitotic states [13, 49], or if they would be prevented to do so by the emergence of conflicting topological constraints. We accordingly took the steered configuration as the starting point of further steering simulations, where we switched off the Hi-C-based target constraints and replaced them with aspecific ones promoting the spatial proximity of all beads at genomic distances smaller than 200 kb (~66 beads).

We found that the hindrance towards establishing these local constraints was minimal. In fact, most of these constraints were satisfied at the end of the steering protocol, as shown in Figure 14.8. As a matter of fact, the lack of significant topological barriers, and the plasticity of the chromosome models allows the condensing chromosomes to segregate neatly, see Figure 14.9, with almost no interchain linking in qualitative accordance with FISH observations [23, 33, 34].

14.4 GENE COREGULATION-COLOCALIZATION STUDY

The physics-based approach presented above allowed for recovering several functionally related features of the human genome by using Hi-C-based structural data.

This fact stimulates an even more challenging question: can one infer the structural organization of the genome by using functionally related data only?

Along these lines, we recently carried out a modeling study [17] that was inspired by the so-called *gene-kissing* hypothesis [50]. This phenomenologically based hypothesis [51] posits that the coexpression of genes on the same or different chromosomes is reflective of their colocalization in space. While not all

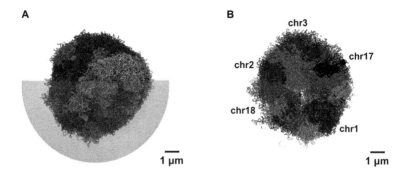

Figure 14.9 Chromosome mitotic recondensation. The chromosome conformations obtained using the using the initial phenomenological placement of ref. [23], and the steering based on the Hi-C data in refs. [7] and [8] were recondensed towards mitotic-like arrangements by means of harmonic constraints between pairs of beads at 200 kb (66 beads). Reproduced with permission from ref. [20].

coregulated sets of genes are necessarily expected to be colocalized, an increasing number of cases where the gene-kissing hypothesis holds have been reported, from bacteria [52] to eukaryotes [53, 54].

In our study of [17], which we revisit here, we explored the relationship between colocalization and coregulation of genes using physical models to test a drastic version of the hypothesis. Specifically, we asked whether it is physically feasible to colocalize *simultaneously* all pairs of genes that are significantly coregulated. A positive answer to this question is not obvious, given the significant frustration expected to arise from conflicting colocalization constraints, and would provide a strong, albeit indirect, element for the more general viability of the hypothesis.

14.4.1 Coregulated gene pairs

As a first step, we identified the pairs of genes of human chromosome 19 (the one with highest gene density) whose transcriptional activity is significantly coregulated.

We followed the strategy of ref. [22] and analyzed a publicly available [55] set of 20,255 expression profiles for the 1,278 genes in human chr19 measured in 591 distinct microarray experiments. The number of expression profiles is about 16-times larger than the number of genes, ensuring a good statistical coverage. However, the expression profile dataset, based on the HG-U133A Affymetrix chip, is rather heterogeneous because it contains data from different human tissues, different cell types, different biological conditions, and is processed with different normalization protocols. To use data from all experiments on equal footing, this heterogeneity had to be accounted for. Following ref. [22], we introduced a coarse-graining procedure of the expression profiles. All expression levels in each experiment were equipartitioned into three discrete states: low, medium, and high, depending on the measured intensity.

Next, to measure the degree of correlation, or covariation, of any gene pair, we computed their mutual information value (MI) overall available (coarse-grained) expression profiles, see Figure 14.10A, i and j

$$M I_{ij} = \sum_{\alpha} \sum_{\beta} \pi_{\alpha\beta} \ln \left(\frac{\pi_{\alpha\beta}}{\pi_{\alpha+}\pi_{+\beta}} \right) \qquad (14.11)$$

where α [β] runs over the 3 discretized expression levels of gene i [j], $\pi_{\alpha\beta}$ is the joint probability that, in a given experiment, the expression levels α and β are respectively observed for probe sets i and j, while the quantities $\pi_{\alpha+} = \sum_{\gamma} \pi_{\alpha\gamma}$ and $\pi_{+\beta} = \sum_{\gamma} \pi_{\gamma\beta}$ are the marginal probabilities to observe expression level α [β] for gene i'[j]. The MI thus provides a statistically founded measure of how the gene expression pattern for gene i is predictable, assuming the knowledge of another pattern j (or, *vice versa*).

To identify the genes with statistically significant MI values, we first grouped all possible distinct gene pairs according to their genomic distances [22]. We used in total 15 groups, each covering a 4 Mb-long interval. Next, using the baseline statistics of ref. [56], we fitted the distribution of the pairwise MI values in each group, using the exponential function $a \, x \exp(-bx)$ expected from the statistical null model, see Figure 14.10B.

Finally, we singled out the statistically significant coregulated pairs in each group as those with atypically large deviations from the null distribution [17], see Figure 14.10C and D for the bin including pairs at distance in the [28:32] Mbp range.

Overall, we identified a total of 1,487 non-redundant probe set pairs, involving 412 distinct genes. As shown in Fig 14.10E, these genes are distributed over the entire chromosome 19, and the gene pairs span genomic distances up to almost the total chromosome length. In view of their statistically significant covariation of expression levels, we regarded these pairs as being genuinely coregulated [22].

14.4.2 Colocalization of coregulated gene pairs

We then enforced the spatial colocalization of these pairs in coarse-grained models of chromosome 19. The model conformations are entirely analogous to those discussed in section 14.2. Specifically, we represented chromosome 19 as a chain of 19,710 beads of thickness $\sigma = 30$ *nm*, and persistence length $l_p = 150nm$, for a total contour length of ~59.13Mb.

Because we are exclusively focusing on chromosome 19, we mimicked the nuclear crowding by placing multiple copies of the chromosome in the periodic cubic simulation box. More precisely, we placed six copies within a box of side equal to 3 μm, so to match the typical nuclear density $0.012bp / nm^3$. The six chromosome copies were initially prepared in the idealized rod-like mitotic arrangement, placed in a random but non-overlapping manner inside the simulation box, and were then let to relax with a standard push-off protocol of $1,200\tau_{LJ}$.

This initially relaxed state, shown in Figure 14.11A, was then subjected to a steering protocol for bringing into contact the coregulated pairs. The steering

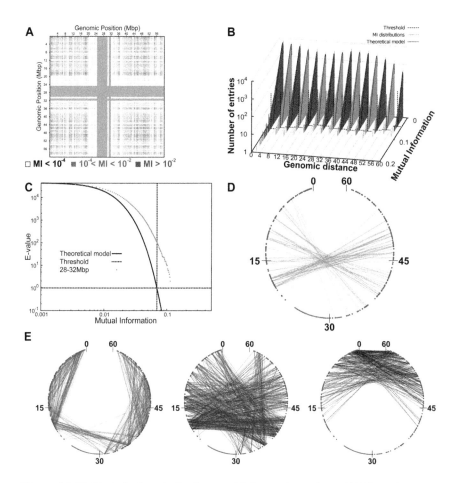

Figure 14.10 Statistical analysis of mutual information values. (A) Mutual information (MI) matrix for any pairs of genes on human chromosome 19. The middle point of each *locus* identifies its position along the chromosome. The centromeric region is indicated as gray stripes, and has no probed gene in it. (B) Histograms of values of mutual information for pairs of genes located at various intervals of their genomic separation. The black lines correspond to fitting the histograms with the theoretical (null case) MI distribution. The vertical black dashed lines correspond to the estimated threshold values. (C) Example of E-value (expected number of false positives) distribution for gene pairs located at genomic separation in the range 28–32 Mbp. Here and in all the distributions at a given interval of genomic separation (see panel B), the threshold for significant MI values is at E-value equal to 1.0. (D) The network of coregulated gene pairs resulting from the analysis at 28–32 Mbp separation is represented on a circularized cartoon sketch of chromosome 19 (The scale is in μm). The significantly high values of Mutual Information correspond to connections (cyan links) between coregulated gene pairs (red dots). (E) The entire network of coregulated gene pairs is represented as 3 sub-networks for different genomic separations between the involved *loci* of 0–20 Mbp (left), 20–40 Mbp (middle) and 40–60 Mbp (right), respectively. Reproduced with permission from ref. [17].

process was carried out by *simultaneously* ramping up an attractive harmonic (spring) interaction between the centers of mass of the significantly coregulated regions.

Attempting the simultaneous colocalization of ~1,500 paired regions is expectedly a frustrating problem. The pairs, in fact, involve only ~500 distinct regions, meaning that on average each region will be steered towards three other *loci*, likely at very large genomic distances. From these considerations alone, we were then expecting that only a subset of the target constraints would be colocalizable, corresponding to a core of spatially coregulated genes common to the heterogeneous ensemble of cell types in datasets.

The steering process, however, had a different outcome from what we had expected. In fact, we found that as much as 80% of the target gene pairs were colocalized at the end of the steering process.

The result is illustrated in Figure 14.11B, which shows that the proportion of satisfied constraints could actually be even higher, by continuing the steering process, which we did not pursue, as this would also introduce artifactual overstretching of the chain. The result is both appealing and thought-provoking, because it suggests that spatial colocalization could be a general and physically viable mechanism to ensure an efficient and robust coregulation of genes. At the same time, it poses the question of whether an analogous colocalization compliance should be expected for an equivalent, but generic polymer system or whether it emerges from specific features of the coregulatory network and of the initial conformation. We addressed these questions by considering various alternative models.

First, to examine the role of the initial conformation, we replaced the relaxed mitotic-like chains with random walks. These walks, which were initially relaxed

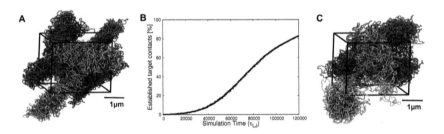

Figure 14.11 Re-shaping of chromosome 19 models under the action of spatial constraints based on gene pair coregulation. (A) The initial rearrangements (before steering) of the six copies of model human chromosome 19 are shown together with the cubic simulation box. Due to the periodic boundary conditions of the simulation the chromosome chains can protrude out from the elementary box. (B) The curve describes the increasing percentage of satisfied gene coregulation spatial constraints during the steered dynamics. The criterion for a contact between two genes is defined on the spatial proximity (<120 nm) of the respective centers of mass of the chromosome stretches hosting them. (C) The rearrangements of chromosome 6 models are shown at the end of the steering process. Chromosome regions involved in the coregulatory network are highlighted in red. Adapted with permission from ref. [17].

to remove steric clashes, showed a negligible steering compliance, see Figure 14.12. This is because the random nature of the walks creates a significant level of intra- and inter-chain entanglement that hinders chain plasticity. The effect is visible in the accompanying snapshot of the end (post-steering) configuration, and confirms the intuition that the ordered relaxed mitotic conformation is an important element for the reconfiguration compliance of the model chromosomes.

For the additional alternative models, we, therefore, maintained the relaxed mitotic-like states as initial configurations and, instead, randomly rewired the network of colocalization constraints. We considered two rewiring protocols.

First, we kept fixed the sequence location of the original target regions, but we reshuffled the pairings. To be as conservative as possible, the random reshuffling was carried out while keeping fixed the number of constraints involving each target region. This reorganized network was found to have about half the colocalization compliance of the original one, see Figure 14.12.

Secondly, we reorganized the network by displacing the target regions in a random (but non-overlapping) manner along the chromosome. The original target pairings were instead preserved. Strikingly, this reshuffled target network

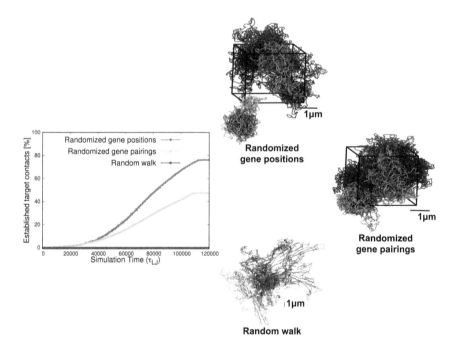

Figure 14.12 Steering process in the randomized cases. The time-evolution of the percentage of satisfied spatial constraints is shown for the 3 randomized variants. The 3 corresponding configurations reached at the end of the steering protocol are shown on the right. Chromosome regions that take part in the pairs of loci to be colocalized are highlighted in red. Adapted with permission from ref. [17].

had about the same compliance of the original network. Figure 14.12 shows, in fact, that about 80% of the target contacts between the displaced corresponding regions are established at the end of the steering process, see Figure 14.11.

Taken together, the results from the two alternative models indicate that the high colocalization compliance of the phenomenological coregulated pairs reflects specific properties of the coregulatory network, while the positioning of the corresponding genes on the chromosomes is less crucial.

Based on the *a posteriori* analysis and inspection of the coregulatory network, we concluded that this specific feature is directly related to the presence of coregulatory cliques, that is groups of genes that are mutually significantly coregulated. It is, in fact, the presence of these cliques that arguably facilitates the gene colocalization, because it guarantees that many target pairings can be simultaneously satisfied by bringing them all together in space.

14.4.3 Testing the coregulation-colocalization hypothesis: macrodomain organization of chromosome 19

The analysis described in the previous subsection shows that the colocalization of significantly coregulated pairs is physically viable, and that this is possible thanks to the abundance of coregulatory cliques.

A more stringent question is whether imposing the colocalization of coregulated pairs alone suffices to induce the correct spatial organization of chromosome 19. To explore this, we compared the contact map of the post-steered conformations of chromosome 19 with the actual Hi-C maps obtained by Dixon et al. [7]. Because the number of colocalization constraints is small compared to the degrees of freedom of the model chromosomes, we examined the accord of the two maps at the level of macro-domains.

Specifically, we used a general (K-medoids) clustering algorithm to partition the chromosomes into up to ten domains. We then compared the overlap of the subdivisions of the steered chromosome contact map with those of the Hi-C one, for an equal number of domains. We found that the overlaps were all statistically significant compared to random partitions for all considered numbers of domains. In particular, the good correspondence of subdivisions into an intermediate number of domains (seven not counting the pre-assigned centromere) is visible in Figure 14.13. The corresponding overlap is 79%, and the probability to observe a match that is equally good or better is less than 3%.

The overall good consistency of the post-steering spatial organization of chromosome 19, which we recall is based exclusively on coexpression data, gives a strong, albeit indirect, element supporting the relevance of colocalization as a general and biologically viable means of achieving gene coregulation *in vivo*.

To further explore how spatial chromosomal arrangement and gene coexpression interrelate on this chromosome, we generated side-by-side Circos plots [57] of significant Hi-C interactions (from ZiNB) and coexpressed genes (see Figure 14.14). These plots reveal a striking correspondence between coexpressed and colocalized regions throughout the chromosome. When inspecting this information in relation to features such as gene expression and gene density,

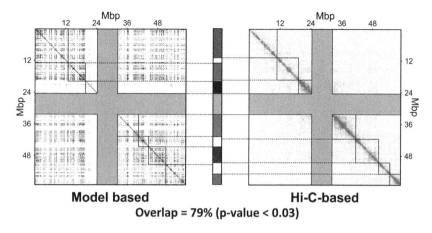

Model based   **Hi-C-based**
Overlap = 79% (p-value < 0.03)

Figure 14.13 Partition in macrodomains of structures obtained using gene colocalization constraints. The contact maps for Chr19 obtained at the end of the steered-MD simulations and the Hi-C interaction map from ref. [7] are shown on the left and right, respectively. The grey bands mark the centromere region. The boundaries of the 7 spatial domains (centromere excluded) are overlaid on the matrices. The overlapping parts of the two partitions account for a large fraction (79%) of the chromosome (centromere excluded). They are visualized in the sketch of chromosome 19 in the center (different colors indicate different domains). Non-overlapping regions are shown in white, and the centromere region in grey. Adapted with permission from ref. [17].

it is evident that the coexpressed genes form an active hub of colocalization. By comparison with our previous radial positioning plots (Figure 14.5), where active regions are shown to be positioned towards the nuclear interior, it is likely that the active hub on chromosome 19 is positioned in the active nuclear center. Intriguingly, as the Circos plots show, spatial colocalization is less prominent around the centromere. This is supported from the previous observation that centromeres tend to be positioned more towards the nuclear periphery [58]. The compatibility of a general colocalization of active regions arising from our genome-wide 3D models further suggests that coregulation might be encoded in long-range, non-local contacts (i.e., >10 Mb) rather than short-range contacts, which seem to be involved in processes organizing TAD structure [8]. The compatibility of these short-range contacts [8] added as constraints after the steering of non-local contacts, suggest that these two types of interactions serve independent mechanisms. The overall emerging view is a multi-level organization of chromatin starting at the level of TADs or subTADs. The figure shows the location of Hi-C interaction points and some aspects of gene regulation for chromosome 19, such as coregulated genes, gene density, gene families, and the location of housekeeping genes. Chromosome 19 contains nearly 55% repetitive sequences, notably due to an unusually high content of short interspersed nuclear elements (SINEs), with Alu repeats constituting 25.8% of the chromosome [59]. The G + C content of the chromosome is also unusually high,

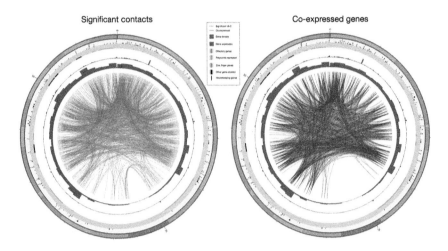

Figure 14.14 Comparison of colocalization networks from Hi-C interactions and gene pairs coregulation. Left: Circos plot illustrating significant interactions (red arcs) on chromosome 19 based on the hESC cell line from [8], and used as constraints in the 3D modeling procedure. Various features along chromosome 19 are plotted. The outer circle illustrates the genomic position on chromosome 19. The purple circular histogram illustrates the relative gene density. The green circular histogram illustrates the relative gene expression of the genes on the chromosome. Colored segments with gray background illustrate genomic positioning of gene cluster classes from [60]. The black segments on the light blue background indicate the positions of housekeeping genes on chromosome 19. Right: The same plot as on the left, but with the 1,487 pairs of coexpressed genes (used as constraints in [17]) indicated in blue.

with an average of 48%, reflecting the high gene density of the chromosome. In fact, chromosome 19 has the highest density of genes of all human chromosomes, where exons cover around 6.4% of the sequence, and protein-coding loci (exons plus introns) span around 50% of the chromosome. Also of interest from an organizational perspective is the prevalence of duplication structures of two types: tandemly clustered gene families and large segmental duplications. Chromosome 19 shows evidence of extensive genomic duplication with 7.35% of the sequence sharing sequence homology to more than one location in the genome. This is predominantly due to an increase in intrachromosomal duplications (6.20% of the sequence). These duplications often contain paralog genes that may lead to similar proteins that cooperate in common pathways and in protein complexes. More than 25% of the genes on chromosome 19 are members of large, well-defined, tandemly clustered gene families. The largest group of such genes on chromosome 19 encodes Krüppel-type (or C2H2) zinc finger transcription factor (ZNF) proteins, with 266 of the approximately 800 total human genes of this type located primarily within 11 large familial clusters. In a recent study on coregulation of paralog genes [60], it was found that paralog gene pairs are enriched for colocalization in the same TAD, share more often common enhancer elements than expected, and have increased contact frequencies over large genomic distances.

Coregulation of genes through colocalization in 3D is compatible with different types of mechanisms. TADs represent genes in relatively close proximity, while another recently described feature is that of coexpression domains, as defined by being regions containing highly coexpressed co-linear genes [61]. These chromosomal domains were found to be more coexpressed both internally in a domain, and between domains. Hi-C interaction maps represent physical proximity, and these contacts can be both tissue-specific or relatively invariant. They found that housekeeping genes are slightly overrepresented for coexpression in domains, as is also observed from our model. Further, we also find that these genes are more likely to be colocalized in 3D. Taken together, our results suggest a 3D colocalization-dependent regulation of active genes in the nuclear center.

14.5 CONCLUSIONS

Based on the work of ref. [20], we discussed how viable coarse-grained three-dimensional models of entire genomes can be obtained by using phenomenological data, and particularly Hi-C interaction maps, as distance restraints in otherwise general and aspecific chromosome models.

The statistical analysis of the phenomenological data is a key step of the process, because it can single out the pairs of genomic *loci* whose contact probability is much enhanced compared to the reference, background value. As we discuss, when this step is performed using appropriate statistical criteria, the significant pairs that are identified are largely compatible, meaning that they can be mostly colocalized simultaneously in a steered-MD approach, without generating conflicting topological constraints. As we discuss with a specific example, this is important also in view of combining data from various sources to improve the spatial resolution of the genomic model.

Besides being relevant *per se*, having access to three-dimensional genomic models can elucidate the tight connection between structure and function. This is not otherwise manifested in the raw Hi-C data. In particular, we showed that the latter encodes the preferential radial placement of gene-rich and gene-poor regions, lamina-associated domains, regions enriched with activating or repressing histone modifications and of loci associated with positive or negative bands of the Giemsa staining technique. That such functionally oriented properties can be recovered from Hi-C data and made evident with physics-based modeling is not obvious *a priori* and is very promising for the perspectives of genome modeling.

Finally, as an exemplary case of the close structure–function relationship, we revisited a recent *in silico* study of the gene-kissing hypothesis on human chromosome 19 [17]. The latter refers to the possibility, actually documented for specific sets of genes, that an effective means of coregulating genes would be to colocalize them in space, where their expression can be concerted locally by shared cellular machinery. As an extreme test of this hypothesis, we applied the same physical approach mentioned, but requiring the proximity of significantly coregulated pairs instead of significant Hi-C-based ones.

The modeling was aimed at verifying two open issues: (i) whether it is at all feasible to colocalize in space simultaneously a large number of coregulated loci and (ii) whether the resulting conformation is compatible with independent Hi-C measurements. We showed that the answer to both questions is affirmative, again a result that is striking and not granted *a priori*.

Overall both types of results reflect positively on the possibility of unveiling new functional/structural features in the future as more detailed phenomenological constraints become available.

REFERENCES

1. Dekker, J., Rippe, K., Dekker, M., and Kleckner, N. Capturing chromosome conformation. *Science* 295, 1306 (2002).
2. Zhao, Z., Tavoosidana, G., Sjölinder, M., Göndör, A., Mariano, P., Wang, S., Kanduri, C., Lezcano, M., Sandhu, K. S., Singh, U., et al. Circular chromosome conformation capture (4C) uncovers extensive networks of epigenetically regulated intra-and interchromosomal interactions. *Nature Genetics* 38(11), 1341 (2006).
3. Simonis, M., Klous, P., Splinter, E., Moshkin, Y., Willemsen, R., de Wit, E., van Steensel, B., and de Laat, W. Nuclear organization of active and inactive chromatin domains uncovered by chromosome conformation capture-on-chip (4C). *Nature Genetics* 38, 1348 (2006).
4. Dostie, J. et al. Chromosome conformation capture carbon copy (5C): A massively parallel solution for mapping interactions between genomic elements. *Genome Res.* 16, 1299 (2006).
5. Lieberman-Aiden, E. Lieberman-Aiden E., van Berkum N.L., Williams L., Imakaev M., Ragoczy T., Telling A., Amit I., Lajoie B.R., Sabo P.J., Dorschner M.O., Sandstrom R., Bernstein B., Bender M.A., Groudine M., Gnirke A., Stamatoyannopoulos J., Mirny L.A., Lander E.S. and Dekker J. Comprehensive mapping of long-range interactions reveals folding principles of the human genome. *Science* 326, 289 (2009).
6. Kalhor, R., Tjong, H., Jayathilaka, N., Alber, F., and Chen, L. Genome architectures revealed by tethered chromosome conformation capture and population-based modeling. *Nature Biotechnology* 30(1), 90–98 (2012).
7. Dixon, J. R., Selvaraj, S., Yue, F., Kim, A., Li, Y., Shen, Y., Hu, M., Liu, J. S., and Ren, B. Topological domains in mammalian genomes identified by analysis of chromatin interactions. *Nature* 485(7398), 376–380 (2012).
8. Rao, S. S., Huntley, M. H., Durand, N. C., Stamenova, E. K., Bochkov, I. D., Robinson, J. T., Sanborn, A. L., Machol, I., Omer, A. D., Lander, E. S., and Aiden, E. L. A 3d map of the human genome at kilobase resolution reveals principles of chromatin looping. *Cell* 159(7), 1665–1680 (2014).
9. Nagano, T., Lubling, Y., Stevens, T. J., Schoenfelder, S., Yaffe, E., Dean, W., Laue, E. D., Tanay, A., and Fraser, P. Single cell Hi-C reveals cell-to-cell variability in chromosome structure. *Nature* 502(7469) (2013).

10. Nagano, T., Lubling, Y., Varnai, C., Dudley, C., Leung, W., Baran, Y., Cohen, N. M., Wingett, S., Fraser, P., and Tanay, A. Cell cycle dynamics of chromosomal organisation at single-cell resolution. *bioRxiv* 094466 (2016).

11. Stevens, T. J., Lando, D., Basu, S., Atkinson, L. P., Cao, Y., Lee, S. F., Leeb, M., Wohlfahrt, K. J., Boucher, W., O'Shaughnessy-Kirwan, A., et al. 3d structures of individual mammalian genomes studied by single-cell Hi-C. *Nature* 544(7648), 59–64 (2017).

12. Nora, E. P., Lajoie, B. R., Schulz, E. G., Giorgetti, L., Okamoto, I., Servant, N., Piolot, T., van Berkum, N. L., Meisig, J., Sedat, J., et al. Spatial partitioning of the regulatory landscape of the x-inactivation center. *Nature* 485(7398), 381 (2012).

13. Naumova, N., Imakaev, M., Fudenberg, G., Zhan, Y., Lajoie, B. R., Mirny, L. A., and Dekker, J. Organization of the mitotic chromosome. *Science* 342(6161), 948–953 (2013).

14. Lupiáñez, D. G., Kraft, K., Heinrich, V., Krawitz, P., Brancati, F., Klopocki, E., Horn, D., Kayserili, H., Opitz, J. M., Laxova, R., et al. Disruptions of topological chromatin domains cause pathogenic rewiring of gene-enhancer interactions. *Cell* 161(5), 1012–1025 (2015).

15. Duan, Z., Andronescu, M., Schutz, K., McIlwain, S., Kim, Y. J., Lee, C., Shendure, J., Fields, S., Blau, C. A., and Noble, W. S. A three-dimensional model of the yeast genome. *Nature* 465(7296), 363 (2010).

16. Wüthrich, K. Protein structure determination in solution by NMR spectroscopy. *Journal of Biological Chemistry* 265(36), 22059–22062 (1990).

17. Di Stefano, M., Rosa, A., Belcastro, V., di Bernardo, D., and Micheletti, C. Colocalization of coregulated genes: A steered molecular dynamics study of human chromosome 19. *Plos Comput. Biol.* 9(3), e1003019 (2013).

18. Paulsen, J., Lien, T. G., Sandve, G. K., Holden, L., Borgan, O., Glad, I. K., and Hovig, E. Handling realistic assumptions in hypothesis testing of 3d co-localization of genomic elements. *Nucleic Acids Research* 41(10), 5164–5174 (2013).

19. Paulsen, J., Rødland, E. A., Holden, L., Holden, M., and Hovig, E. A statistical model of chia-pet data for accurate detection of chromatin 3d interactions. *Nucleic Acids Research* 42(18), e143–e143 (2014).

20. Di Stefano, M., Paulsen, J., Lien, T. G., Hovig, E., and Micheletti, C. Hi-C-constrained physical models of human chromosomes recover functionally-related properties of genome organization. *Scientific Reports* 6, 35985 (2016).

21. Paulsen, J., Sekelja, M., Oldenburg, A. R., Barateau, A., Briand, N., Delbarre, E., Shah, A., Sørensen, A. L., Vigouroux, C., Buendia, B., et al. Chrom3d: three-dimensional genome modeling from Hi-C and nuclear lamin-genome contacts. *Genome Biology* 18(1), 21 (2017).

22. Belcastro, V., Siciliano, V., Gregoretti, F., Mithbaokar, P., Dharmalingam, G., Berlingieri, S., Iorio, F., Oliva, G., Polishchuck, R., Brunetti-Pierri, N., and di Bernardo, D. Transcriptional gene network inference from a massive dataset elucidates transcriptome organization and gene function. *Nucleic Acids Research* 39, 8677 (2011).

23. Bolzer, A., Kreth, G., Solovei, I., Koehler, D., Saracoglu, K., Fauth, C., Müller, S., Eils, R., Cremer, C., Speicher, M. R., and Cremer, T. Three-dimensional maps of all chromosomes in human male fibroblast nuclei and prometaphase rosettes. *PLoS Biology* 3(5), e157 (2005).

24. Kremer, K. and Grest, G. S. Dynamics of entangled linear polymer melts: A molecular-dynamics simulation. *J. Chem. Phys.* 92, 5057 (1990).

25. Finch, J. T. and Klug, A. Solenoidal model for superstructure in chromatin. *Proc. Natl. Acad. Sci. USA* 73, 1897 (1976).

26. Rosa, A. and Everaers, R. Structure and dynamics of interphase chromosomes. *Plos Comput. Biol.* 4, e1000153 (2008).

27. Rosa, A., Becker, N. B., and Everaers, R. Looping probabilities in model interphase chromosomes. *Biophys. J.* 98, 2410 (2010).

28. Vettorel, T., Grosberg, A. Y., and Kremer, K. Statistics of polymer rings in the melt: A numerical simulation study. *Phys. Biol.* 6, 025013 (2009).

29. Rosa, A. and Everaers, R. Ring polymers in the melt state: The physics of crumpling. *Phys. Rev. Lett.* 112, 118302 (2014).

30. Grosberg, A., Rabin, Y., Havlin, S., and Neer, A. Crumpled globule model of the three-dimensional structure of DNA. *Europhys. Lett.* 23, 373 (1993).

31. Mirny, L. A. The fractal globule as a model of chromatin architecture in the cell. *Chromosome Res.* 19, 37–51 (2011).

32. Bystricky, K., Heun, P., Gehlen, L., Langowski, J., and Gasser, S. M. Long-range compaction and flexibility of interphase chromatin in budding yeast analyzed by high-resolution imaging techniques. *Proc. Natl. Acad. Sci. USA* 101, 16495 (2004).

33. Cremer, T. and Cremer, C. Chromosome territories, nuclear architecture and gene regulation in mammalian cells. *Nature Rev. Genet.* 2, 292 (2001).

34. Branco, M. R. and Pombo, A. Intermingling of chromosome territories in interphase suggests role in translocations and transcription-dependent associations. *PLoS Biol.* 4, e138 (2006).

35. Misteli, T. Beyond the sequence: Cellular organization of genome function. *Cell* 128(4), 787–800 (2007).

36. Solovei, I., Kreysing, M., Lanctôt, C., Kösem, S., Peichl, L., Cremer, T., Guck, J., and Joffe, B. Nuclear architecture of rod photoreceptor cells adapts to vision in mammalian evolution. *Cell* 137(2), 356–368 (2009).

37. Müller, S. and Almouzni, G. Chromatin dynamics during the cell cycle at centromeres. *Nature Reviews Genetics* 18(3), 192–208 (2017).

38. Plimpton, S. Fast parallel algorithms for short-range molecular dynamics. *J. Comp. Phys.* 117, 1 (1995).

39. Valet, M. and Rosa, A. Viscoelasticity of model interphase chromosomes. *The Journal of chemical physics* 141(24), 12B651_1 (2014).

40. Fraser, J., Rousseau, M., Shenker, S., Ferraiuolo, M. A., Hayashizaki, Y., Blanchette, M., and Dostie, J. Chromatin conformation signatures of cellular differentiation. *Genome biology* 10(4), R37 (2009).

41. Bau, D. and Marti-Renom, M. A. Structure determination of genomic domains by satisfaction of spatial restraints. *Chromosome Res.* 19, 25–35 (2011).
42. Nagano, T., Lubling, Y., Stevens, T. J., Schoenfelder, S., Yaffe, E., Dean, W., Laue, E. D., Tanay, A., and Fraser, P. Single-cell Hi-C reveals cell-to-cell variability in chromosome structure. *Nature* 502(7469), 59–64 (2013).
43. Kalhor, R., Tjong, H., Jayathilaka, N., Alber, F., and Chen, L. Genome architectures revealed by tethered chromosome conformation capture and population-based modeling. *Nature Biotechnology* 30(1), 90–98 (2012).
44. Federico, C., Cantarella, C. D., Di Mare, P., Tosi, S., and Saccone, S. The radial arrangement of the human chromosome 7 in the lymphocyte cell nucleus is associated with chromosomal band gene density. *Chromosoma* 117(4), 399–410 (2008).
45. Boyle, S., Gilchrist, S., Bridger, J. M., Mahy, N. L., Ellis, J. A., and Bickmore, W. A. The spatial organization of human chromosomes within the nuclei of normal and emerin-mutant cells. *Human Molecular Genetics* 10(3), 211–220 (2001).
46. Tanabe, H., Müller, S., Neusser, M., von Hase, J., Calcagno, E., Cremer, M., Solovei, I., Cremer, C., and Cremer, T. Evolutionary conservation of chromosome territory arrangements in cell nuclei from higher primates. *Proceedings of the National Academy of Sciences* 99(7), 4424–4429 (2002).
47. Kind, J., Pagie, L., Ortabozkoyun, H., Boyle, S., de Vries, S. S., Janssen, H., Amendola, M., Nolen, L. D., Bickmore, W. A., and van Steensel, B. Single-cell dynamics of genome-nuclear lamina interactions. *Cell* 153(1), 178–192 (2013).
48. Le Dily, F., Baù, D., Pohl, A., Vicent, G. P., Serra, F., Soronellas, D., Castellano, G., Wright, R. H., Ballare, C., Filion, G., et al. Distinct structural transitions of chromatin topological domains correlate with coordinated hormone-induced gene regulation. *Genes & Development* 28(19), 2151–2162 (2014).
49. Sikorav, J. and Jannink, G. Kinetics of chromosome condensation in the presence of topoisomerases: a phantom chain model. *Biophysical Journal* 66(3 Pt 1), 827 (1994).
50. Spilianakis, C. G., Lalioti, M. D., Town, T., Lee, G. R., and Flavell, R. A. Interchromosomal associations between alternatively expressed loci. *Nature* 435, 637–645 (2005).
51. Cavalli, G. Chromosome kissing. *Curr. Opin. Genet. Dev.* 17, 443 (2007).
52. Trussart, M., Yus, E., Martinez, S., Bau, D., Tahara, Y. O., Pengo, T., Widjaja, M., Kretschmer, S., Swoger, J., Djordjevic, S., et al. Defined chromosome structure in the genome-reduced bacterium *Mycoplasma pneumoniae*. *Nature Communications* 8 (2017).
53. Fanucchi, S., Shibayama, Y., Burd, S., Weinberg, M. S., and Mhlanga, M. M. Chromosomal contact permits transcription between coregulated genes. *Cell* 155(3), 606–620 (2013).

54. Cubeñas Potts, C., Rowley, M. J., Lyu, X., Li, G., Lei, E. P., and Corces, V. G. Different enhancer classes in *Drosophila* bind distinct architectural proteins and mediate unique chromatin interactions and 3d architecture. *Nucleic Acids Research* 45(4), 1714–1730 (2017).
55. Parkinson, H., Kapushesky, M., Shojatalab, M., Abeygunawardena, N., Coulson, R., Farne, A., Holloway, E., Kolesnykov, N., Lilja, P., Lukk, M., Mani, R., Rayner, T., Sharma, A., William, E., Sarkans, U., and Brazma, A. Arrayexpress-a public database of microarray experiments and gene expression profiles. *Nucleic Acids Res.* 35 (2007).
56. Goebel, B., Dawy, Z., Hagenauer, Z., and Mueller, J. An approximation to the distribution of finite sample size mutual information estimates. *IEEE International Conference on Communications* 2, 1102 (2005).
57. Krzywinski, M., Schein, J., Birol, Connors, J., Gascoyne, R., Horsman, D., Jones, S. J., and Marra, M. A. Circos: An information aesthetic for comparative genomics. *Genome Research* 19(9), 1639–1645 (2009).
58. Weierich, C., Brero, A., Stein, S., von Hase, J., Cremer, C., Cremer, T., and Solovei, I. Three-dimensional arrangements of centromeres and telomeres in nuclei of human and murine lymphocytes. *Chromosome research* 11(5), 485–502 (2003).
59. Grimwood, J., Gordon, L. A., Olsen, A., Terry, A., Schmutz, J., Lamerdin, J., Hellsten, U., Goodstein, D., Couronne, O., Tran-Gyamfi, M., et al. The DNA sequence and biology of human chromosome 19. *Nature* 428(6982), 529–535 (2004).
60. Ibn-Salem, J., Muro, E. M., and Andrade-Navarro, M. A. Co-regulation of paralog genes in the three-dimensional chromatin architecture. *Nucleic Acids Research* 45(1), 81–91 (2017).
61. Soler-Oliva, M. E., Guerrero-Martnez, J. A., Bachetti, V., and Reyes, J. C. Analysis of the relationship between coexpression domains and chromatin 3d organization. *PLoS Computational Biology* 13(9), e1005708 (2017).

Index

Printed and bound by CPI Group (UK) Ltd, Croydon, CR0 4YY

24/10/2024

01778304-0016